Nucleic Acids
in
Plants

Volume I

Editors

Timothy C. Hall, Ph.D.

Professor
Department of Horticulture
University of Wisconsin
Madison, Wisconsin

Jeffrey W. Davies, Ph.D.

Head
Virology Department
John Innes Institute
Norwich, England

CRC PRESS, INC.
Boca Raton, Florida 33431

Library of Congress Cataloging in Publication Data

Main entry under title:

Nucleic acids in plants.

  Bibliography: p.
  Includes indexes.
  1. Nucleic acids. 2. Botanical chemistry.
3. Plant viruses. I. Hall, Timothy C. II. Davies,
Jeffrey W.
QK898.N8N83   581.8'732   78-25602
ISBN 0-8493-5291-6 (Volume I)
ISBN 0-8493-5292-4 (Volume II)

© 1979 by CRC Press, Inc.

International Standard Book Number 0-8493-5291-6 (Volume I)
International Standard Book Number 0-8493-5292-4 (Volume II)

Library of Congress Card Number 78-25602
Printed in the United States

PREFACE

These are exciting times in the biological sciences. Each day brings new discoveries that help to explain how nucleic acid sequence and structure molds the form and function of every living organism. The intelligence that distinguishes man from other creatures has always made him curious about himself and the universe around him. In retrospect, one can see the gathering concepts of biology stimulated by Linnaeus who initiated the classification of living things in a logical fashion, and accelerated by Darwin's recognition of evolution as the process of slow change and diversification. Mendel, whose work was nearly overlooked, pioneered the science of genetics. The studies of Avery and of Hershey elegantly demonstrated that DNA is the carrier of genetic information. The transition to the era of molecular biology was made by the brilliant insight of Watson and Crick. They correctly deduced the structure of DNA, and succinctly showed how it could provide a simple chemical code for genetic information while also serving as its own template for replication from generation to generation.

It is upon these foundations that the rapid succession of todays discoveries are built. Almost all of the innovative work of the 1960s flowed from studies on microorganisms because of their short life cycles and their relatively simple genetic content. Fortunately, results from these studies can now be used towards advancing animal and plant biology.

Our ambition in the organization of this book was to explore the current status of knowledge about nucleic acids in plants. We wanted the reader to be able to learn how this research is being undertaken. Therefore, we asked the contributing authors to include details of approaches and methods. Where feasible, they have provided protocols that can be followed by those who wish to repeat results, extend data, make improvements, or use them in new applications. We have done our best to accentuate work in plants. Results from studies on animals and microorganisms have been used only as they underlie techniques in plant research or as they point to new directions applicable to plant studies.

The title *Nucleic Acids in Plants*, rather than *Nucleic Acids of Plants,* was chosen deliberately. We felt that it was essential to include information on the nucleic acids of plant viruses and viroids. Strictly, these are nucleic acids that are *in*, not *of* plant cells. Viral nucleic acids are not rarities in plant cells; their presence is widespread in natural plant populations. Viruses and viroids have evolved (and continue to evolve) in adaptation with the biosynthetic systems of plant cells. Therefore, they represent appropriate model systems for the scientist whose ultimate interest is the more complicated metabolism of proteins and nucleic acids of the plant cell. Consequently, much topical research on nucleic acids in plants follows from techniques developed in connection with viral studies.

We greatly appreciate the efforts of all our contributors. We were fortunate in attracting young and active researchers to describe their study areas. They have endeavored to make their chapters readable and to reflect their enthusiasm for the subject. We learned that for the active researcher it is truly hard to find the time to write a review. Inevitably, some of the data that were novel when submitted will be almost out of date by the time this book is published; despite the dedicated cooperation of CRC Press in expediting its appearance. The impression that this is regrettable is tempered by the realization that this is merely a symptom of the rapid progress being made in nucleic acid research. Indeed, we hope that this book will stimulate its readers to participate in the rapid advancement of knowledge concerning nucleic acids in plants.

T. C. Hall
J. W. Davies

THE EDITORS

Timothy C. Hall, Ph.D., is a Professor of Horticulture at the University of Wisconsin, Madison. He is a member of the American Society for Microbiology, the Biochemical Society (U.K.), and is chairman of the Recombinant DNA Committee of the American Society of Plant Physiologists. Dr. Hall received his B.Sc. in 1962 and his Ph.D. in 1965 from the University of Nottingham, England. Following a year at the University of Minnesota as Louis W. and Maud Hill Foundation Research Fellow in the Horticultural Science Department, Dr. Hall was appointed to the University of Wisconsin faculty in 1966. During 1977 he undertook sabbatical research-study at the Université Louis Pasteur, Strasbourg, France.

Throughout his career Dr. Hall has been interested in fostering the use of molecular biological approaches towards the enhancement of crop plant productivity. He has published many scientific articles in the fields or virology, cold stress physiology, protein synthesis, and nucleic acids in plants.

Jeffery W. Davies, Ph.D., is the Head of the Virology Department of the John Innes Institute, Norwich, England.

Dr. Davies received his B.Sc. in 1963, and Ph.D. in 1966, at the University of Nottingham, England, where he began his research into plant protein synthesis, using protoplasts and cell-free extracts. From 1966 to 1970 he was Lecturer in Molecular Biology at the University of Edinburgh. During this time he developed an interest in the structure and translation function of RNA, especially of viruses. After an E.M.B.O. Fellowship in Leiden, The Netherlands, he was appointed as Assistant Scientist in the Biophysics Laboratory at the University of Wisconsin, Madison, where he continued this research from 1972 to 1976 with plant viruses.

Dr. Davies is a member of the Institute of Biology, a member of the Biochemical Society, the Society for General Microbiology, and the American Society for Microbiology. In 1975, he was elected as a Fellow of the Linnean Society of London.

CONTRIBUTORS

Wayne M. Becker, Ph.D.
Professor of Botany
University of Wisconsin
Madison, Wisconsin

Edwin T. Bingham, Ph.D.
Professor of Agronomy
University of Wisconsin
Madison, Wisconsin

Jeffery W. Davies, Ph.D.
Head, Virology Department
John Innes Institute
Norwich, England

Elizabeth Dickson, Ph.D.
Assistant Professor
Laboratory of Genetics
The Rockefeller University
New York, New York

Robert B. Goldberg, Ph.D.
Associate Professor of Biology
University of California
Los Angeles, California

Timothy C. Hall, Ph.D.
Professor of Horticulture
University of Wisconsin
Madison, Wisconsin

Roger Hull, Ph.D.
Research Scientist
Department of Virology
John Innes Institute
Norwich, England

Leslie C. Lane, Ph.D.
Assistant Professor of Plant Pathology
University of Nebraska
Lincoln, Nebraska

Christopher J. Leaver, Ph.D.
Lecturer
Department of Botany
University of Edinburgh
Edinburgh, Scotland

Jeff Schell, Ph.D.
Professor
Department of Genetics
State University
Gent, Belgium

Krishna K. Tewari, Ph.D.
Professor of Biochemistry
Department of Molecular Biology and
  Biochemistry
University of California
Irvine, California

Virginia Walbot, Ph.D.
Assistant Professor of Biology
Washington University
St. Louis, Missouri

Jacques-Henry Weil, Sc.D.
Professor of Biochemistry
Universite Louis Pasteur
Strasbourg, France

Milton Zaitlin, Ph.D.
Professor of Plant Pathology
Cornell University
Ithaca, New York

TABLE OF CONTENTS

Volume I

Volume II

III. PLANT VIRUS NUCLEIC ACIDS

IV. VIROIDS, PLASMIDS, AND GENETIC ENGINEERING

Section I
Plant DNA

PLANT GENOME ORGANIZATION AND ITS RELATIONSHIP TO CLASSICAL PLANT GENETICS*

Virginia Walbot and Robert Goldberg

TABLE OF CONTENTS

---

* In this chapter, "sequence" frequently refers to a region of DNA rather than to a known nucleotide series.

I. INTRODUCTION

A. A Brief History of Plant Genetics

As Sturtevant[1] notes, the examination of expression and inheritance of traits in plants has led to many important discoveries in genetics. The classic experiments of Mendel in the 1860s, considered the first demonstration of segregation, were, as any school child knows, performed with humble garden peas. At the turn of the century, two notable plant biologists, Correns and Tschermak, completed a similar set of experiments demonstrating the principles of segregation, again with garden peas, but without knowledge of Mendel's results. After the rediscovery and popularization of Mendel's laws at the turn of the century, a number of plants became experimental tools for the verification and extension of Mendel's results. Cases of non-Mendelian inheritance were also found in plant materials. A remarkable finding early in the 20th century concerned the permanent heterozygosity and balanced lethality of *Oenothera* species in which a ring composed of all of the chromosomes in a fixed array is characteristic of meiotic figures. The haploid chromosome sets of such ring-forming chromosomes are named Renner complexes in honor of the botanist who first deduced the nature of the structural heterozygosity. A succession of other eminent botanists including DeVries and Cleland also devoted much of their life work to this complex phenomenon.

The experiments of Johannsen on pure line selection at the turn of the century foretold many of the problems still encountered in plant breeding programs today. Johannsen selected plants for high and low seed weight over several generations until a number of inbred lines were established which bred true for average seed weight. However, within an inbred line, a considerable variation in seed weight persisted despite recurrent selection for seed weight differences using seeds at the extremes of the distribution of that inbred line. These observations that pure bred lines generate a variety of phenotypes provided the first major modification of the simple pattern of Mendelian inheritance; namely, that environmental influences and many subtle genetic components contribute to whole plant characters such as seed weight or productivity. From these and other observations arose the concept of the polygene, a group of hundreds of genes which contribute quantitatively to a character rather than in a qualitative manner characteristic of simple Mendelian traits. Polygenes were presumed to contribute in a step-wise manner to the building of such macrotraits as body weight, longevity, and productivity, and these gene groups were presumed to be especially subject to environmental factors.

The principles of Mendelian inheritance were firmly established, however, during the subsequent years of the 20th century with the caveat that environmental impact on gene expression was an important variable in plant development. The salient features of Mendelian inheritance include: (1) the particulate nature of the gene—each gene is a discrete hereditary unit that can be mapped to a particular location; (2) genes on different chromosomes show independent assortment at meiosis while genes on the same chromosome are linked; (3) the chromosome contains a group of genes in a fixed linear order; (4) gene regulation is a property of each Mendelian unit except in the exceptional cases where an entire chromosome is repressed, i.e., the Barr body; and (5) in eukaryotes the genes coding for enzymes of a particular biochemical pathway are not contiguous as is found in prokaryotes.

Additional knowledge concerning the behavior of genes was provided by the cytogenetic examination of chromosomes. The chromosome content, including chromosome number and morphology, was seen to be a species characteristic. Chromosomes could also be differentiated into regions of euchromatic and heterochromatic staining propeties. Specialized heterochromatic regions, the centromeres, through which the chromosomes were attached to the spindle apparatus, were noted as a general, although not universal, feature of chromosome morphology. As gene mapping progressed in both *Drosophila* and *Zea mays*, it became clear that the heterochromatic regions, especially the centromeric regions, contained few, if any, genes. In fact, increases in heterochromatin were often associated with gene repression.

Another important concept to emerge from classical genetics is that of the orderliness of the genome and the tenacity with which order is preserved. Chromosome structure and gene order are not easily perturbed so that the conclusion may be drawn that it is either exceedingly difficult to disrupt the structure of the genetic material or that the structure itself is preserved in evolution for significant reasons. The specialized heterochromatic regions of the centromeres are a case in point. No transcription is measurable from such DNA, yet the existence of the centromere segregation is essential in most species for the orderly segregation of the chromosomes in mitosis and meiosis.

In this chapter, it will be seen that an orderliness exists in the organization of the genetically active DNA and that this order is apparently conserved in a wide variety of organisms. The current theories of genome organization and gene regulation are a direct outgrowth of the heritage from classical genetic and cytogenetic observations. Current theories attempt to explain, at a molecular level, these observations and to formulate a set of testable predictions concerning the organization of the genetic, regulatory, and structural DNA sequences of the genome.

B. Overview of Current Concepts of Genome Organization

The nuclear genome of eukaryotes contains most of the genes expressed by a species. In addition to coding for RNA and protein components of the cell, the nuclear genome also contains sufficient informational DNA to regulate the expression of these gene products and structural DNA required for the maintenance of chromosome morphology and function. This chapter examines the extent of our current knowledge of the organization of the informational DNA for genetic and regulatory functions and structural DNA sequences of plant chromosomes, and relates our understanding of DNA sequence organization to chromosome and species evolution.

One of the primary goals of a description of genome organization in eukaryotes is to explain those features of eukaryotic gene expression which differ from prokaryotic organisms. The current status of gene regulation in eukaryotes has been recently reviewed by Lewin.[2] The most striking difference between eu- and prokaryotic genomes is the massive increase in DNA content of eukaryotes; any model of genome organization must deal realistically with this.

Comparison of the DNA renaturation kinetics of prokaryotic DNAs reveals that these DNAs are composed primarily, if not exclusively, of DNA sequences present once per genome; that is, the organism is haploid at every locus and very little sequence homology exists between different segments of the genome. The consequence of this simple second-order renaturation kinetic profile is that the number of genes can be calculated with reasonable accuracy assuming a fixed size of genes. Eukaryotic DNA when reassociated provides a much more complex pattern. A variable but often large component of the DNA reassociates relatively rapidly, indicative of a high level of reiteration in the genome; that is, the organism is effectively polyploid for those sequences.[3] These sequences may be genes such as those for histones or ribosomal RNAs which are reiterated several hundred times in the genome. Other of these repetitive sequences do not appear to code for specific proteins, but they may play some role in the regulation of gene expression. One aspect of these repetitive sequences is that they are not perfectly matched in reassociation experiments; there has been some sequence divergence within the family of related but not identical sequences. Another major component of the genome is those sequences which reassociate as if present only once per haploid genome. This fraction presumably contains Mendelian factors, i.e., most of the genes.[3] In eukaryotes it is not possible, therefore, to calculate the total number of possible genes solely from the content of "unique" DNA since some genes may be found in families repeated many times in each genome.

A significant difference in the organization of prokaryotic vs. eukaryotic genomes is in the distribution of genes of a single biochemical pathway. In prokaryotic organisms, the genes of a pathway are coordinately regulated because they usually share a common set of regulatory sequences preceeding the gene cluster. Probably the best-studied example of such coordinate expression and regulation is found in the lac operon of *Escherichia coli*. In eukaryotes, however, enzymes of a particular biochemical pathway or characteristic of a specialized differentiated state, insofar as they have been mapped, do not share a contiguous location on the genetic map of that organism. Genes of a biochemical pathway are dispersed throughout the genome, although coordinate regulation is still accomplished. Hence, any model of genome organization in eukaryotes must seek to explain how the diverse dispersed genes can be recognized by regulatory molecules; clearly this recognition is unlikely to lie in the gene sequence itself, since this is a unique sequence of nucleotides.

The major requirements of a theory of gene organization and regulation in eukaryotes are that it explain the large size of eukaryotic genomes, the dispersed nature of the genes, and the presence of repetitive DNA. The model proposed by Britten and Davidson[5] (discussed most recently in Davidson et al.)[6] is the primary current theory and the basis for most experimental work in the field of eukaryotic genome organization. The key feature of their theory is a model of gene and regulatory DNA sequence spacing in which genes, the unique copy component of the genome, are each preceded by one or more regulatory DNA sequences drawn from the repetitive DNA families. Coordinate gene regulation would, therefore, be accomplished by genes of a pathway sharing a similar spectrum of repetitive regulatory sequences. This theory predicts that repetitive and unique copy DNA must be interspersed in the genome of eukaryotes, a proposal borne out in most genomes examined.[7,8] Thus far the genomes of several plants have been examined in great detail and a cursory study has been made of 20 or more plants. The experimental methodology used in these studies and the experimental results will be discussed in the next section.

II. ORGANIZATION OF PLANT GENOMES

A. Current Status of Eukaryotic DNA Sequence Organization

The chromosomes of all eukaryotes examined have been shown to have interspersed repetitive and single copy DNA sequences.[7,8] Two general patterns of repetitive and single copy DNA sequence arrangement have been identified.[9,10] These patterns are schematically represented in Figure 1. Genomes in which a major fraction of the DNA (> 60%) consists of moderately repetitive sequences, 200 to 400 nucleotide pairs (NTP) in length, interspersed with single copy DNA at intervals less than 1000 to 4000 NTP.

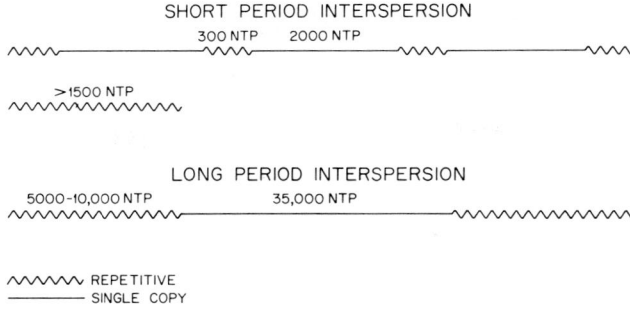

FIGURE 1.   A schematic representation of the long- and short-period interspersion patterns.

are said to be arranged in a short-period interspersion pattern.[9] Coinciding with short interspersion are variable amounts of long repetitive sequences, extending for at least 1500 NTP and not interspersed with single-copy DNA. The exact chromosomal length of long repetitive DNA is not known, but could easily be 50,000 NTP or longer. Short period interspersion has been found in chromosomes of representative species of all major animal phyla and flowering plants and as such appears to be highly conserved evolutionarily.

Genomes which lack short repetitive sequences but contain long stretches of repeated DNA (> 4000 NTP), interspersed with single-copy sequences at long intervals (> 10,000 NTP), are said to be organized in a long-period interspersion pattern.[10] To date, only a few eukaryotes, with small genome sizes (< 0.4 pg), have been demonstrated to have this interspersion pattern.[10-13] While there may prove to be exceptions, a correlation appears to exist between genome size and the general organization of chromosomal DNA (i.e., large genomes — short interspersion and small genomes — long interspersion). It has been proposed that organisms with long-period interspersion evolved from ancestors with larger genomes and DNA sequences arranged in the short period pattern.[12] More experimentation is needed to test this hypothesis.

A summary of the patterns of DNA sequence organization found in eukaryotic genomes investigated to date is presented in Table 1. The phylogenetic generality of DNA sequence interspersion in eukaryotic chromosomes implies a fundamental role in genomic activity. However, what functional significance long- and short-period interspersion has in relation to transcriptional processes, gene regulation, and chromosomal structure in eukaryotes is not presently understood.

B. Methods Employed in Analyzing DNA Sequence Organization

1. DNA Preparation

An essential prerequisite for the study of DNA sequence arrangement is a large quantity (> 20 mg) of highly purified, undegraded DNA. For studies of plant genome organization this is a difficult, although not impossible, goal to attain. Generally, plant DNA extractions are laborious and the results (purity, yield, and/or single-strand fragment size) not always ideal. This is primarily due to large quantities of polysaccharides, pigments, and nucleases in plant cells; a low nuclear/cellular volume ratio (i.e., large vacuolated cells with small nuclei); and cell walls which are difficult to break. Two DNA extraction procedures, however, have proven very effective in the extraction of DNA from plant tissues. One is a modified Marmur procedure[32] with an additional polysaccharide removal step[33,34] and the other a slightly modified urea-phosphate-hydroxyapatite (MUP) method developed by Britten.[35,36] These procedures are outlined

TABLE 1

The Pattern of DNA Sequence Organization in Eukaryotic Genomes

Kingdom	Organism	Dominant interspersion pattern	Ref.
Plant			
	Tobacco	Short	14
	Pea	Short	15—17
	Soybean	Short	18
	Cotton	Short	19
	Broad bean	Short	20
	Wheat	Short	21
	Rye	Short	22
Animal			
	Xenopus	Short	9
	Sea urchin	Short	23
	Oyster	Short	24
	Surf clam	Short	24
	Jellyfish	Short	24
	Sea hare	Short	25
	Rat	Short	26
	Horseshoe crab	Short	24
	Human	Short	27
	Silk moth	Short	28
	Housefly	Short	12
	Silverfish	Short	29
	Drosophila	Long	10,11
	Honey bee	Long	12
	Blowfly	Long	29
	Chironomus	Long	30
Fungal			
	Achlya	Long	13
Protist			
	Dictyostelium	Short	31

in Tables 2 and 3.

The degree of purity of plant DNA (total cellular or organellar) obtained by these techniques is similar. Both generate excellent DNA, with identical T_m and hyperchromicity values, and no contaminating polysaccharide, protein, or RNA molecules. The effort involved, the total quantity of DNA isolated, and the DNA fragment length differ significantly for these two approaches. The MUP procedure is definitely the method of choice for plant DNA extraction. A MUP DNA isolation is simple, rapid, and yields maximum amounts of plant DNA (e.g., 100 μg/g of expanded tobacco leaf material). Due to shearing, though, the average single strand fragment length of MUP isolated DNA is generally less than 10,000 nucleotides (NT). While this fragment length is more than adequate for most DNA sequence organization studies, longer lengths may be desired in some instances, for example, electron microscopic investigation of DNA arrangement of genomes with long-period interspersion or fine-scale sequence organization of long repetitive sequences. For longer fragment lengths, DNA prepared by the modified Marmur procedure or any analogous method should be used. The modified Marmur procedure involves considerably more time and effort and results in lower DNA yields (e.g., 10 μg/g of expanded tobacco leaf material), but the single strand fragment length is usually longer (generally > 30,000 NT).

Whichever procedure is used, it is essential to evaluate each DNA preparation for

TABLE 2

Modified-Marmur Extraction Procedure for Plant DNA

1. Grind plant material (leaves, roots, embryos, etc.) to a fine powder in a prechilled mortar ($-20°C$) with the aid of liquid nitrogen.[a]
2. Thaw leaf powder in the presence of one volume of TE buffer containing 4% SDS. The temperature should be raised to 40°C to facilitate cellular lysis.[b]
3. Collect the cellular lysate by passing the thawed leaf slurry through five layers of fine-mesh cheesecloth.
4. Repeat Steps 1 to 3 on the "pulp" in order to completely rupture cell walls.
5. Digest proteins by adding 100 μg/mℓ of self-digested pronase B and incubating for 2 hr at 37°C.
6. Adjust the cellular lysate to 1 M sodium perchlorate and extract remaining proteins with a 25:24:1 solution of phenol:chloroform:isoamyl alcohol. Repeat until interface is free of denatured protein and then extract one time with a 24:1 chloroform:isoamyl alcohol solution.
7. Precipitate nucleic acids by adding two volumes of cold 95% ethanol, spool DNA onto glass rods, and resuspend in TN buffer (See Table 3) at 200 to 500 μg/mℓ.[c]
8. Hydrolyze RNA by digesting with 100 μg/mℓ RNase A and 30 units/mℓ RNase T_1 for 2 hr at 37°C.[d]
9. Repeat Steps 5 to 7.
10. Adjust DNA solution to 1 M NaCl. Add CTAB (cetyltrimethylammonium bromide, Eastman) to 2% using a 10% stock solution. Pellet the white polysaccharide precipitate in the centrifuge.[e]
11. Lower the NaCl concentration to 0.4 M and collect precipitated DNA by spooling on glass rods.[f]
12. Wash precipitated DNA three times with 70% ethanol - 0.1 M sodium acetate, pH 6.0, and two times with 95% ethanol. Dry the DNA, resuspend in TN buffer, and store over chloroform at 4°C.[g]

[a] Ignited and washed sand can be used to facilitate pulverization. This procedure can also be used for organelle isolation by beginning at Step 2 and eliminating Steps 3 and 4.
[b] TE buffer is 0.2 M Tris-HCl, 0.2 M EDTA, pH 9.0. The pH of the lysate should be monitored and adjusted to pH 8.0 if necessary. A low cellular lysate pH (<4) can cause DNA degradation.
[c] Winding DNA in small amounts on several glass rods will facilitate resuspension. If DNA is unspoolable, pellet precipitated nucleic acids in a centrifuge tube, resuspend in TN buffer at a moderate DNA concentration (500 μg/mℓ), adjust NaCl to 0.15 M, and repeat precipitation and spooling steps. DNA should now spool. This procedure can be repeated several times until all spoolable material is collected. It is not uncommon to have highly pigmented, spoolable material at this stage. Note that polysaccharides also precipitate with ethanol, are viscous in solution, and spool onto glass rods.
[d] Adjust NaCl to 0.15 M. Preincubate RNase A at 90°C in TN buffer, pH 5.0 for 15 min. After cooling, adjust pH to 7.6, and add RNase T_1.
[e] All steps with CTAB should be at room temperature since CTAB precipitates at low temperatures. If no precipitate appears, the absence of polysaccharides cannot be inferred. Some polysaccharides are insoluble in CTAB at high ionic strength while others are soluble at both low and high ionic strengths and not precipitated with CTAB.[34] The latter will be eliminated by CTAB precipitation of the DNA (Step 11).
[f] DNA is now a CTAB salt and insoluble at low salt concentrations.
[g] This step removes CTAB and converts DNA back into the Na$^+$ salt form.

TABLE 3

MUP Extraction Procedure for Plant DNA[a]

1. Grind plant material (leaves, roots, embryos, seedlings, etc.) to a fine powder with the aid of liquid nitrogen in a cold mortar ($-20°C$).[b]
2. Thaw cell powder in the presence of three volumes of MUP buffer to 10 to 15°C. This is easily accomplished by placing the mortar in a partially filled sink with circulating tap water (20 to 40°C). Stir the slurry continuously with a pestle.[c]
3. Repeat Steps 1 and 2 several times in order to completely break plant cells.[d]
4. Extract protein with an equal volume of 25:24:1 phenol:chloroform:isoamyl alcohol.[e]
5. Reprocess interface by adding one volume of MUP buffer and repeating Steps 3 and 4.
6. Pass the combined cellular lysate over a hydroxyapatite (HAP) column operated at room temperature. The amount of HAP in the column should be at least 1/9 the volume of the pooled cellular lysate in MUP buffer. To speed up the procedure air pressure should be used (1 to 5 psi).[f]

7. Elute RNA, protein, polysaccharides, etc. from the HAP with UP buffer. Elution should be continued until the A_{260} is zero (generally 25 to 75 column volumes). During elution the column should be occasionally stirred.[f]
8. Remove urea from the HAP by eluting with 5 to 10 column volumes of 0.01 *M* PB.
9. Elute DNA with 0.5 *M* PB while simultaneously stirring the HAP. Use 1 column volume of 0.5 *M* PB at a time to keep the eluted DNA as concentrated as possible.
10. Concentrate the DNA by dialysis against 0.3 *M* sodium acetate, pH 6.0 and ethanol precipitation or by sedimentation for 12 hr in a Beckman 60 rotor (to pellet molecules $\geq$ 4S [Svedberg units]).
11. Resuspend DNA at 200 to 400 $\mu g/ml$ in 0.05 *M* NaCl, pH 7.6 (TN buffer) and store over chloroform at 4°C.[g]

[a]   A modified Britten MUP procedure.[35,36]
[b]   This protocol can be used with excellent success for organelle DNA isolation. In this case, purified organelles are directly suspended in MUP buffer and lysed by freezing and thawing several times.
[c]   MUP buffer consists of 8 *M* urea, 0.24 *M* phosphate buffer (PB), 0.01 *M* EDTA, and 2% SDS. PB is an equimolar mixture of sodium monobasic and dibasic phosphate. This solution should be prefiltered.
[d]   When isolating total cellular DNA, this step is essential for maximum yields.
[e]   This is easily accomplished by stirring the mixture in beaker with a magnetic stirrer for 15 to 30 min.
[f]   A "fast" flow rate lot of the HAP should be used. Generally BioRad non-DNA grade is excellent for this purpose. However, each lot of HAP should be tested for its DNA binding properties under standard conditions (0.12 *M* PB) before being purchased. Not all lots of HAP "behave" normally. HAP should be packed in 0.5 *M* PB, washed until no A_{260} is eluted, and equilibrated with UP buffer. UP buffer consists of 8 *M* urea, 0.24 *M* PB. Before chromatographing the cellular lysate, the HAP should be stirred.
[g]   Typically, plant DNA isolated by this procedure will have a hyperchromicity of 28 to 30% (as percent of final absorbance), A_{260}/A_{230} ratio of 2.3, A_{260}/A_{280} ratio of 1.85, T_m of 85 to 88°C in 0.12 *M* PB (depending on plant species), and a modal single-strand fragment length of 8000 to 10,000 nucleotides. Generally 1 mg of DNA can be isolated from 10 g of leaf material. Higher yields can be obtained when using embryos or seedlings due to greater cell number per equivalent fresh weight. On the other hand, callus cultures generally yield small amounts of DNA. This procedure has proven successful in DNA isolation from tobacco, pea, soybean, pinto bean, cotton, pumpkin, *Achlya, Chlorella,* and yeast in our laboratory.

purity and size before using in DNA sequence organization studies. This will save effort and eliminate possible misinterpretation of reassociation data due to impurities or size errors. The T_m and hyperchromicity values should be measured by melting the DNA in the spectrophotometer. No "foot" (i.e., shallow absorbance rise) should be present below the melting region (an indication of RNA or single-stranded DNA) and the hyperchromicity value should be 28 to 30% (as percent of the final absorbance value). It may also be desirable to melt the DNA in tetraethylammonium chloride (TEACl) solution which has been demonstrated to eliminate the effects of base composition on T_m[37] and reveal undetected contaminants.[38] The DNA should be subjected to isopycnic CsCl centrifugation to check for possible satellite DNA losses as well as contamination by RNA and polysaccharides. The latter can easily be detected in a Schlieren pattern of DNA at equilibrium in CsCl. The single strand fragment length of newly prepared DNA (and periodically stored DNA) should be measured by analytical band sedimentation in the analytical ultracentrifuge,[39] sucrose density gradient centrifugation,[9] or electron microscopy.[40,41]

2. DNA Reassociation Experiments

The genome of any organism can be characterized with respect to DNA content (basic or haploid amount), representation of DNA repetition classes (quantity of repetitive and single copy DNA as well as average reiteration frequency of repeated DNA classes), and sequence complexity (total number of NTP in nonrepeating sequences) by following the reassociation of complementary DNA sequences over a range of Cot values [Cot is simply the product of the initial concentration of single-strand DNA (M nucleotide) and the time (seconds) of reassociation] that is, constructing a Cot curve which describes the reassociation kinetics of the DNA.[3] The mathematics and theory of DNA reassociation kinetics have been presented in papers by Britten and Kohne,[3,42,43] Wetmur and Davidson,[44] and Britten-Davidson and their associates.[45,46] This subject has also been elegantly reviewed by Britten et al.,[36] Davidson,[47] and Wetmur.[48]

The technical aspects of performing and analyzing DNA reassociation experiments have been reviewed in detail by Britten et al.[36] This paper should be closely scrutinized before attempting to analyze a genome with respect to its reassociation kinetics. Briefly, short DNA fragments (200 to 400 NT in length) are denatured and allowed to reassociate into DNA/DNA duplexes at a specific experimental annealing criterion (usually 60°C, 0.18 M Na$^+$). Short fragments are necessary to insure that each DNA sequence has an equal chance of colliding with its complement in the renaturation reaction and forming a stable DNA/DNA duplex.[3] This is especially important when investigating the reassociation kinetics of DNA sequences from organisms with short period interspersion. The use of long DNA fragments (> 1000 NT) will cause extensive network formation due to interspersed repetitive and single copy DNA sequences. In fact, network formation in renatured samples of long DNA is a direct test for short interspersion (see below). The criterion used in the annealing reaction should allow for a maximum reassociation rate while setting a specific level at which stable duplexes can form between related but not identical sequences of a repetitive DNA family. This is generally 20°C below the T_m of similar-sized DNA in the salt used for the annealing reaction. Since 1% nucleotide sequence mismatching lowers the T_m of DNA/DNA duplexes by 1°C,[36] up to 20% mismatching between related repetitive DNA sequences will be observed under such conditions. This is within the range of mismatching (15%) contained in duplexes derived from the renaturation of the most highly divergent plant repetitive sequences (see Figure 10 and References 14 and 18). It must be emphasized that results obtained with DNA/DNA reassociation are strongly dependent on the criterion at which annealing is conducted. It is critical, therefore, that the same annealing criterion be maintained for a given set of experiments.

The degree of reassocation, after annealing, can be determined by hydroxyapatite (HAP) fractionation,[3] measurement of hypochromicity in the spectrophotometer,[3,44] or resistance to single-stranded specific nucleases (generally S-1 nuclease of *Aspergillus oryzae*).[36,45] The HAP assay, diagrammatically presented in Figure 2, measures the fraction of DNA molecules which contain a HAP-bindable duplex region. At a 60°C, 0.18 M Na$^+$ criterion the minimum duplex length measured by HAP is approximately 30 NTP.[36] On the other hand, the S-1 nuclease and hypochromicity assays measure the fraction of DNA nucleotides which are based paired at each point in the renaturation reaction.

Generally the fraction of DNA fragments which bind to HAP and the fraction of DNA nucleotides which are based paired (e.g., resistant to S-1 nuclease) will be similar only at the beginning and end of a reassociation reaction. This is because each randomly sheared DNA molecule in the annealing reaction has a distinct nucleotide sequence at its 3′- and 5′-end.[47] Hence, HAP-bindable DNA molecules will contain significant amounts of single-stranded DNA flanking base paired regions (overlap). Early in the renaturation reaction, the amount of overlap is approximately 55% of the length of the annealed DNA (i.e., 45% of HAP-bound DNA will be S-1 nuclease sensitive).[45] Because of overlap, DNA sequence interspersion, and the fact that DNA fragments bind to HAP as a result of the renaturation of the most repetitive sequences along their length, the HAP assay cannot be used to measure the genomic content of DNA sequence repetition classes (i.e., highly repetitive, middle repetitive, and single-copy

DNA REASSOCIATION ANALYSIS USING HYDROXYAPATITE

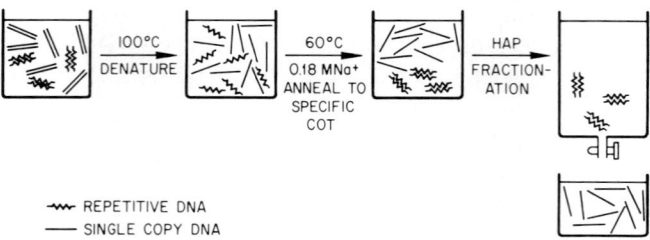

--⌁⌁⌁-- REPETITIVE DNA
──── SINGLE COPY DNA

FIGURE 2.   A schematic representation of a DNA reassociation experiment using hydroxyapatite chromatography.

DNA). This assay will generally increase the observed fraction of repetitive DNA while decreasing the observed fraction of single-copy DNA. This effect is especially significant for genomes organized in a short-period interspersion pattern. For the actual proportion of DNA sequence repetition classes in a genome, the S-1 nuclease or optical renaturation assay must be used.

a. HAP Reassociation Kinetics of Soybean DNA

An experiment which measures the reassociation kinetics of short DNA fragments of one plant DNA, the soybean plant (*Glycine max* Linnaeus cultivar 'Dare'®), analyzed by HAP chromatography, is presented in Figure 3a. A least squares analysis of these data, using a computer program written by Britten et al.,[36] suggests the presence of at least three second-order DNA components, each renaturing with distinct rate constants. It follows that these components contain sequences which differ in concentration and hence average reiteration frequency. The soybean DNA Cot curve is very similar to most other higher plant HAP reassociation curves, in that the most slowly renaturing DNA fragments (i.e., those fragments containing only single-copy sequences) represent a minor fraction of the total DNA population.[14,49,50] There are exceptions, however, such as cotton DNA[19] and as more experimentation on plant DNA is carried out, others will undoubtedly be revealed.

An analysis of the reassociation kinetics of whole or or unfractionated DNA can provide estimates of the relative range of DNA repetition classes and their kinetic characteristics (e.g., fraction of DNA fragments, second-order rate constants, average reiteration frequency, etc.). However, in order to directly demonstrate the presence of discrete repetition classes and to kinetically describe each class precisely, further analysis is required. A more detailed description of a genome can be obtained by preparing specific kinetic fractions of the DNA, by annealing and HAP chromatography, and then measuring the reassociation kinetics of each fraction individually (i.e., constructing a "minicot" curve).[36] In practice, minicot curves provide more meaningful information if labeled DNA fractions are prepared and then reassociated in the presence of an excess of unlabeled total DNA. From these types of data, it is not only possible to reveal discrete repetition classes, but the relative reiteration frequencies of these classes can be directly computed (since the rate constants are in whole DNA, see Reference 47). In addition, the reassociation kinetics of the most slowly reassociating or enriched single-copy fraction in the whole DNA provides a more accurate measurement of the basic genome size of the organism.

The HAP reassociation kinetics of three kinetic fractions of soybean ^{3}H-DNA, in the presence of an excess of unlabeled total DNA, are presented in Figure 3b. These experiments directly demonstrate the presence of at least four DNA classes in the soy-

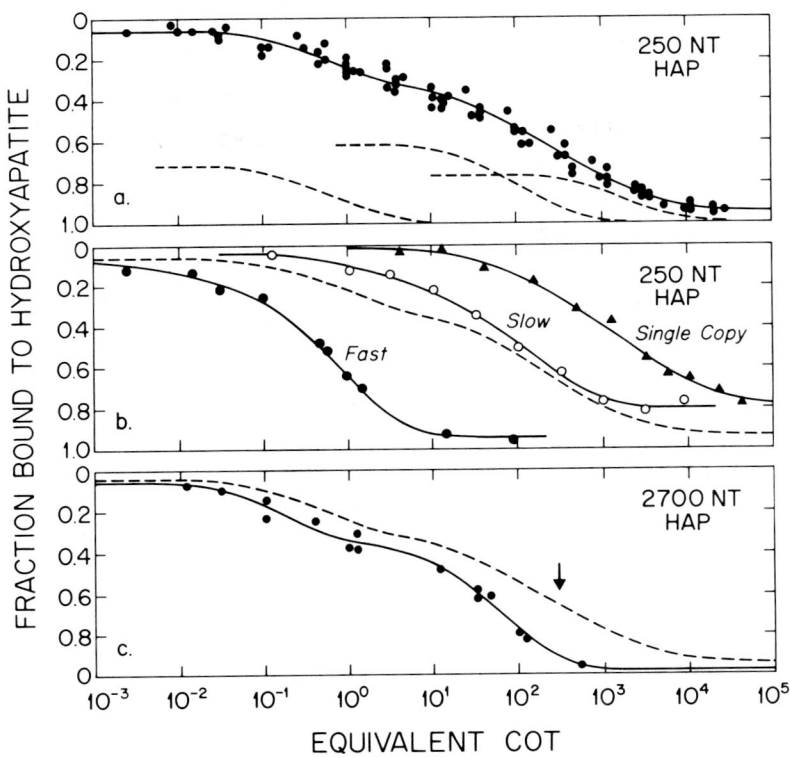

FIGURE 3. Hydroxyapatite reassociation kinetics of soybean DNA. DNA samples were reassociated at a criterion equal or equivalent to 0.18 M Na$^+$, 60°C. The fraction of fragments containing duplex regions was assayed by HAP chromatography at 60°C, in 0.12 M phosphate buffer.[14] The DNA length in nucleotides is listed for each curve. (a) Reassociation of 250 NT total DNA fragments. The solid curve portrays a least squares solution of the renaturation data using three second-order components represented by the dashed curves: 28% with a K of 1.48 M^{-1} sec^{-1} (fast), 38% with a K of 0.0096 M^{-1} sec^{-1} (slow), and 23% with a K of 0.00051 M^{-1} sec^{-1} (single copy). Approximately 7% of the DNA bound to HAP at the earliest Cot measured (10^{-3}; foldback and very fast) while 4% failed to bind at Cot 5 × 10^4. For this solution the rate of each component was fixed at a value obtained from the ressociation of labeled kinetic fractions, enriched for each component, in the presence of unfractionated DNA (see Figure 3b). (b) Reassociation of isolated kinetic components of soybean DNA. Three kinetic fractions of ^{3}H-DNA were prepared by reassociation and HAP fractionation.[36] The preparative history and percent of total DNA for each fraction were fast: Cot 10^{-5} UNB, Cot 1 BD, 22%; slow: Cot 20 UNB, Cot 75 BD, 7%; single copy: Cot 20,000 BD, Cot 2500 UNB, 17%. BD and UNB refer to DNA bound and unbound to HAP after reassociation to the indicated Cot. Renaturation of 250 NT ^{3}H-fast DNA sequences in the presence of 833-fold excess of unlabeled total DNA fragments (●). The solid curve represents the best least squares solution of the data yielding two second-order components: 12% with a K of 148 M^{-1} sec^{-1} (very fast) and 77% with a K of 1.42 M^{-1} sec^{-1} (fast). An additional 7% of the tracer bound to HAP at Cot 10^{-3} (foldback). Renaturation of isolated 250 NT ^{3}H-slow DNA fragments in the presence of a 7000-fold excess of unlabeled 250 NT total DNA fragments (○). The solid curve represents the best least squares solution for two second-order components: 16% with a K of 0.62 M^{-1} sec^{-1} (residual fast) and 60% with a K of 0.0096 M^{-1} sec^{-1} (slow). Renaturation of 250 NT ^{3}H-single-copy sequences in the presence of a 152-fold excess of total 250 NT DNA fragments (▲). The solid curve represents the best two component least squares solution: 28% with a K of 0.015 M^{-1} sec^{-1} (residual slow) and 50% with a K or 0.00051 M^{-1} sec^{-1} (single copy). The dashed curve portrays the renaturation of the driver DNA. (c) Reassociation of 2700 NT DNA fragments. The solid curve represents the best two component least squares solution to the renaturation data: 32% with a K of 6.94 M^{-1} sec^{-1} (fast, see Figure 7a) and 63% with a K of 0.018 M^{-1} sec^{-1} (slow plus covalently linked nonreassociated single-copy DNA, see Figure 7b). An additional 5% of the fragments bound to HAP at Cot 10^{-3} (foldback and very fast). The dashed curve portrays the reassociation of 250 NT DNA fragments for comparison. A Cot of 300 is indicated by the arrow.[18] (From Goldberg, R. B., *Biochem. Genet.*, 16, 45, 1978. With permission.)

bean genome (the fast fraction contains two second-order components, fast and very fast) termed very fast, fast, slow, and single copy. The average reiteration frequency of DNA sequences represented in these classes is 290,000, 2800, 19, and 1, respectively. It should be noted that each class of repeated DNA may contain families with a range of reiteration frequencies, the average of which is indicated in these reassociation experiments. This fine scale heterogeneity of repetition frequencies would not be detected in these experiments, but could be detected by analyzing the reassociation kinetics of cloned DNA fragments from each repetitive class.[51] From the single-copy rate constant (0.00051 M^{-1} sec^{-1}) and that of similar sized fragments of *Escherichia coli DNA* (0.22 M^{-1} sec^{-1}) the haploid genome size of soybean can be computed to be 1.91 pg.

b. S-1 Nuclease Reassociation Kinetics of Soybean DNA

In order to demonstrate the difference between the HAP and S-1 nuclease renaturation assays, measurements of the S-1 nuclease reassociation kinetics of short soybean DNA fragments are presented in Figure 4. Mild S-1 nuclease conditions were used in this experiment which digest >98% of the single-stranded DNA, but leave even extensively mismatched duplexes intact.[52] As explained above, S-1 nuclease measurements of reassociation kinetics provide a direct estimate of the genomic content of DNA sequence repetition classes. The least squares analysis of these data indicates that the single-copy and repetitive sequence content of the soybean genome is 35 to 40% and 60 to 65%, respectively. It should be noted that these values differ significantly from those obtained by the HAP reassociation assay. This suggests that the single-copy and repetitive sequences in soybean chromosomes are arranged in a short interspersion pattern. In addition, the large difference in the fraction of DNA molecules containing a slow repetitive sequence (HAP assay) and the actual genomic content of slow sequences (S-1 nuclease assay) suggest that slow repetitive sequences are interspersed among single-copy DNA. Direct proof of this will be presented below. A summary of the DNA sequence components in the soybean genome and their kinetic characteristics is presented in Table 4.

3. DNA Sequence Organization Analysis — A Hypothetical Experiment

The renaturation kinetics of short DNA fragments (< 400 NT) provides only limited information regarding the arrangement of DNA sequences in eukaryotic chromosomes. Using these data alone, it is not possible to ascertain the extent of DNA sequence interspersion nor the overall genome organization pattern. Although there are many diverse approaches to the study of DNA sequence arrangement, these are based on a comparison of the HAP reassociation kinetics of DNA fragments differing in length.[9,53] The results of two hypothetical experiments, comparing the renaturation of 300 NT and 2700 NT DNA fragments from genomes with distinct patterns of sequence organization, are presented in Figure 5.

The theoretical renaturation kinetics of DNA fragments from a genome with no interspersion (i.e., repetitive and single-copy DNA on different chromosomes) or a long period interspersion pattern are portrayed in Figure 5a. A diagrammatic representation of HAP bound and nonbound DNA molecules at two Cot values is also presented. These structures are a simplistic example of the type of molecules which would be visualized directly in the electron microscope. The renaturation kinetics demonstrate that all repetitive sequences reassociate by Cot 10 while no single-copy sequences have reacted (at both fragment lengths). Cot 10 HAP binding, therefore, establishes an experimental criterion which distinguishes between the renaturation of repetitive and single-copy sequences. Notice that the fraction of 300 NT and 2700 NT DNA fragments binding to HAP at Cot 10 is identical, but that the renaturation rates differ—the

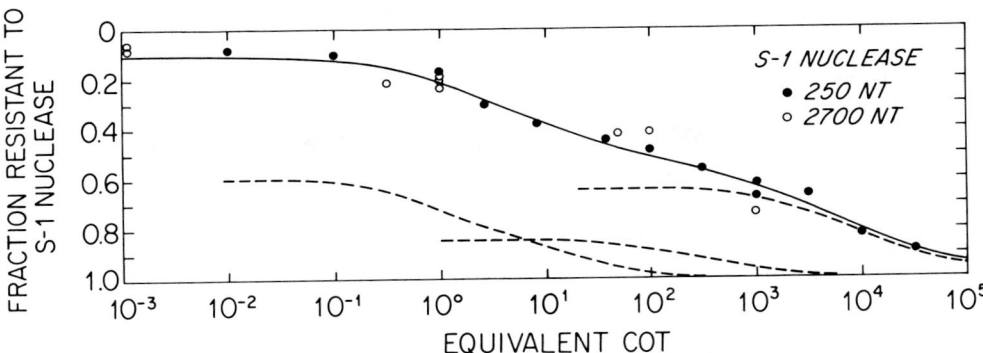

FIGURE 4.   S-1 nuclease reassociation kinetics of soybean DNA. Samples of 250 NT DNA fragments were reassociated in 0.3 M NaCl—0.01 M Pipes, pH 6.7, directly treated with S-1 nuclease according to the mild conditions of Britten et al.,[52] and the S-1 nuclease resistance determined by HAP chromatography at 60°C.[18,24] The solid line through the data points represents the best three component least squares solution to the 250 NT data. This analysis uses the equation:

$$\frac{C}{C_0} = \left[\frac{1}{1 + KCot} \right]^{0.44} \tag{1}$$

which describes the rate at which DNA nucleotides become S-1 nuclease resistant in a renaturation reaction.[45,47] For this solution the second-order rate constants of the fast, slow, and single-copy components were set at those values derived from the HAP reassociation kinetics (Figure 3). The dashed curves portray the renaturation of each component: 40%, fast; 13%, slow; and 37%, single copy. An additional 9% of the nucleotides were S-1 nuclease resistant at Cot 10^{-3} (foldback and very fast). It should be noted that the percentage of each component in the least squares analysis can be varied to some degree with less than a 10% increase in root mean square error. The range of genomic values for the fast, slow, and single-copy components within this domain of error are 28 to 40%, 13 to 25%, and 33 to 45%, respectively. The average of these values are presented in Table 4.[18] (From Goldberg, R. B., *Biochem. Genet.,* 16, 45, 1978. With permission.)

TABLE 4

Sequence Components of the Soybean Genome

Component	Fraction of genome[a]	K[b]	Cot$_{1/2}$	Average no. copies[c]	Complexity[d]
Foldback[e]	0.05				
Very fast	0.03	148	0.007	290,000	194
Fast	0.34	1.42	0.704	2,800	2.2×10^5
Slow	0.19	0.0096	104	19	1.8×10^7
Single copy	0.39	0.00051	1961	1	7.1×10^8

a   These values represent best estimates derived from the S-1 nuclease reassociation kinetics, HAP reassociation kinetics, and sequence organization pattern of soybean DNA.

b   Second-order rate constant (M^{-1} sec^{-1}) for 250 NT fragments in whole DNA at a 60°C, 0.18 M Na$^+$ criterion. The Cot$_{1/2}$ is the reciprocal of the K value.

c   Computed from the ratio of repetitive and single copy rate constants in whole DNA.[14,18,47]

d   Genomic complexity as expressed in NTP. Calculated from the genome size of 1.81×10^9 NTP, the genomic fraction of each component, and the average number of copies per genome.[14,47]

e   Foldback refers to DNA sequences arranged as tandem reverse repeats and which bind to HAP as a result of intramolecular base pairing. This reaction is not second order and occurs instantaneously. All plant DNA examined have been shown to have foldback sequences (cf. References 14 and 19).

longer fragments annealing faster by a factor of three. This rate increase is exactly the square root of the ratio of long to short DNA fragments. From the results of Figure 5a, several experimental predications can be made:

1. The hyperchromicity and S-1 nuclease resistance of duplexes bound to HAP at Cot 10 will be the same regardless of length.
2. Electron microscopic examination of long and short Cot 10 bound molecules will reveal primarily linear duplexes.
3. The renaturation kinetics of long ^{3}H-DNA fragments which were bound to HAP at Cot 10, subsequently sheared to 300 NT, and renatured in the presence of short, unlabeled, total DNA, will reveal only repetitive sequences. It is also possible, however, that both repetitive and single-copy sequences will be observed in this "playback" experiment with repetitive sequences being in the vast majority. The former result is predicted for a genome with no interspersion, while the latter for one with a long-period interspersion pattern. The percent single-copy DNA revealed in the "playback" experiment will be a function of the average spacing interval between single-copy and repetitive sequences in the long interspersion pattern (i.e., longer intervals, less observed single-copy DNA and shorter intervals, more single copy).

Using the results of Figure 5a and those of other experiments suggested by data in Figure 5a, it is possible to derive the following conclusions regarding the arrangement of repetitive and single-copy DNA sequences in this hypothetical genome:

1. The populations of 300 NT and 2700 NT DNA contain fragments entirely comprised of repetitive or single-copy sequences (i.e., minimal or no interspersion). This follows from the absence of quantitative differences in HAP binding at Cot 10.
2. The difference in renaturation rate of long and short DNA fragments is due solely to the effect of fragment length on rate of reassociation (see Reference 44).
3. Repetitive and single-copy sequences are not interspersed or they are very long (> 5 to 10,000 NT) and arranged in a long interspersion pattern. The spacing intervals between repetitive and single-copy sequences in this pattern can be determined directly from electron microscopic examination of very long DNA molecules (> 50,000 NT) renatured to a repetitive Cot or estimated from the results of a "playback" experiment (see References 10, 11, and 13).

A strikingly different result is presented in Figure 5b for a hypothetical genome with a short interspersion pattern. At Cot 10, 50% of the 300 NT DNA fragments bind to HAP. On the other hand, at this Cot value all of the 2700 NT fragments are HAP bound. Since only repetitive sequences renature by Cot 10, this difference can only be explained by the presence of repetitive and single-copy DNA sequences on the same 2700 NT DNA molecule. It is also important to observe that the renaturation rate of repetitive sequences in this experiment is the same for both long and short DNA fragments. Several experimental predications can be made from these data:

1. The hyperchromicity and S-1 nuclease resistance of long and short DNA molecules bound to HAP at Cot 10 will be different. The long molecules will display significantly less hyperchromicity and S-1 nuclease resistance. This is a direct result of the presence of long regions of unreassociated single-copy DNA covalently linked to repetitive duplexes. The hyperchromicity and S-1 nuclease resistance values will

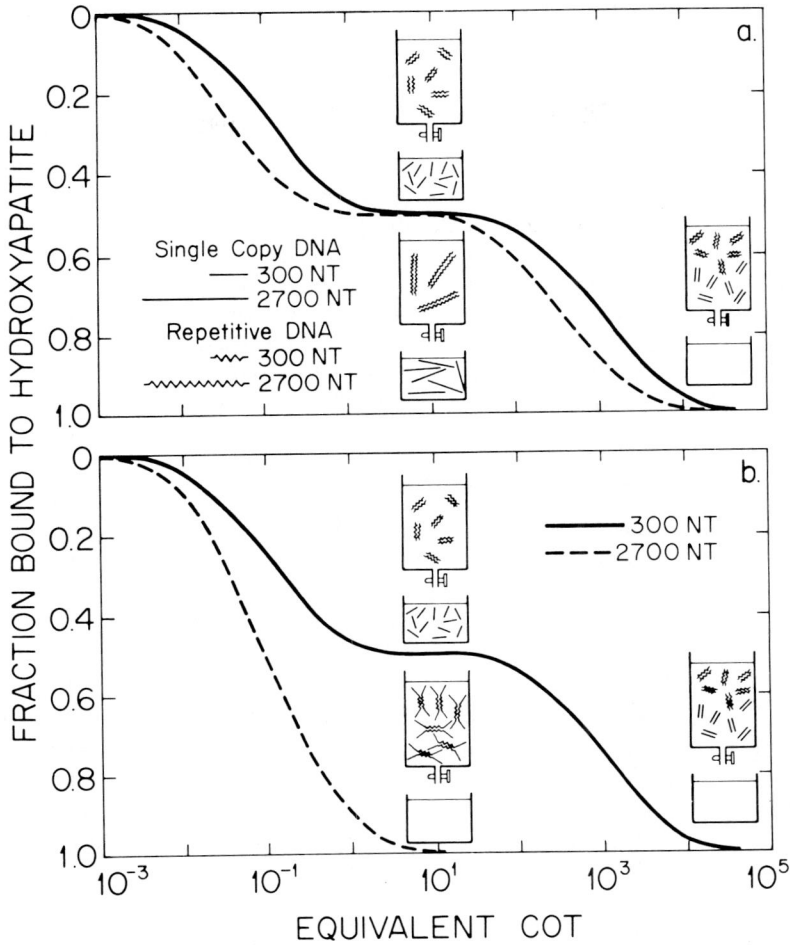

FIGURE 5. A hypothetical DNA sequence organization experiment. Theoretical HAP reassociation curves were constructed for 300 NT and 2700 NT DNA fragments from a hypothetical genome with 1 pg of DNA (haploid). In this genome 50% of the DNA is repetitive and 50% single copy. The repetitive class is homogeneous, all sequences having a reiteration frequency of 10,000. The second-order rate constants for the repetitive and single copy components, at the 300 NT fragment length, were set at 10 and 0.001 M^{-1} sec^{-1}, respectively. Two cases were considered. One in which the genome had no interspersion or a long period interspersion pattern and the other in which the genome was arranged in a short-period interspersion pattern. The insets represent the HAP fractionation of renatured DNA molecules at the indicated Cot values. The double-stranded (HAP-bound) and single-stranded (nonbound) DNA molecules represent simplified structures which would be visualized in the electron microscope. (a) No interspersion or long-period interspersion pattern. Repetitive and single-copy sequences were assumed to be on different chromosomes or >50,000 NTP in length and distributed on the same chromosome. The second-order rate constants for the repetitive and single-copy components, at the 2700 NT fragment length, were set at 30 and 0.003 M^{-1} sec^{-1}, respectively. This is a threefold rate increase and exactly the square root of the ratio of longer to shorter fragment lengths $[(^{2700})^{1/2}]$. A rate increase of this magnitude is expected, in the absence interspersion, for DNA fragments comprised entirely of one sequence class.[44] (b) Short-period interspersion. In this case it was assumed that the entire genome was arranged in a short interspersion pattern and no long repeats were present. Hence, all single-copy sequences are 2700 NTP in length or less and contiguous to 300 NTP interspersed repetitive sequence elements. Only one DNA component, repetitive, is able to reassociate in this case. The rate constant for the reacting repetitive sequences was set at 10 M^{-1} sec^{-1}, exactly that of the repetitive component at the 300 NT fragment length.

be approximately one third of that obtained for the shorter DNA molecules (if all repetitive DNA sequences are interspersed and 300 NTP in length; see Reference 9).

2.   Electron microscopic examination of HAP-bound, Cot 10, long DNA molecules will reveal complex double fork or H-structures and few linear duplexes. These structures will contain a short interspersed repetitive duplex flanked by four single-stranded, single-copy tails. Some DNA molecules will contain two or more duplexes separted by single-stranded DNA. From such structures, the spacing intervals between repetitive and single-copy DNA sequences in the short period interspersion pattern can be determined (see Reference 54).

3.   Shearing Cot 10 bound 2700 NT ³H-DNA molecules to 300 NT and analyzing their renaturation kinetics in the presence of total unlabeled DNA ("playback") will reveal sequences with two reiteration frequencies, repetitive and single copy, in the same proportion as that obtained for the short DNA fragments. This experiment provides direct evidence for single-copy sequences contiguous to interspersed repeats.

4.   The amount of DNA binding to HAP at Cot 10 will increase linearly with the fragment length and then plateau. At the inflection point of this "R-binding" curve, the DNA fragment length is equal to the single-copy spacing interval between two interspersed repeats (see References 9 and 23).

From the hypothetical experiment presented in Figure 5b, the following features of DNA sequence arrangement are revealed:

1.   The size of interspersed repetitive sequences is 300 NT in length. This conclusion derives from the similarity of renaturation rates of 300 NT and 2700 NT DNA fragments. Since the rates are the same, the size of the sequences which are reassociating must also be the same.

2.   All single-copy sequences in this genome are interspersed with short repetitive sequences. This logically follows from the observation that all of the long DNA fragments bind to HAP at Cot 10.

3.   Single-copy sequences range up to 2700 NT in length. Since this experiment does not reveal the single-copy spacing intervals between interspersed repeats, the average genomic length could be less than this value (but not more).

4.   Few, if any, long repetitive sequences not interspersed with single-copy DNA are present in this genome. If there were a significant fraction of long repetitive sequences, the renaturation rate of the 2700 NT DNA fragments would be faster.

While the results presented in Figure 5 are hypothetical and represent extreme cases, they serve to illustrate how experiments comparing the renaturation and HAP binding of long and short DNA fragments can be used to ascertain the pattern of DNA sequence organization. In reality, eukaryotic genomes are much more complex, and a variety of approaches are necessary in order to establish a complete and quantitative profile of the arrangement of repetitive and single-copy DNA (e.g., S-1 nuclease resistance, hyperchromicity measurements, EM, etc.). What follows is a series of experiments which describe the organization of DNA sequences in one plant genome.

4. DNA Sequence Organization Analysis — The Soybean Genome
a. Reassociation Kinetics of Long DNA Fragments
The HAP reassociation kinetics of 2700 NT fragments of soybean DNA are presented in Figure 3c. The least squares analysis indicates the presence of two second-order components, 32% of the sequences renaturing with a K (rate constant) of 6.92

M^{-1} sec^{-1} (fast) and 63% with a K of 0.018 M^{-1} sec^{-1} (slow). We can infer from these results that some repetitive sequences are interspersed with single-copy DNA while others are not appreciably interspersed at this fragment length.

It is instructive to compare these results to those obtained from the renaturation of 250 NT DNA fragments (Figure 3a and dashed curve in Figure 3c) and to the hypothetical experiments of Figure 5. This will allow the general features of DNA sequence organization in the soybean genome to be revealed. Recall that four major sequence classes have been identified in soybean DNA: very fast, fast, slow, and single copy (Figures 3a and 3b and Table 4). Over the range of Cot values required for the renaturation of very fast and fast DNA sequences (up to Cot 10), there is minimal difference in the amount of long or short DNA fragments binding to HAP. Hence at Cot 3, 31% of 250 NT and 38% of 2700 NT DNA fragments are HAP bound. This result suggests that very fast and fast DNA sequences are not appreciably interspersed among other sequence classes in the soybean genome (i.e., slow and single copy) nor between each other (compare to Figure 5a). If fast sequences are not interspersed, we can use the 250 NT renaturation data to predict that approximately 30% of 2700 NT DNA fragments should renature with a rate of 4.67 M^{-1} sec^{-1} $[(\frac{2700}{250})^{1/2} \times 1.42\ M^{-1}$ sec$^{-1}]$. The observed results are entirely consistent with these predictions.

In contrast to these results, there is a significant difference in the fraction of 250 NT and 2700 NT DNA fragments binding to HAP over the range of Cot values necessary for the reassociation of slow sequences (Cot 10 to 1000). At two Cot values, 300 and 1000, 67 and 77% of short DNA fragments bind to HAP. On the other hand, at Cot 300, 90% of the long DNA fragments bind to HAP, while over 96% do so at Cot 1000. Since the vast majority of duplexes which form at these Cot values are a result of the renaturation of slow DNA sequences, this difference in HAP binding can only be explained by the interspersion of slow and single-copy sequences in the soybean genome (compare to Figure 5b).

From the renaturation of long DNA fragments, we can deduce several facts about the arrangement of DNA classes in the soybean genome. First, very fast and fast DNA sequences are not significantly interspersed among slow or single-copy sequences. Second, repetitive sequence classes differing in average reiteration frequency are not significantly interspersed. Third, slow and single-copy sequences are arranged in a short-period interspersion pattern. Fourth, single-copy DNA sequences average 2700 NT or less. Fifth, at least 70% of the single-copy sequences in soybean are included in the short-period interspersion pattern. This value was computed from the differences in 250 NT and 2700 NT HAP binding at Cot 300 (23%), the amount of single-copy self-reassociation which could occur at Cot 300 (7%), and the fraction of 250 NT fragments which are entirely single copy (23%). Hence, $\frac{23\% - 7\%}{23\%} = 70\%$. More direct evidence for these features of the soybean genome will be presented below.

b. Hyperchromicity and S-1 Nuclease Resistance of Renatured DNA Fragments

An independent confirmation of repetitive and single-copy sequence interspersion can be obtained from a comparison of the hyperchromicity and S-1 nuclease resistance of long and short DNA fragments renatured to a Cot value at which only repetitive sequences react (see discussion of Figure 5). If repetitive and single-copy sequences are arranged in a short-period interspersion pattern, these measurements should yield different values for renatured long and short DNA molecules. This disparity will be due to the presence of unreassociated single-copy DNA linked to repetitive duplexes at the long fragment length.

Melting profiles of duplexes formed by annealing soybean 250 NT and 6000 NT DNA fragments to Cot 100 and HAP fractionation are presented in Figure 6. A melt-

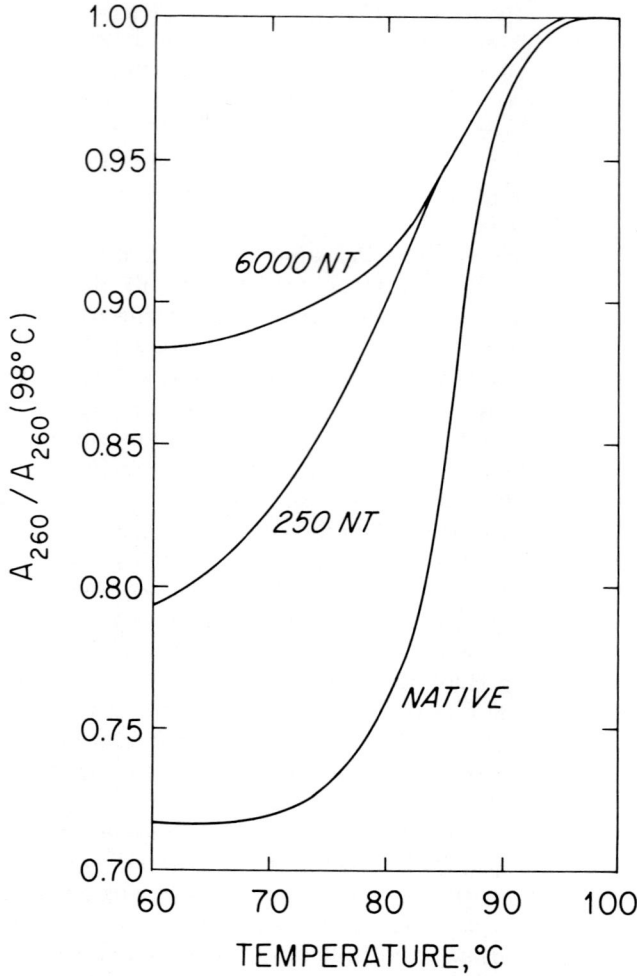

FIGURE 6.   Melting profiles of renatured DNA. Soybean 250 NT
and 6000 NT DNA fragments were denatured, individually reassociated to Cot 100 at 60°C in 0.12 *M* PB, and fractionated over HAP
at the same experimental criterion. DNA molecules containing a repetitive duplex region were eluted from HAP with 0.5 *M* PB, dialyzed
into 0.12 *M* PB, and melted in the spectrophotometer. A melt of native soybean DNA was included as a reference standard. The T_m and
hyperchromicity values measured in these experiments were 250 NT
fragments, 79°C and 0.209 (hyperchromicity as percent of final absorbance); 6000 NT fragments, 82°C, 0.118; native DNA, 86°C,
0.271. An approximate correction can be made for the effects of fragment length on T_m by using the relationship ΔT_m = [650], where N is
the fragment length.[36] The corrected T_m values are 250 NT, 81.6°C;
6000 NT, 82.1°C. Since 1% base sequence mismatching lowers the T_m
of native DNA by 1°C,[36] soybean repetitive sequences average 4%
sequence divergence. This assumes that the average % GC of renatured repetitive sequences is similar to that of total soybean DNA.
This is a reasonable assumption since no satellite is demonstrated by
centrifuging soybean DNA to equilibrium in CsCl. We cannot exclude
the possibility, however, of a "hidden" satellite.

TABLE 5

Duplex Content of Renatured Long DNA Fragments

Cot	1	100
Fraction of molecules bound to HAP[a]	0.34	0.77
Fraction of nucleotides resistant to S-1 nuclease[b]	0.26	0.42
Average duplex content of HAP bound molecules[c]	0.77	0.55

[a] Values obtained from the reassociation curve presented in Figure 3c.
[b] Values obtained from the 2700 NT renaturation data presented in Figure 4.
[c] Computed by dividing the fraction of nucleotides resistant to S-1 nuclease by the fraction of molecules bound to HAP.

ing profile of native soybean DNA is also presented as a reference. At Cot 100, only 1% of 250 NT duplexes and 4% of 6000 NT duplexes could result from the reassociation of single-copy sequences. Several conclusions can be derived from the measurements made in this experiment. First, the T_m of soybean repetitive duplexes is approximately 4°C lower than that of native DNA, after normalization for the effect of fragment length on thermal stability.[36] While this result indicates the presence of divergent repetitive sequences, it also suggests that a large fraction of soybean repetitive sequences is relatively nondivergent and/or has a higher GC content than the bulk of the DNA. This derives from the fact that similar experiments with other eukaryotic DNAs have revealed an average of 10 to 15% divergence of repetitive DNA sequences.[14,24] Second, the observed T_m (uncorrected for fragment length) of long and short repetitive duplexes differ (82 vs. 79°C). Since duplexes containing identical sequences have been melted in each case, this difference is probably the result of duplex length effects on T_m[36] and suggests that a large fraction of long repetitive sequences are contained in soybean DNA ($\geq$ 6000 NT). Finally, the hyperchromicity of Cot 100 long and short duplexes is less than that of native DNA. This result indicates the presence of single-stranded regions flanking repetitive duplexes at both fragment lengths.[9] However, the hyperchromicity of the renatured 6000 NT DNA fragments (0.118) is significantly less than that of the renatured 250 NT DNA fragments (0.209). Thus, much longer single-strand stretches are contiguous to repetitive duplexes at the long fragment length. This is a direct result of the short period interspersion of repetitive and single-copy sequences in the soybean genome (see discussion of Figure 5b).

The hyperchromicity experiments presented in Figure 6 provide confirmatory evidence for repetitive and single-copy sequence interspersion in soybean DNA, but not for which repetitive class is contained within the short interspersion pattern. Evidence has previously been presented which suggests that single-copy sequences are interspersed with low-frequency repetitive DNA (slow) but not significantly with moderately (fast) or highly repetitive (very fast) sequences (Figure 3). If this is indeed the case, long DNA fragments binding to HAP as a result of the renaturation of fast and very fast sequences (< Cot 10) should be primarily double stranded, while those binding at higher Cot values (> 10) should have a significantly lower duplex content due to the presence of extensive regions of unreassociated single copy DNA.

Table 5 presents measurements of the average duplex content of 2700 NT DNA fragments bound to HAP after renaturation to Cot 1 and 100. These data were computed from the HAP reassociation experiments presented in Figure 3 and the S-1 nuclease reassociation experiments presented in Figure 4. At Cot 1, approximately 80% of the nucleotides of HAP-bound DNA molecules are double stranded. In contrast, only 55% of the nucleotides in DNA molecules bound to HAP at Cot 100 are in duplex structures. These results are in accord with predictions made from the renaturation kinetics and support the notion that fast and very fast sequences are not appreciably interspersed in soybean DNA, but that slow and single-copy DNA are arranged in a short-period interspersion pattern. Very fast and fast DNA sequences are most likely, therefore, organized as long stretches of repeated DNA.

c. A Kinetic Test for DNA Sequence Interspersion

The experiments presented to this point have enabled us to deduce that some repetitive sequences in soybean chromosomes are interspersed with single-copy DNA (slow sequences) while others are not (very fast and fast). However, in order to directly determine the arrangement of specific repetition classes, a different strategy must be employed.

Two kinetic fractions of 3000 NT ^{3}H-DNA fragments were isolated by renaturation and HAP fractionation. One fraction contained sequences which renatured between Cot 0.1 and 1.0. Molecules of DNA binding to HAP at these Cot values should contain duplex regions due to the renaturation of fast DNA sequences as well as any single-stranded slow or single-copy sequences linked to slow duplexes (note that any interspersed sequences will be significantly enriched for in this fractionation). The other fraction contained DNA fragments which did not bind to HAP at Cot 30. Included in this fraction should be molecules with interspersed slow and single-copy sequences plus any fragments entirely comprised of slow or single-copy DNA. Both ^{3}H-DNA fractions were sheared to 250 NT and then separately renatured in the presence of excess unlabeled total DNA ("playback" experiment). The results of these reassociation experiments are presented in Figure 7.

The DNA sequence representation in the Cot 0.1 to 1.0 fraction is presented in Figure 7a. The least squares analysis clearly demonstrates that the large majority of renaturing DNA sequences are derived from one kinetic class, reacting with a rate constant consistent with those of fast sequences. Since this experiment is sensitive enough to detect less than 5% of the sequences in total DNA, we can conclude that the fast DNA class is not significantly interspersed with slow or single-copy sequences in the soybean genome (see discussion of Figure 5a).

The renaturation kinetics of the second kinetic fraction is presented in Figure 7b. In contrast to the previous results, the reassociation kinetics reveals the presence of two second-order components with rate constants equivalent to those of slow and single-copy DNA sequences. This result substantiates, therefore, that slow and single-copy DNA sequences are arranged in a short period interspersion pattern in the soybean genome (see discussion of Figure 5b).

d. Size Distribution of Repetitive DNA Sequences

The very fast and fast DNA classes constitute almost 40% of the nucleotide sequences in soybean chromosomes (Table 4). Since sequences of these classes are not significantly interspersed with other sequence components (i.e., slow and single copy), a large fraction of soybean DNA should consist of long repeats, having a tandem or clustered sequence arrangement.[7] In order to account for the extensive interspersion of single-copy sequences at a 2700 NT fragment length, however, short repetitive sequences must also be present. Two types of procedures can be used to determine the size distribution of repetitive DNA sequences. One method utilizes gel filtration chro-

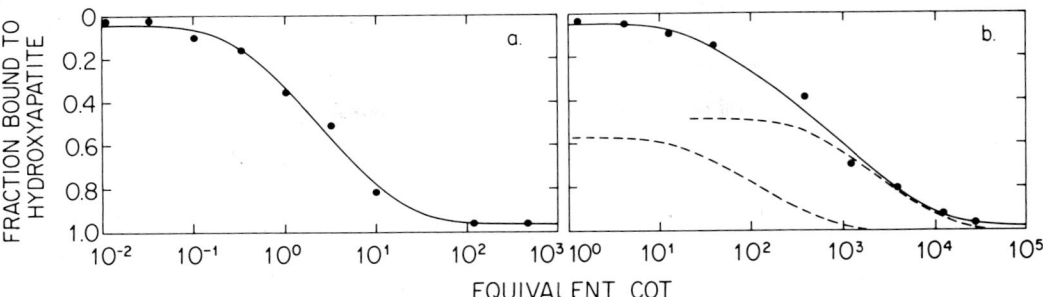

FIGURE 7. Reassociation of kinetic fractions isolated from long soybean DNA. Kinetic fractions of soybean DNA were prepared by reassociating 3000 NT ^{3}H-DNA fragments at 60°C in 0.12 M PB and subsequent HAP fractionation at the same criterion. The reassociation history and percent of total DNA for each fraction were: fast — Cot 10^{-1} UNB, Cot 1 BD, 22%; slow — Cot 30 UNB, 36%. After fractionation, the single-strand fragment length of the fast fraction was 2800 NT and the slow 2500 NT. These fractions contain sequences represented in each component of the 2700 NT reassociation curve presented in Figure 3c. Each fraction was sheared in the Virtis 60 homogenizer to 250 NT[36], and the DNA sequence representation measured by reassociation in the presence of an excess of 250 NT unlabeled total DNA (mass ratio 416/1). (a) Reassociation kinetics of the fast fraction (Cot 0.1 to 1). The solid line through the data points represents the best least squares solution for one second-order component. For this solution K is equal to 0.44 M^{-1} sec^{-1}. (b) Reassociation kinetics of the slow fraction (Cot >30). The solid curve through the data points is a least squares solution for two second-order components. For this solution the second-order rate constants were set at values corresponding to the renaturation of 250 NT fragments of slow and single-copy DNA (Figure 3a and 3b and Table 4). The dashed curves represent elements of the overall solution: 45% with a K of 0.0096 M^{-1} sec^{-1} and 52% with a K of 0.00051 M^{-1} sec^{-1}.[18]

matography to separate S-1 nuclease-resistant repetitive duplexes.[24,52] The other method directly measures the size of repetitive duplexes in the electron microscope, using conditions which distinguish between double- and single-stranded regions of DNA.[10,40,41,54]

i. Agarose Gel Filtration

To determine the size distribution of sequences in each class of soybean repetitive DNA, agarose gel filtration profiles were constructed for S-1 nuclease-resistant repetitive duplexes formed from the renaturation of 6000 NT DNA fragments to Cot < 10^{-4} (instantaneous HAP binding), Cot 1, and Cot 100. From these gel filtration profiles, the distribution of repetitive sequence lengths contained within the foldback (very fast and fast) as well as slow DNA classes were determined.[25] The results of these experiments are presented in Figure 8.

The gel filtration profile presented in Figure 8a indicates that most of the foldback mass (> 70%) is excluded from the agarose column. Hence, the mass average size of duplexes (stems; see Reference 55) contained within foldback structures of soybean DNA exceeds 1500 NTP. Note, however, that duplexes of foldback DNA are also included in the agarose column. While these represent only a small portion of the mass, a large number of molecules are probably represented. It is most likely, therefore, that foldback structures in soybean DNA have a range of sizes. This notion has recently been supported by direct electron microscopic examination.[56]

The size distribution of Cot 1 S-1 nuclease-resistant duplexes is presented in Figure 8b. The length distribution of very fast and fast DNA sequences (dashed curve) can be estimated by subtracting the foldback contribution (Figure 8a) from the total Cot 1 duplex profile. Over 60% of very fast and fast duplexes are excluded from the gel filtration column (minimum duplex length 1500 NTP). This result is consistent with a

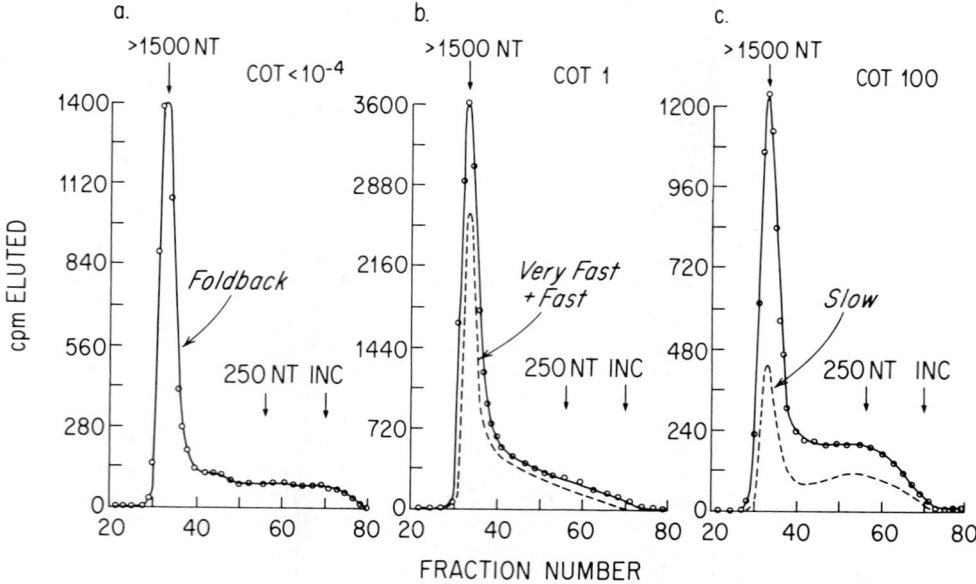

FIGURE 8.   Size distribution of soybean repetitive DNA sequences. DNA fragments 6000 NT in length were reassociated in 0.05 M Pipes (pH 6.7) - 0.18 M NaCl at 60°C to Cot < 10^{-4} (a), Cot 1 (b), or Cot 100 (c) and single-stranded fragments removed by mild treatment with S-1 nuclease.[24,52] S-1 nuclease-resistant duplexes were harvested by HAP chromatography. Approximately 6, 22, and 42% of the genome was resistant to S-1 nuclease at Cot < 10^{-4}, 1, and 100, respectively. These values are in good agreement with those predicted from a knowledge of the second-order rate constants of components of soybean DNA (Table 4), the S-1 nuclease reassociation kinetics of soybean DNA (Figure 4), and the effect of length on reassociation rate; $[K_{6000} = K_{250}(^{6000})^{1/2}]$. The components of the soybean genome which are resistant to S-1 nuclease at the annealed Cot values are foldback (10^{-4}), very fast (10^{-4} to 1), and slow (1 to 100). The size distribution of S-1-resistant repetitive duplexes was determined by Agarose A-50 chromatography (Biorad, 100 to 200 mesh) in 0.12 M PB. The position of the exclusion (>1500 NT), inclusion, and 250 NT peaks are given by the arrows. (a) The size distribution of foldback sequences. Greater than 70% of the foldback sequences are excluded from the Agarose A-50 column. This value was obtained by estimating the area under the exclusion peak in relation to the total curve. (b) The size distribution of very fast and fast repetitive sequences. The solid curve represents the distribution of Cot 1 S-1-resistant duplexes. The distribution of very fast and fast DNA sequences can be estimated by subtracting the Cot 10^{-4} duplex contribution (Figure 8a) from the total Cot 1 duplex profile. The dashed curve represents this distribution. At least 60% of very fast and fast duplexes exceed 1500 nucleotide pairs in length. (c) The size distribution of slow repetitive sequences. The distribution of Cot 100 S-1-resistent duplexes is portrayed by the solid curve. The dashed curve represents the distribution of slow duplexes reassociated by Cot 100. This distribution was estimated by subtracting the Cot 1 duplex contribution (Figure 8b) from the total Cot 100 duplex profile. Approximately 5% of these duplexes could be single copy. Approximately 60% of the slow duplexes chromatographed with a modal size distribution of 250 nucleotide pairs, while 30 to 40% have a minimum duplex length of 1500 nucleotide pairs.[18] (From Goldberg, R. B., *Biochem. Genet.*, 16, 45, 1978. With permission.)

clustered and/or tandem arrangement of sequences in these DNA classes. Duplexes which are included in the column probably result from the heterogeneous size distribution in the original 6000 NT DNA population.

Figure 8c presents the size distribution of Cot 100 S-1-resistant repetitive duplexes. Notice that the majority of mass is excluded from the column, a result consistent with a large fraction of long repetitive DNA in soybean chromosomes. The size distribution of slow DNA sequences (dashed curve) was constructed by subtracting the Cot 1 repetitive duplex contribution from the total Cot 100 profile. A distinctly different distribution of sequence lengths is contained within the slow DNA class. Approximately 60% of the slow duplexes are included in the agarose column and have a modal size

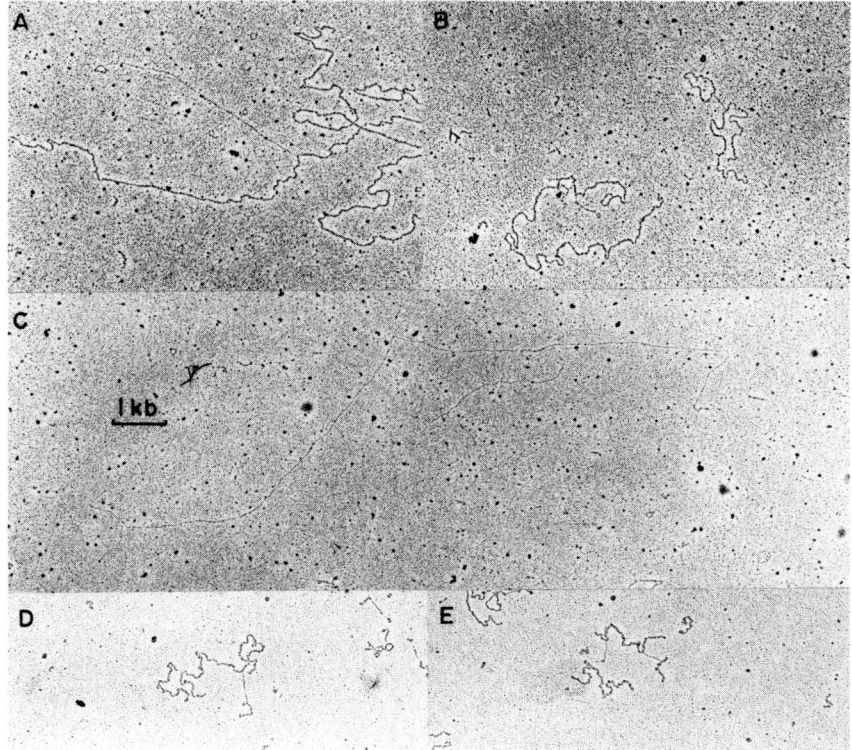

FIGURE 9.   Electron micrographs of reassociated long soybean DNA fragments. Three kinetic fractions of 7000 NT DNA fragments were prepared by reassociation and HAP fractionation. The reassociation history of each DNA preparation was: foldback — Cot 10^{-5} BD; fast
— Cot 0.01 UNB, Cot 1 BD; slow — Cot 5 UNB; Cot 100 BD. Each fraction was extracted
with a 24:23:1 solution of phenol:chloroform:isoamyl alcohol solution and dialyzed into 0.01
M PB, 0.001 M EDTA (pH 7.6). The DNA was subsequently treated with gene 32 protein
and prepared for electron microscopy according to the procedure of Pellegrini et al.[57] (A)
Foldback fraction: long stem—small loop; (B) foldback fraction: short stem—large loop; (C)
fast fraction; (D) slow fraction: double fork; (E) slow fraction: molecule with 2 double forks
and interspersed single-copy DNA.

of 250 NTP. This is the exact size expected for repetitive sequences contained within a
short interspersion pattern.[9] Another 30 to 40% of the duplexes are excluded from the
agarose column. This suggests that some low-frequency repetitive sequences are also
organized as long repeats in soybean chromosomes.

ii. Electron Microscopic Examination of Renatured DNA*

The size distribution of soybean repetitive sequences was also studied in the electron
microscope using a new spreading procedure employing gene 32 DNA binding protein.[41,57] This procedure produces unambiguous contrast between regions of double-
and single-stranded DNA. As such it is far superior to the standard Kleinschmidt-form-
amide method for studies of DNA sequence organization.[40]

Representative DNA structures containing foldback, fast, and slow duplexes are presented in Figure 9. Notice the striking contrast between single-stranded (thick) and

*   In collaboration with Dr. Maria Pellegrini.

duplex (thin) DNA regions. The foldback structures presented in Figures 5, 9A, and 9B contain both single-stranded loops and double-stranded stems. While size hetero-geneity clearly exists in the stem and loop regions, long-stem duplexes constitute the bulk of the mass. The molecules presented in Figure 9C represent linear duplexes con-taining fast DNA sequences. As predicted from the other experiments, little single-stranded DNA is present in these molecules and the duplex lengths are very long. This confirms that fast DNA sequences are organized as long repeats and are not signifi-cantly interspersed in soybean chromosomes. Finally, representative slow duplex struc-tures are shown in Figures 5, 9D, and 9E. The structure in 9D is a double-fork or H-structure containing a short interspersed repetitive sequence element and four single-stranded tails of single-copy DNA. This structure is exactly the type predicted from the short-period interspersion of slow and single-copy sequences in soybean DNA.[54] The molecule in Figure 9E contains two double forks and interspersed single-copy DNA.

e. Summary of Soybean DNA Sequence Organization

The major features of the soybean genome revealed in these experiments are

1. Approximately 60% of the nucleotide sequences in soybean are repetitive and 40% single copy.
2. A large fraction of soybean DNA (45 to 50%) is composed of long repetitive DNA sequences organized in a tandem or clustered sequence arrangement. In-cluded in this fraction are most sequences of the very fast and fast DNA classes as well as some slow sequences.
3. Repetitive sequence classes differing significantly in reiteration frequency are not interspersed among each other. Thus, there is little interspersion of fast and slow DNA sequences or very fast and fast DNA sequences. Note that repetitive se-quences from families with different reiteration frequencies could be inter-spersed. However, their reiteration frequency would have to be similar (e.g., 750 copies and 2500 copies).
4. Approximately 50% of the soybean genome is arranged in a short period inter-spersion pattern of 250 NTP slow sequences and single-copy DNA. At least 70%, and most likely all single-copy sequences, are represented in the short period pattern. The maximum single-copy spacing interval between two interspersed slow sequences is approximately 2700 NTP. The actual average size of soybean single-copy DNA is most likely 1500 NTP or equivalent to the number-average size of polysomal messenger RNA in soybean.[58,59]
5. The sequence arrangement of soybean DNA is similar to other eukaryotes with comparable genome sizes. The unique aspects of soybean DNA sequence orga-nization are the large fraction of long repeats and the apparent absence of repet-itive sequence class interspersion.

C. Comparative Aspects of Plant Genome Organization
1. A Comparison of Several Plant Genomes

Extensive studies of the arrangement of repetitive and single-copy DNA sequences have been completed for the genomes of eight flowering plants, representing both mon-ocotyledonous and dicotyledonous species. Diverse approaches were used in these ex-periments, including HAP binding measurements, kinetic analysis of long and short DNA renaturation hyperchromicity, S-1 nuclease, and electron microscopic studies of renatured DNA. These experiments have provided a quantitative and detailed profile of DNA sequence arrangement in higher plant chromosomes. The major conclusion derived from these studies is that the general and qualitative features of plant DNA sequence arrangement are similar to those of animal chromosomes. Thus, the short-

TABLE 6

Comparative Aspects of Plant Genome Organization[a]

Plant	Genome size[b]	Genomic fraction of single-copy DNA	Genomic complexity[c]	Repetitive sequence classes (no. copies per genome)[d]	Repetitive sequence classes interspersed with single-copy DNA at fragment lengths < 4000 NT[e]	Fraction of repetitive DNA in 200 to 400 NT sequence elements	Fraction of single copy DNA interspersed at fragment lengths < 4000 NT	Minimum genomic fraction period interspersion pattern	Ref.
Nicotiana tabacum (tobacco)	1.65 pg 1.5 × 10^9 NTP[f]	0.45	6.2 × 10^8 NTP	250 15,000	Yes ?	0.35	>0.80	0.55	14
Glycine max (soybean)	1.97 pg 1.8 × 10^9 NTP	0.39	7.0 × 10^8 NTP	19 2,800 290,000	Yes No No	0.11	>0.70	0.40	18
Pisum sativum (pea)	0.42 pg 3.9 × 10^8 NTP	0.45	1.8 × 10^8 NTP	90 3,600	Yes ?	0.55	>0.90	0.70	15—17
Gossypium hirsutum (cotton)	0.80 pg 7.3 × 10^8 NTP	0.68	5.0 × 10^8 NTP	130 125,000	Yes ?	0.80[g]	>0.80	0.60	19
Secale cereale (rye)	7.4 pg 6.8 × 10^9 NTP	0.30	2.0 × 10^9 NTP	850 38,000 600,000	? ? ?	?	>0.80	?	22,60
Triticum aestivum (wheat)	5.7 pg 5.1 × 10^9 NTP	0.25	1.4 × 10^9 NTP	250 20,000	? ?	0.80[h]	>0.65	0.75	21,61
Vicia faba (broad bean)	48 pg 4.4 × 10^10 NTP	0.20	9.7 × 10^9 NTP	30 3,600 110,000	? Yes ?	0.75	>0.65	0.80	20
Achlya ambisexualis[i] (water mold)	0.04 pg 3.7 × 10^7 NTP	0.85	3.1 × 10^7 NTP	70 2,200	No No	<0.05	<0.03[i]	None[i]	13

[a]   Constructed similar to that of Davidson et al.[7] to enable a direct comparison to be made of plant and animal DNA sequence organization.
[b]   Kinetic estimate of genome size. Due to uncertainties in evaluating the ploidy of plant cells, this is the most reliable estimate of plant genome size.
[c]   (Genome size) × (fraction of single-copy DNA); fraction of single-copy DNA estimated from S-1 nuclease or optical reassociation data.
[d]   Refers to average number of copies of each repetitive sequence class. More repetition classes may be present in some species, but due to methods employed they were not detected. It is important to note that each class contains a range of repetition frequencies, the average of which is indicated.
[e]   NT refers to single-strand fragment length in nucleotides.
[f]   Refers to nucleotide pairs.
[g]   Average size of repetitive sequences is 1250 NTP.
[h]   Range of repetitive sequence lengths is 400 to 800 NTP.
[i]   Oomycete fungus; included for comparative purposes.
[j]   No short-period interspersion is found in Achlya. Approximately 3% of single-copy sequences are contiguous to repetitive DNA at a 5000 NT fragment length. Hence, Achlya has a very long period interspersion pattern.

period interspersion pattern has been found in all plant species investigated to date and as such has been phylogenetically conserved across kingdom lines. This implies that short-period interspersion evolved early in eukaryotic evolution and that it performs a fundamental, but presently unknown, role in eukaryotic genomic activity.

The major results extracted from plant DNA sequence organization studies are presented in Table 6. Also included in this table are data regarding the organization of DNA in the chromosomes of the water mold *Achlya*. These measurements reveal the striking contrast between higher plant and fungal DNA sequence organization. Some general features of higher plant DNA sequence arrangement are briefly discussed below.

a. Repetitive DNA Content

Higher plant genomes contain a large percentage of repetitive DNA. For example, 60% of the DNA in soybean chromosomes is repeated and 75% is repeated in wheat. These quantities are generally higher than those found in the chromosomes of animals with similar genome sizes.[7,8] Why plant cells require a large quantity of DNA sequences which are repeated remains a mystery. Regardless of their functional role, however, the genomic content of reiterated DNA appears to be a major feature distinguishing plant and animal genomes.

b. Genomic Complexity

The genomic or single-copy complexity of higher plant DNA varies over almost two orders of magnitude. The single-copy complexity of two legumes, pea and broad bean, is 1.8×10^8 and 9.7×10^9 NTP, respectively. Assuming the average size of single-copy DNA to be 1500 NTP, this is enough potential information for 1.5×10^5 and 6.5×10^6 diverse structural genes. Since only a small fraction of the single-copy DNA codes for messenger RNA (see below), we are confronted with two problems. First, why do higher plants with virtually the same level of biological complexity (e.g., two legume species) have vastly different genomic complexities? Second, what biological role (if any) does the vast majority of single-copy sequences have in plant cells? A similar paradox exists for the single-copy DNA in animal chromosomes.[7,8] Thus, as Davidson et al. point out, single-copy complexity appears to be a measure of potential and not actual amount of genetic information available to an organism.[7]

c. Repetitive DNA Sequence Classes

The chromosomes of all higher plants studied have a very heterogeneous spectrum of DNA repetition classes. Unfortunately, not enough kinetic fractionation experiments have been performed to accurately assess the extent of repetitive DNA sequence class heterogeneity. All plant genomes investigated appear, however, to have a similar range of DNA repetition classes. Generally middle repetitive DNA, either low frequency repetitive DNA (10 to 100 copies per sequence), moderately repetitive DNA (100 to 1000 copies per sequence), or both are the dominant repetitive classes in plant genomes. Some plants have more highly repetitive DNA sequences (> 10,000 copies per sequence), but these usually represent a minor fraction of the total DNA.

The low frequency or moderately repetitive DNA classes contribute sequences which are interspersed with single-copy DNA. Generally only a fraction of these classes are represented in the short-period interspersion pattern, the remainder being organized as long repetitive sequences not contiguous to single-copy DNA. The interspersion of DNA sequences from different repetition classes (e.g., moderate and low frequency) remains an unresolved and important question. Detailed experiments with soybean DNA failed to reveal repetitive sequence class interspersion (see Figure 7). However, indirect evidence in a number of plant genomes including tobacco,[14] pea,[15,17] and mung bean[17] suggests that this may be a possibility. Because of the high proportion of repetitive DNA, a resolution of the fine scale arrangement of repetitive DNA classes may provide clues for the functional role of repeated DNA in plant cells.

d. Size of Repetitive DNA Sequences

The length of individual repetitive DNA sequences in plant chromosomes is heterogeneous. Both long (> 1500 NTP) and short (200 to 400 NTP) repetitive sequences are present in plant chromosomes. In most plant genomes, the short repetitive sequences are interspersed with single-copy DNA in the short-period interspersion pattern. However, the size range of interspersed repeats is generally greater than that found in animal chromosomes (see Reference 19). Because of limitations of DNA size, the range of long repetitive DNA sequence lengths cannot be determined. However, recent experiments involving electron microscopic examination of renatured repetitive DNA from soybean chromosomes reveal sequences at least 15,000 NTP in length.[56]

Long repetitive DNA sequences constitute a major fraction of the repeated DNA in plant chromosomes (25 to 80%). Generally, these sequences are in the majority. The quantity of long repetitive DNA sequences in plant chromosomes usually exceeds the amount present in animal chromosomes. Since the quantity of short repetitive DNA is sufficient to enable each single-copy sequence to be contiguous to an interspersed repeat, the long repetitive sequences probably represent the additional repeated DNA in plant chromosomes.

The number of families of short interspersed repeated DNA sequences can be computed from a knowledge of the genome size, genomic fraction of interspersed repeats, average reiteration frequency, and average length in the chromosomes. These calculations indicate that there are approximately 1000 famiies in cotton,[19] 4000 in tobacco,[14] and 40,000 in soybean.[18] Clearly, plant chromosomes contain a large number of diverse repetitive sequences, each repeated an average of 20 to 300 times and all contiguous to single-copy DNA. What functional role these families of repetitive sequences perform in plant cells remains a major, unresolved question.

e. Single-copy DNA

Most higher plant genomes studied contain less than 50% single-copy DNA. Some, like broad bean and wheat, have only minor fractions of single-copy DNA. Despite the small genomic quantity, however, the number of individual single-copy sequences greatly exceeds the number of estimated structural genes (see below).

In all plant species examined, over 70% of the single-copy sequences are 4000 nucleotides in length or less and terminate in an interspersed repetitive element. While small amounts of single-copy DNA may be interspersed at longer genomic distances or not interspersed at all, the vast majority of higher plant single-copy sequences are contained within the short period interspersion pattern. In most plant chromosomes, the average size of single-copy DNA appears to be 1000 to 2000 NTP, in close agreement with the average size of messenger RNA in plant cells.

Up to the present time, no higher plant genome has been shown to be organized in a long-period interspersion pattern in which mostsingle-copy sequences are 10,000 nucleotides in length or longer. This is probably due to the fact that all plant genomes investigated have relatively large amounts of DNA (> 0.5 pg). It is reasonable to predict that a plant with a small genome size (0.1 to 0.2 pg) will have mostly long single-copy DNA sequences and a long-period interspersion pattern. For example, the DNA of the water mold *Achlya,* a representative of the fungal kingdom which has very small genomes (< 0.1 pg), is arranged in a very long interspersion pattern.[31]

2. Sequence Divergence in Families of Repeated DNA

The arrangement of long and short repeated sequences in plant chromosomes suggests a fundamental difference in their function and evolutionary history. The organization of short repetitive sequences is consistent with the hypothesis that interspersed repetitive DNA may regulate contiguous structural genes.[5,6] On the other hand, long repetitive sequences (which are highly complex) have organizational properties analogous to those of multigene families (ribosomal RNA genes, histone genes, etc.). In addition, short repetitive sequences could conceivably be derived from long repeats by chromosomal translocation and evolutionary divergence.[3,62-64]

The divergence of sequences contained within families of long and short repeated DNA has been examined in three plant species. These measurements were made by annealing long DNA to a Cot at which only repetitive sequences react, digesting unreassociated DNA with S-1 nuclease and fractionating the resistant duplexes by agarose gel filtration. Fractions containing repetitive duplexes of different length were then melted in the spectrophotometer in order to measure the precision of base pairing. The results of one such experiment with soybean DNA are shown in Figure 10, and these data are compared to similar measurements with tobacco and pea DNA (Table 7).

In all plants examined, reassociated long repetitive duplexes have a thermal stability close to that of native DNA indicating relatively little sequence divergence. A strikingly different result is obtained with the short repetitive duplexes. These molecules are highly mismatched, indicating that a high degree of sequence divergence occurs within families of short interspersed repetitive DNA. This correlation between organizational

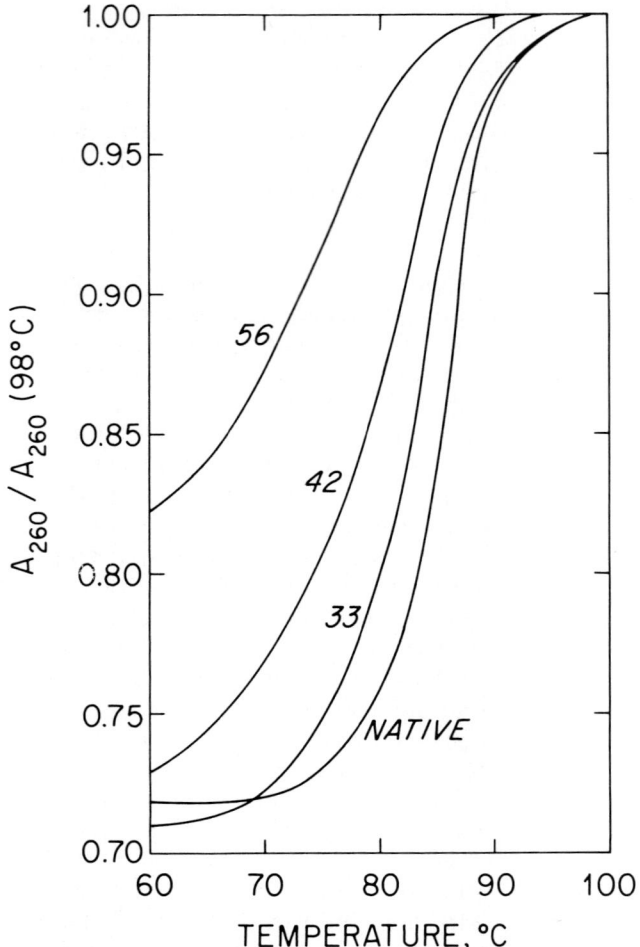

FIGURE 10.   Thermal denaturation profiles of Cot 100 S-1-resistant duplexes. Specific fractions of the Agarose A-50 eluate portrayed in Figure 8c were melted in the spectrophotometer. The number designates the specific elution fraction melted. A melt of native soybean DNA is included as a reference. The observed T_m and hyperchromicity values measured were native DNA — 86°C, 0.282; Fraction 33 (exclusion peak, > 1500 nucleotide pairs) — 83°C, 0.289; Fraction 56 (inclusion peak, 250 nucleotide pairs) — 70°C, 0.232. After correcting for the effects of fragment length on T_m,[36] the T_m values are Fraction 33, 83.4°C; Fraction 42, 79.7°C; Fraction 56, 72.6°C. From the corrected T_m values it can be computed that the average percent base sequence divergence in each melted fraction is Fraction 33, 2.6%; Fraction 42, 6.3%; Fraction 56, 13.4%.

pattern (i.e., length of repetitive DNA) and degree of base sequence divergence has also been demonstrated in a number of diverse animal species and is probably a general property of eukaryotic repeated DNA.[52,62,64]

Since members of a family of long repeated DNA have almost 100% sequence homology, they could be recent additions to plant chromosomes (by an overreplication of an existing sequence), highly conserved evolutionarily, or both. At least one family of long repetitive DNA, the ribosomal RNA genes, has been shown to be highly con-

TABLE 7

Thermal Stability of Long and Short Repetitive Sequences in Several Plant Species

Plant	T_m[a]			% Divergence[b]	
	Native DNA	Long duplexes	Short duplexes	Long duplexes	Short duplexes
Tobacco	87.5	87.5	71.0	0	16.5
Soybean	86.0	83.0	73.0	3	13
Pea	87.5	85.0	78.5	2.5	9

[a] Measured in 0.12 M PB. Long duplexes were excluded from an Agarose A-50 column ($\geqslant$ 1500 NTP) while short duplexes were those included and had a modal size of 200 to 400 NTP. The short repetitive duplexes contain interspersed repeated sequences. See legend to Figure 10 for details.
[b] The average percent divergence was calculated by assuming that a 1°C reduction in native T_m, after correction for the effects of fragment length on T_m, equals 1% nucleotide sequence divergence.[36]

served in the plant kingdom and eukaryotes in general.[65] On the other hand, heterologous reassociation experiments with animal DNA suggest that precisely paired repetitive sequences are species specific and therefore of recent origin.[66] Whether short interspersed repeats are derived from long repetitive DNA has not yet been established. However, recent experiments with muskmelon DNA suggest that this may be the case.[67]

III. RELATIONSHIP OF GENOME ORGANIZATION TO PLANT GENETICS

A. How Many Genes Are There?

1. Eukaryotic Genome Size

A major constraint on models of eukaryotic genome organization is the requirement to account for the large size of eukaryotic genomes. As Sparrow et al.[68] have pointed out, the amount of DNA calculated as haploid DNA content in free-living organisms spans a range from 0.007 pg for an average bacterium to 100 pg or more for some plants and salamanders. Thus, the eukaryotes with the highest DNA content contain 10^5 times more DNA than most bacteria. The range of DNA content for major plant groups is shown in Table 8. The minimal amount of DNA for a particular phylum or class among the animal groups examined appears to increase with evolutionary complexity,[69] although within each division there are species which have a haploid DNA content considerably higher than the minimum. Within a major group such as a phylum or class, the distribution of DNA amount may approximate a logarithmic normal curve so that most species are concentrated at the lower end of the distribution.[70]

Not only does DNA content vary among plants but also the number and size of the chromosomes. In plants, there is a general correlation between DNA content and chromosome number so that ferns of high DNA content have chromosome numbers of several hundred and are certainly polyploid. Primitive and slowly evolving groups such as the psilopsids and ferns and the more thoroughly studied cases of salamanders[68] and the fishes often have a very high DNA content.[71,72]

The high DNA content of many eukaryotes is often summarized as the "C" value paradox. The paradox lies in the negative correlation between advances in organismal organization and DNA content. Although the minimal amount of DNA required to specify a particular taxon appears to increase with increasing levels of complexity as already discussed, it is clear that many organisms at a particular level of organization

TABLE 8

Lowest, Modal, and Highest Values for Nuclear or Cellular DNA Content in Major Groups of Microorganisms and Plants[a]

Group	Number spp.	Lowest value		Median value[b]		Highest value	
Bacteria	36	*Chlamydia trachomatis*	0.0012	*Staphylococcus aureus*	0.015	*Aerobacter aerogenes*	0.057
Cyanophyceae	3	*Anacystis nidulans*	0.012	(No other records)		*Oscillatoria princeps*	17.0
Fungi	10	*Saccharomyces cerevislae*	0.048	*Aspergillus nidulans*	0.088	*Ustilago maydis*	0.38
Unicellular algae	14	*Chlorella ellipsoidea*	0.080	*Euglena gracilis*	5.8	*Gonyaulax polyedra*	400.0[d]
Bryophyta	4	*Marchantia polymorpha*	0.9	*Sphagnum* sp.	2.6	*Riccia* sp.	8.6
				Mnium sp.	7.8		
Spore bearing vascular plants	16	*Selaginella kraussiana*	5.1	*Pteridium aquilinum*	44.0[d]	*Tmesipteris* sp.	630.0[d]
Gymnosperms	13	*Ephedra fragilis*	8.4	*Picea mariana*	68.0	*Pinus resinosa*	140.0
Angiosperms	36	*Arabidopsis thaliana*	2.0	*Clematis jackmannii*	28.0	*Lilium longiflorum*	180.0
						Sprekelia formosissima	350.0[d]

[a]   From Sparrow et al., 1972. Nucleotides of DNA in billions ($\times 10^9$).

[b]   Due to the small size of the samples, median values are more informative than mean or mode.

[c]   Since recorded values for angiosperms include a disproportionate number of species with large chromosomes, this value is probably higher than the actual mean or mode.

[d]   These species are highly polyploid.

From Stebbins, G. L., in *Evolutionary Biology*, 1976. With permission from Plenum Press.

have far more DNA than would seem to be required (e.g., compare DNA content of leguminous plants in Table 6). Stebbins[73] has attempted to analyze this paradox by correlating the intrageneric variation in DNA content with phylogenetic specialization. It might be hypothesized that more complex organisms or the organisms at the termini of phylogenetic lines might contain either considerably less DNA, since all "junk" DNA would be discarded, or they might contain more DNA required for complex regulation. Analysis of DNA content vs. phylogenetic specialization suggests both increases and decreases of DNA in the plant groups thus far examined.[74-78] On the other hand, most animal groups show a decreasing DNA content with specialization (see review by Hinegardner).[79]

The amount of DNA per nucleus is positively correlated with several aspects of cell life. There is a positive correlation between the length of the cell cycle and the nuclear DNA content in plants.[76,80] Increasing DNA content can be accompanied either by a longer S period or a longer G_1 period; typically, increases in either S or G_1 account for the increase in cell cycle of a high nuclear DNA content species compared to closely related low DNA content species. The contribution with regard to S is likely a function of the increased amount of time required to synthesize a larger nuclear genome during replication; however, since eukaryotic cell nuclear DNA is replicated simultaneously at many sites, there is no *a priori* reason for an absolute requirement for a longer S in order that more DNA can be replicated. The increase in G_1 period in some species may relate to another aspect of cell economy, namely that the ratio of nuclear volume and nuclear DNA content is well correlated with cell volume.[68] Cells, in general, maintain a fixed nuclear to cytoplasmic volume ratio. Division is initiated when cell volume exceeds a specific size; hence cells with a high nuclear DNA content will generally reach a larger size prior to division. Most cell growth is during G_1 of the cell cycle, and G_1 may be proportionately extended in species with a high nuclear DNA content.

From the data presented on genome size, it is not possible to draw many conclusions regarding the functional significance of extremely large genomes. Some plant groups have such an extraordinarily large DNA content that the only reasonable explanation may be a high degree of polyploidization in the absence of selection for a reduction in genome size. Perhaps it is more profitable to focus on the majority of plants with moderate 1 to 10-pg genome size. The number of potential genes in organisms of this level of genome complexity has been calculated both genetically and using DNA reassociation kinetics as described in the next section.

2. Calculation of Gene Number from Genetic Data, DNA Reassociation, and DNA/ RNA Hybridization Kinetics

Calculation of the exact number of genes in eukaryotes has been attempted using several kinds of evidence; this field has been recently reviewed by Bishop.[81] In bacteria such as *E. coli*, in which a substantial fraction of the genome has been genetically mapped and for which a precise genome size has been measured, it has recently been determined that there are approximately 2500 structural genes.[82] In eukaryotes the situation is considerably more complex since unique and repetitive DNA fractions exist, less precise estimates of genome size are available, and genome sizes are much larger.

Most of the current estimates of eukaryotic genome size using genetic evidence are based on studies in *Drosophila* involving three parameters: mutation rates, calculation of the number of (potential) complementation groups on individual chromosomes, and determination of the frequency of lethal mutations. These three lines of evidence point to a gene number of approximately 5000 in this species.[83] However, such estimates must be accepted with caution due to the difficulty of scoring events. For example, it is known that null alleles at some loci do not result in lethality. Consequently, the *Drosophila* genetic estimates should be regarded as minimal.

Another concept that has confused calculation of gene number is polygenes. Polygenes, a term coined by Mather in the 1940s, contribute to the quantitative variation of traits in contrast to Mendelian unit characters that form a discontinuous distribution. Polygenes are usually assumed to have individual small effects on macro traits such as body size, plant yield, photosynthetic efficiency, body shape, etc. Polygenes are analyzed by biometrics and are each assumed to consist of hundreds or possibly thousands of different genes. Thus, the large amounts of unique-copy DNA of eukaryotes might be required for these characters. This prevailing view of polygenes has been recently challenged by Thompson[84] who has reanalyzed the continuous distribution reported for various polygene traits. Thompson assumes that polygenes are only one or two Mendelian characters of variable heritability and variable "weight" affecting the phenotype. With this assumption, continuous variation is possible even with only two alleles at a single locus in some cases. Even in complex cases, it is usually possible to describe most of the natural variation in terms of a few genes with major responsibility for the determination of phenotype, while the large number of genes responsible for the general viability of the organism will have a minor cumulative impact on trait development. Thompson's in-depth analysis of wing vein development in *Drosophila* demonstrates that the major share of the variance in a very complex developmental pattern can be determined by only 2 to 3 loci and, furthermore, that each major contributive gene has a specific developmental and biochemical role in the construction of the pattern.

Thompson's analysis illustrates two important facts. First, all genes must be ultimately responsible for a specific gene product with a specific role in the organism. There is no gene for yield, for example, but many specific enzymes may contribute to a high yield. Second, macro traits of plants or animals will be affected to a small extent by the summation of each of the specific genes required for cell and organismal survival. These genes, called the "housekeeping genes", code for enzyme of basic cell processes such as the Krebs cycle, cell division, cell growth, maintenance of viable mitochondria and chloroplasts, etc. Different alleles at housekeeping gene loci will contribute in a variable fashion to the final size and growth rate of the organism. In fact, loci at control points in metabolic pathways may contribute to relatively large quantitative changes in organismal fitness.[85] However, these housekeeping genes are the same for all traits examined, although the relative contribution of a specific gene may change slightly if a different phenotypic character is examined; thus the polygenes constitute only a relatively small group of perhaps several thousand genes essential to the general health of the organism and are the same for each major trait of the organism.

The concept of housekeeping genes gains further support from DNA/RNA hybridization experiments using messenger RNA and either cDNA (DNA complementary to the mRNA population) or single-copy DNA.[86,87] These experiments unambiguously establish the number of structural genes expressed in a given cell type. By comparing the expression of structural genes in different tissues and developmental stages estimates can be made of the total amount of genetic material expressed into mRNA in an organism's entire life cycle.[88] In addition, knowing the number of diverse messenger RNAs which are shared by different tissues provides estimates of the number of structural genes that perform housekeeping functions.[88,89]

In all organisms and cell types investigated so far only a small fraction of the genome complexity is transcribed into messenger RNA. For example, in leaves of the tobacco plant only 5% of the single-copy DNA is transcribed into messenger RNA actively translated on leaf polysomes.[90] This amount of DNA is the equivalent of only 27,000 diverse structural gene transcripts. Surprisingly, during the entire embryonic development of the sea urchin only 35,000 diverse structural genes are transcribed and perhaps in the whole life cycle less than 50,000.[88] Of this number, approximately 1000 to 2000 are shared by different cell types, while the remainder are characteristic of a given cell type. These experiments indicate, therefore, that eukaryotes have many fewer genes than are potentially available to them in their chromosomal DNA and that some of these genes perform housekeeping functions. The main point, however, is that genome size and genomic complexity in no way provide estimates of the total number of genes in eukaryotic cells.

B. Polyploidization, Heterozygosity, and Evolution

Current theories in evolutionary genetics place a high premium on heterozygosity and the contribution of heterozygosity to overall fitness. The traditional view holds that increased chromosomal heterozygosity results in higher polymorphism at individual loci.[91,92] In this view the obligate structural heterozygotes, in which a restricted set of chromosomal segregation patterns are found, would tend to maintain the highest level of heterozygosity. This would be true, in particular, for genomes such as that of *Oenothera* in which the entire chromosome complement is involved.[93] In such cases, the entire genome is free from recombination. This interpretation, however, overlooks the important fact that heterozygosity is fixed when restricted segregation is imposed; consequently, heterozygosity will be maintained at a fixed level that may be high, low, or intermediate compared to systems with outcrossing and/or recombination.

Fixed heterozygosity is usually thought to be advantageous due to increased biochemical diversity and homeostatic ability in a variable environment.[85,94] Polymorphism is exceedingly high in plants and is greatest for proteins regulating flux through biochemical pathways as might be expected from the prediction that heterozygosity promotes biochemical homeostasis.[85]

Population genetics has examined primarily loci that exhibit Mendelian inheritance. These genes are present only once per haploid chromosome set; in plants known to be tetra- or polyploid, the dosage of each Mendelian factor is considered to be correspondingly increased. The Mendelian factors have been determined, for the most part, to be in the single-copy fraction of the genome. In practice, it is not routinely possible to distinguish a diploid from a tetraploid organism solely on the basis of the reassociation kinetics of the single-copy component, that is, it is usually not possible to confidently conclude that the presumed single-copy fraction is reassociating as if present twice or four times per cell. In a recently constructed autotetraploid, each single-copy sequence will, of course, actually be present four times per cell. However, over evolutionary time, autotetraploid unique sequences may diverge in nucleotide sequence and no longer behave as if present four times per genome. Many higher plant species are actually allotetraploids, the result of hybridization of closely related species. The single-copy sequences of such allotetraploids are typically present only once per haploid cell if the strictest definition of single copy is applied. However, during reassociation there may be considerable cross-hybridization between similar, nearly identical single-copy sequences from the two closely related progenitor species. Consequently, the reassociation behavior of the single-copy fraction may be intermediate between that expected for a diploid or tetraploid species.

Another way to examine the dosage of particular gene sequences is to survey the extent of heterozygosity at particular loci. Selander[95] has recently reviewed the literature on enzyme polymorphism in plants and animals. The composite picture for plants is that within a population 46% of the loci will exhibit polymorphism and any one individual will be heterozygous at 17% of loci examined. This extent of polymorphism presumably reflects the nucleotide divergence of formerly identical and highly similar sequences to sequences of less similarity. At some loci at which no heterozygosity can be detected at the protein level, there may still be a nucleotide "memory" of polyploidization or other forms of gene duplication. For example, in a recent provocative study, Ferris and Whitt[96] examined the level of heterozygosity of tetraploid Catostomidae, a group of fishes in which tetraploidization occurred approximately 50 million years ago. They found more than a 50% loss of duplicate gene expression at the 30 loci examined. The fate of the nonexpressed gene(s) is not known. Such genes could have evolved new functions, become silent, or are regulated so that they appear only at specific stages of the life cycle or in specific tissues not assayed in the investigation. Thus, although measurable protein heterozygosity is high, it does not necessarily reflect the actual content of similar but divergent nucleotide sequences in the genome. This conclusion makes it simpler to understand the observation of Allard and Kahler[97] that self-fertilizing plant species can maintain a high level of genetic variation over the population level; the variation can meet or even exceed that found in outcrossing species. The maintenance of variation at the protein level will be a reflection of the need for gene products in both self-compatible and incompatible species; selective silencing of loci in self-compatible and incompatible species is likely a random process. The genomes of each type would be expected to contain essentially identical levels of diversity in nucleotide sequence over short time ranges.

C. Prospects for Future Research

Over the past few years, a large amount of exciting and new information has been obtained regarding DNA sequence arrangement in plant chromosomes. This knowledge has been gained through a concerted effort by many laboratories and aided by technical innovations which have allowed experimentation with plant DNA that was not previously possible. We now understand the general manner in which genetic material is arranged in plant chromosomes and have obtained some insight concerning the number and molecular arrangement of plant genes.

However, we have only scratched the surface. Large gaps of knowledge exist concerning the arrangement of plant chromosomal DNA. In particular, we do not understand what significance patterns of genome arrangement have in relation to normal cellular processes. This is largely due to the lack of information about DNA expression in plant cells. To be sure, if a functional meaning is to be given to the pattern of sequence organization found in plant genomes, we must have information concerning which DNA sequences are transcribed in plant nuclei and the extent of differential transcription in plant development. Definitive studies on the arrangement and expression of specific gene sequences in plant chromosomes can then be more meaningfully performed with the goal of understanding the basic mechanisms controlling gene function in plant cells.

We believe that plant molecular geneticists are rapidly approaching the goal of comprehending in molecular terms classical genetic phenomena. Notable in this regard will be the unraveling of the molecular basis of controlling element function in higher plants. Clearly, this is one of the most genetically well defined and intriguing genetic systems in higher eukaryotes.[98,99]

IV. ADDENDUM

Since writing this chapter several papers have been published on eukaryotic DNA sequence organization.

1. Plants
    a. Soybean[18]
    b. Parsley[100]
    c. Pea[38]
    d. Mung bean[38]
2. Animals
    a. Rat[101]
    b. Pig[102]
    c. Human[103]
    d. Mouse[104]
    e. Crab[105]
3. Fungal
    a. *Achyla*[13]

We would also like to point out that eukaryotic DNA sequence organization studies are now being performed with cloned DNA fragments. Experiments using recombinant DNA technology will add a new dimension to our knowledge of how DNA sequences are arranged in plant chromosomes.

ACKNOWLEDGMENT

Research results reported here were supported in part by grants from the National Science Foundation to each author.

REFERENCES

1. **Sturtevant, A. H.,** *A History of Genetics,* Harper & Row, New York, 1965.
2. **Lewin, B. L.,** *Gene Expression, Vol. 2,* John Wiley & Sons, New York, 1975.
3. **Britten, R. J. and Kohne, D. E.,** Repeated sequences in DNA, *Science,* 161, 529, 1968.

4. **Lewin, B. L.,** *Gene Expression, Vol. 1,* John Wiley & Sons, New York, 1975.

5. **Britten, R. J. and Davidson, E. H.,** Gene regulation in higher cells — a theory, *Science,* 165, 349, 1969.

6. **Davidson, E. H., Klein, W. H., and Britten, R. J.,** Sequence organization in animal DNA and a speculation on hnRNA as a coordinate regulatory transcript, *Dev. Biol.,* 55, 69, 1977.

7. **Davidson, E. H., Galau, G. A., Angerer, R. C., and Britten, R. J.,** Comparative aspects of DNA sequence organization in metazoa, *Chromosoma,* 51, 253, 1975.

8. **Angerer, R. C. and Hough-Evans, B. R.,** Sequence organization of eukaryotic DNA, in *Hormone Action, Vol. 1,* O'Malley, B. and Birnbaumer, L., Eds., Academic Press, New York, 1977, chap. 1.

9. **Davidson, E. H., Hough, B. R., Amenson, C. S., and Britten, R. J.,** General interspersion of repetitive with nonrepetitive elements in the DNA of *Xenopus, J. Mol. Biol.,* 77, 1, 1973.

10. **Manning, J. E., Schmid, C. W., and Davidson, N.,** Interspersion of repetitive and non-repetitive DNA sequences in the *Drosphila melanogaster* genome, *Cell,* 4, 141, 1975.

11. **Crain, W. R., Eden, F. C., Pearson, W. R., Davidson, E. H., and Britten, R. J.,** Absence of short period interspersion of repetitive and non-repetitive sequences in the DNA of *Drosophila melanogaster, Chromosoma,* 56, 309, 1976.

12. **Crain, W. R., Davidson, E. H., and Britten, R. J.,** Contrasting patterns of DNA sequence arrangement in *Apis mellifera* (honeybee) and *Musca domestica* (housefly), *Chromosoma,* 59, 1, 1976.

13. **Hudspeth, M. E. S., Timberlake, W. E., and Goldberg, R. B.,** DNA sequence organization in the water mold, *Achlya, Proc. Natl. Acad. Sci. U.S.A.,* 74, 4332, 1977.

14. **Zimmerman, J. L. and Goldberg, R. B.,** DNA sequence organization in the genome of *Nicotiana tabacum, Chromosoma,* 59, 227, 1977.

15. **Goldberg, R. B.,** unpublished data, 1977.

16. **Thompson, W. F.,** Sequence organization in pea DNA, *Carnegie Inst. Washington Yearb.,* 75, 356, 1976.

17. **Murray, M. G., Cuellar, R. E., and Thompson, W F.,** Studies on DNA sequence organization, *Biochemistry,* 17, 5781, 1978.

18. **Goldberg, R. B.,** DNA sequence organization in the soybean plant, *Biochem. Genet.,* 16, 45, 1978.

19. **Walbot, V. and Dure, L. S.,** Developmental biochemistry of cotton seed embryogenesis and germination. VII. Characterization of the cotton genome, *J. Mol. Biol.,* 101, 503, 1976.

20. **Taylor, W. C. and Bendich, A. J.,** personal communication, 1977.

21. **Flavell, R. B. and Smith, D. B.,** Nucleotide sequence organization in the wheat genome, *Heredity,* 37, 231, 1976.

22. **Smith, D. B. and Flavell, R. B.,** Nucleotide sequence organization in the rye genome, *Biochim. Biophys. Acta,* 474, 82, 1977.

23. **Graham, D. E., Neufeld, B. R., Davidson, E. H., and Britten, R. J.,** Interspersion of repetitive and nonrepetitive DNA sequences in the sea urchin genome, *Cell,* 1, 127, 1974.

24. **Goldberg, R. B., Crain, W. R., Ruderman, J. V., Moore, G. P., Barnett, T. R., Higgens, R. C., Gelfand, R. A., Britten, R. J., and Davidson, E. H.,** DNA sequence organization in the genomes of five marine invertebrates, *Chromosoma,* 51, 225, 1975.

25. **Angerer, R. C., Davidson, E. H., and Britten, R. J.,** DNA sequence organization in the mollusc *Aplysia californica, Cell,* 6, 29, 1975.

26. **Bonner, J., Garrand, W. T., Holmes, D. S., Sevall, J. S., and Wilkes, M.,** Functional organization of the mammalian genome, *Cold Spring Harbor Symp. Quant. Biol.,* 38, 303, 1974.

27. **Schmid, C. W. and Deininger, P. L.,** Sequence organization of the human genome, *Cell,* 6, 345, 1975.

28. **Efstratiadis, A., Crain, W. R., Britten, R. J., Davidson, E. H., and Kafatos, F. C.,** DNA sequence organization in the lepidopteran *Antheraea pernyi, Proc. Natl. Acad. Sci. U.S.A.,* 73, 2289, 1976.

29. **French, C. and Manning, J. E.,** personal communication, 1977.

30. **Wells, R., Royer, H. D., and Hollenberg, C. P.,** Non *Xenopus*-like DNA sequence organization in the *Chironomus tentans* genome, *Mol. Gen. Genet.,* 147, 45, 1976.

31. **Firtel, R. A. and Kindle, K.,** Structural organization of the genome of the cellular slime mold *Dictyostelium discoideum:* Interspersion of repetitive and single copy DNA, *Cell,* 5, 401, 1975.

32. **Marmur, J.,** A procedure for the isolation of DNA from microorganisms, *J. Mol. Biol.,* 3, 208,

33. **Katterman, F.,** Purification of DNA by cetyltrimethylammonium bromide as a potential source of error for the determination of 5-methylcytosine residues in DNA, *Anal. Biochem.,* 63, 156, 1975.

34. **Darby, G. K., Jones, A. S., Kennedy, J. F., and Walker, R. T.,** Isolation and analysis of the nucleic acids and polysaccharides from *Clostridium welchii, J. Bacteriol.,* 103, 159, 1970.

35. **Britten, R. J., Pavich, M., and Smith, J.,** A new method for DNA purification, *Carnegie Inst. Washington Yearb.,* 68, 400, 1970.

36. **Britten, R. J., Graham, D. E., and Neufeld, B. R.,** An analysis of repeating DNA sequences by reassociation, in *Methods in Enzymology,* Vol. 29E, Grossman, L. and Moldave, K., Eds., Academic Press, New York, 1974, 363.

37. **Melchoir, W. B. and Von Hippel, P. H.,** Alteration of the relative stability of dA · dT and dG · dC base pairs in DNA, *Proc. Natl. Acad. Sci. U.S.A.,* 70, 298, 1973.

38. **Murray, M. G., Preisler, R. S., and Thompson, W. F.,** Contaminants affecting plant DNA reassociation, *Carnegie Inst. Washington Yearb.,* 1976, 240.

39. **Studier, F. W.,** Sedimentation studies of the size and shape of DNA, *J. Mol. Biol.,* 11, 373, 1965.

40. **Davis, R. W., Simon, M., and Davidson, N.,** Electron microscopic heteroduplex methods for mapping regions of base sequence homology in nucleic acids, in *Methods in Enzymology,* Vol. 21 (Part D), Grossman, L. and Moldave, K., Eds., Academic Press, New York, 1971, 413.

41. **Wu, M. and Davidson, N.,** Use of gene 32 protein staining of single-strand polynucleotides for gene mapping by electron microscopy: application of the ϕ8Od$_3$ ilvsu$^+$7 system, *Proc. Natl. Acad. Sci. U.S.A.,* 72, 4506, 1975.

42. **Britten, R. J. and Kohne, D. E.,** Nucleotide sequence repetition in DNA, *Carnegie Inst. Washington Yearb.,* 65, 78, 1967.

43. **Britten, R. J. and Kohne, D. E.,** Repeated nucleotide sequences, *Carnegie Inst. Washington Yearb.,* 66, 73, 1967.

44. **Wetmur, J. G. and Davidson, N.,** Kinetics of renaturation of DNA, *J. Mol. Biol.,* 31, 349, 1968.

45. **Smith, M. J., Britten, R. J., and Davidson, E. H.,** Studies on nucleic acid reassociation kinetics: reactivity of single-stranded tails in DNA-DNA renaturation, *Proc. Natl. Acad. Sci. U.S.A.,* 72, 4805, 1975.

46. **Britten, R. J. and Davidson, E. H.,** Studies on nucleic acid reassociation kinetics: empirical equations describing DNA reassociation, *Proc. Natl. Acad. Sci. U.S.A.,* 73, 415, 1976.

47. **Davidson, E. H.,** *Gene Activity in Early Development,* Academic Press, New York, 1977, chap. 6.

48. **Wetmur, J. G.,** Hybridization and renaturation kinetics of nucleic acids, *Annu. Rev. Biophys. Bioeng.,* 5, 337, 1976.

49. **Bendich, A. J. and McCarthy, B. J.,** DNA comparisons among barley, oats, rye, and wheat, *Genetics,* 65, 545, 1970.

50. **Flavell, R. B., Bennett, M. D., Smith, J. B., and Smith, D. B.,** Genome size and proportion of repeated nucleotide sequence DNA in plants, *Biochem. Genet.,* 4, 257, 1974.

51. **Scheller, R. H., Thomas, T. L., Lee, A. S., Klein, W. H., Niles, W. D., Britten, R. J., and Davidson, E. H.,** Clones of individual repetitive sequences from sea urchin DNA constructed with synthetic Eco RI sites, *Science,* 196, 197, 1977.

52. **Britten, R. J., Graham, D. E., Eden, F. C., Painchaud, D. M., and Davidson, E. H.,** Evolutionary divergence and length of repetitive sequences in sea urchin DNA, *J. Mol. Evol.,* 9, 1, 1976.

53. **Britten, R. J. and Smith, J.,** A bovine genome, *Carnegie Inst. Washington Yearb.,* 68, 378, 1970.

54. **Chamberlin, M. E., Britten, R. J., and Davidson, E. H.,** Sequence organization in *Xenopus* DNA studied by the electron microscope, *J. Mol. Biol.,* 96, 317, 1975.

55. **Schmid, C. W., Manning, J. E., and Davidson, N.,** Inverted repeat sequences in the *Drosophila* genome, *Cell,* 5, 159, 1975.

56. **Pelligrini, M. and Goldberg, R. B.,** submitted for publication, 1979.

57. **Pelligrini, M., Manning, J., and Davidson, N.,** Sequence arrangement of the rDNA of *Drosophila melanogaster, Cell,* 10, 213, 1977.

58. **Goldberg, R. B.,** unpublished data, 1977.

59. **Key, J. L. and Siflow, C.,** The occurrence and distribution of poly(A) RNA in soybean, *Plant Physiol.,* 56, 364, 1975.

60. **Ranjekar, P. K., Lafontaine, J. G., and Palotta, D.,** Characterization of repetitive DNA in rye (*Secale cereale*), *Chromosoma,* 48, 427, 1974.

61. **Smith, D. B. and Flavell, R. A.,** Characterization of the wheat genome by renaturation kinetics, *Chromosoma,* 50, 223, 1975.

62. **Britten, R. J. and Davidson, E. H.,** DNA sequence arrangement and preliminary evidence on its evolution, *Fed. Proc. Fed. Am. Soc. Exp. Biol.,* 35, 2151, 1976.

63. **Eden, F. C., Graham, D. E., Davidson, E. H., and Britten, R. J.,** Exploration of long and short repetitive sequence relationships in the sea urchin genome, *Nucleic Acids Res.,* 5, 1553, 1977.

64. **Galau, G. A., Chamberlin, M. E., Hough, B. R., Britten, R. J., and Davidson, E. H.,** Evolution of repetitive and nonrepetitive DNA, in *Molecular Evolution,* Ayala, F. J., Ed., Sinauer Associates, Sunderland, Mass., 1976, chap. 12.

65. **Matsuda, K. and Siegel, A.,** Hybridization of plant ribosomal RNA to DNA: The isolation of an RNA component rich in ribosomal RNA cistrons, *Proc. Natl. Acad. Sci. U.S.A.,* 58, 673, 1967.

66. **Rice, N.,** Thermal stability of reassociated repeated DNA from rodents, *Carnegie Inst. Washington Yearb.,* 69, 472, 1970.

67. **Bendich, A. J. and Taylor, W. C.,** Sequence arrangement in satellite DNA from the muskmelon, *Plant Physiol.,* 59, 604, 1977.

68. **Sparrow, A. H., Price, H. J., and Underbrink, A. G.,** A survey of DNA content per cell and per chromosome of prokaryotic and eukaryotic organisms: Some evolutionary considerations, *Brookhaven Symp. Biol.,* 23, 451, 1974.

69. **Davidson, E. H. and Britten, R. J.,** Organization, transcription, and regulation in the animal genome, *Q. Rev. Biol.,* 48, 555, 1973.

70. **Bachmann, K., Goin, O. B., and Goin, C. J.,** Nuclear DNA amounts in vertebrates, *Brookhaven Symp. Biol.,* 23, 419, 1974.

71. **Pedersen, R. A.,** DNA content, ribosomal gene multiplicity, and cell size in fish, *J. Exp. Zool.,* 177, 65, 1971.

72. **Hinegardner, R. and Rosen, D. E.,** Cellular DNA content and the evolution of teleost fishes, *Am. Nat.,* 106, 621, 1972.

73. **Stebbins, G. L.,** Chromosome, DNA and plant evolution, in *Evolutionary Biology,* Vol. 9, Hecht, M. K., Steere, W. C., and Wallace, B., Eds., Plenum Press, New York, 1976, 1.

74. **Rees, H. and Jones, R. N.,** The origin of the wide species variation in nuclear DNA content, *Int. Rev. Cytol.,* 32, 53, 1972.

75. **Kadir, Z. B. A.,** DNA values in the genus *Phalaris* (Graminae), *Chromosoma,* 45, 379, 1974.

76. **Jones, R. N. and Brown, L. M.,** Chromosome evolution and DNA variation in *Crepis, Heredity,* 36, 91, 1976.

77. **Price, H. J.,** Evolution of DNA content in higher plants, *Bot. Rev.,* 42, 27, 1976.

78. **Bennett, M. D.,** Nuclear DNA content and minimum generation time in herbaceous plants, *Proc. R. Soc. London Ser. B,* 181, 109, 1972.

79. **Hinegardner, R.,** Evolution of genome sizes, in *Molecular Evolution,* Ayala, F. J., Ed., Sinauer Associates, Sunderland, Mass., 1976, chap. 11.

80. **Van't Hof, J.,** Relationships between mitotic cycle duration, S period duration and the average rate of DNA synthesis in the root meristem cells of several plants, *Exp. Cell Res.,* 39, 48, 1965.

81. **Bishop, J. O.,** The gene numbers game, *Cell,* 2, 81, 1974.

82. **Hahn, W. E., Pettijohn, D. E., and Van Ness, J.,** One strand equivalent of the *E. coli* genome is transcribed: Complexity and abundance classes of mRNA, *Science,* 197, 582, 1977.

83. **Judd, B. H., Shen, M. W., and Kaufman, T. C.,** The anatomy and function of a segment of the X chromosome of *Drosophila melanogaster, Genetics,* 71, 139, 1972.

84. **Thompson, J. N.,** Analysis of gene number and gene regulation in polygenic systems, in *Stadler Symp.,* Vol. 9, University of Missouri Press, Columbia, 9, 1977.

85. **Johnson, G. B.,** Enzyme polymorphism and metabolism, *Science,* 184, 28, 1974.

86. **Galau, G. A., Britten, R. J., and Davidson, E. H.,** A measurement of the sequence complexity of polysomal messenger RNA in sea urchin embryos, *Cell,* 2, 9, 1974.

87. **Bishop, J. O., Morton, J. G., Rosbash, M., and Richardson, M.,** Three abundance classes in HeLa cell messenger RNA, *Nature (London),* 230, 199, 1974.

88. **Galau, G. A., Klein, W. H., Davis, M. M., Wold, B. J., Britten, R. J., and Davidson, E. H.,** Structural gene sets active in embryos and adult tissues of the sea urchin, *Cell,* 7, 487, 1976.

89. **Hastie, N. D. and Bishop, J. O.,** The expression of three abundance classes of messenger RNA in mouse tissues, *Cell,* 9, 761, 1976.

90. **Goldberg, R. B., Hoschek, G., Kamalay, J. C., and Timberlake, W. C.,** *Cell,* 14, 123, 1978.

91. **Cleland, R. E.,** *Oenothera. Cytogenetics and Evolution,* Academic Press, New York, 1972.

92. **Darlington, C. D.,** *Evolution of Genetic Systems,* Oliver & Boyd, London, 1958.

93. **Levin, D. A.,** Genetic correlates of translocation heterozygosity in plants, *BioScience,* 25, 724, 1975.

94. **Fincham, J. S. F.,** Heterozygous advantage as a likely general mechanism for enzyme polymorphisms, *Heredity,* 28, 387, 1972.

95. **Selander, R. K.,** Genic Variation in Natural Populations, in *Molecular Evolution,* Ayala, F. J., Ed., Sinauer Associates, Sunderland, Mass., 1976, chap. 2.

96. **Ferris, S. D. and Whitt, G. S.,** Loss of duplicate gene expression after polyploidization, *Nature (London),* 265, 258, 1977.

97. **Allard, R. and Kahler, A.,** Allozyme polymorphisms in plant populations, *Stadler Symp.,* Vol. 3, University of Missouri Press, Columbia, 9, 1971.

98. **Peterson, P. A.,** Basis for diversity of controlling elements in maize, *Mol. Gen. Genet.,* 149, 5, 1976.

99. **McClintock, B.,** The states of a gene locus in maize, *Carnegie Inst. Washington Yearb.,* 66, 20, 1968.

100. **Kiper, M. and Herzfeld, F.,** DNA sequence organization in the genome of *Petroselinum sativum (Umbelliferae), Chromosome,* 65, 335, 1978.

101a. **Pearson, W. R., Wu, J. R., and Bonner, J.,** Analysis of rat repetitive DNA sequences, *Biochemistry,* 17, 51,78.

101b. **Wilkes, M. M., Pearson, W. R., Wu, J. R., and Bonner, J.,** Sequence organization of the rat genome by electron microscopy, *Biochemistry,* 17, 60, 1978.

102. **Arvedimento, U. E., Aquaviva, A. M., and Varrone, S.,** Sequence organization of porcine DNA, *Nucleic Acids Res.,* 3, No. 10, 2491, 1976.

103. **Deininger, P. L. and Schmid, C. W.,** An electron microscope study of the DNA sequence organization of the human genome, *J. Mol. Biol.,* 106, 773, 1976.

104. **Ginelli, E., DiLernia, R., and Corneo, G.,** The organization of DNA sequences in the mouse genome, *Chromosoma,* 61, 215, 1977.

105. **Holland, C. A. and Skinner, D. S.,** The organization of the main component DNA of a crustacean genome with a paucity of middle repetitive sequences, *Chromosoma,* 63, 233, 1977.

STRUCTURE AND REPLICATION OF CHLOROPLAST DNA

K. K. Tewari

TABLE OF CONTENTS

I. INTRODUCTION

The idea that extranuclear cell organelles could contain some of their own genetic information originated from the experiments of Bauer[1] and Correns.[2] Bauer[1] found that reciprocal crosses between the green and white germ layers of *Pelargonium zonale* gave rise to green, variegated, and white offspring. These results were in contrast to those expected from Mendelian inheritance, in which one would have obtained a uniform F_1 generation. Interestingly, the green F_1 plants and the green shoots of variegated plants gave no F_2 segregation, demonstrating that they were true breeding. Furthermore, Bauer observed that when the white germ layers of variegated plants were selfed with the white shoots of other variegated plants, the few seedlings that were produced from such a cross were white and died soon after germination. The data showed that the green and white plastid characteristics were both true breeding and formed a mixture of the two plastid types when crossed. At the same time, Correns[2] was studying another variegated plant, *Mirabilis jalapa*. It was found that green shoots of variegated plants produced green offspring and white shoots gave rise to white offspring. Furthermore, the type of offspring obtained from crossing was exactly the same as from selfing, for it was entirely dependent upon the color of the female shoot. The pollen from different colored shoots had absolutely no effect. These two examples illustrate the two types of non-Mendelian inheritance, biparental and uniparental-maternal, respectively. Since these initial discoveries, non-Mendelian inheritance of plastids and mitochondria has been recognized in other higher plants, *Chlamydomonas reinhardi*, yeast, *Neurospora*, and higher animals, and has been the subject of intensive studies and modern symposia.[3-14]

In 1968, 60 years after the original experiments of Bauer and Correns, the genetic autonomy of chloroplasts could be understood in molecular terms with the conclusive demonstration of the presence of DNA in chloroplasts. The last 10 years have been the most explosive years in the understanding of the molecular biology of chloroplasts and the combined efforts of biologists, geneticists, biochemists, and molecular biologists have brought chloroplasts into the forefront of biological studies. This review is restricted to only those studies which pertain to the structure, replication, and transcription of chloroplast DNA.

II. PHYSICOCHEMICAL PROPERTIES OF CHLOROPLAST DNA

A. Isolation of Chloroplast DNA

Chloroplast (ctDNA) has been isolated from a number of organisms. However, extensive studies on ctDNA have been carried out only in *Euglena gracilis, Chlamydomonas reinhardi,* and in certain higher plants. Therefore, these organisms will form the basis of this review. A detailed description of a general method of isolating ctDNA is given below. This method has been successful in isolating ctDNA from practically all higher plants.[15]

In a typical isolation, 1 kg of leaves is homogenized in 4 ℓ of ice-cold Buffer A containing 0.3 M mannitol/0.05 M Tris/0.003 M ethylenediaminetetraacetic acid (EDTA)/0.001 M mercaptoethanol/0.1% bovine serum albumin, pH 8.0, with two 5 s bursts in a Waring® blender at medium power. The homogenates are filtered through two layers of cheesecloth and four layers of Miracloth® (Calbiochem) and are centrifuged for 10 min at 40 × g at 4°C to remove nuclei. The supernatant is centrifuged at 1020 × g for 15 min at 4°C and the resulting crude chloroplast pellet is suspended in 200 mℓ of Buffer A. MgCl$_2$ (0.01 M) and DNAase I (50 μg/mℓ) are added and the suspension is incubated for 1 hr at 4°C. At the end of the incubation, 600 mℓ of Buffer B containing 0.3 M sucrose/0.05 M Tris/0.02 M EDTA, pH 8.0, is added and the suspension is centrifuged at 1500 × g for 15 min at 4°C. The pellet is washed twice by suspending it in 600 mℓ of Buffer B and centrifuging for 15 min at 1500 × g. The final pellet is suspended in 48 mℓ of Buffer C containing 0.05 M Tris/0.02 M EDTA, pH 8.0. Pronase (200 μg/mℓ) and 12 mℓ of Buffer C containing 10% sodium sarkosyl is added and the suspension is incubated for ½ hr at 37°C. The ctDNA is isolated from this lysate by extracting with an equal volume of phenol buffered with 0.1 M Tris, pH 12.0. The aqueous phase is reextracted with an equal volume of phenol. Two volumes of 95% alcohol are added to the aqueous phase, the mixture is kept in a freezer overnight, and the precipitate is collected by centrifugation and dissolved in 5 mℓ of Buffer C. The solution is incubated for 2 hr at 37°C with 50 μg/mℓ of RNase and 50 units/mℓ of RNase T$_1$ followed with 200 μg/mℓ of Pronase and a further incubation of 2 hr. At the end of incubation, two phenol extractions are carried out and the aqueous phase is dialyzed against a total volume of 10 ℓ of SSC with five changes in 24 hr.

The method described above yields pure ctDNA from higher plants uncontaminated with nuclear and mitochondrial DNA. The success of the method depends upon the fact that higher plant chloroplasts can be obtained in relative intact condition, therefore, it is possible to remove contaminating nuclear DNA by treating with DNase. The contaminating mitochondria are quantitatively removed during washing with Buffer B. However, the ctDNA from *Euglena* and *Chlamydomonas* cannot be obtained by using DNase because the physical methods used to obtain a cell-free homogenate from these organisms invariably result in broken chloroplasts. The isolation of ctDNA from these organisms is, therefore, carried out by isolating the DNA from cell fractions enriched for chloroplasts by differential or gradient centrifugations (without the DNase treatment) and separating the ctDNA in CsCl density gradients.

B. Buoyant Density of Chloroplast DNA

DNA isolated from chloroplasts of *Chlamydomonas* has been found to have a density of 1.695 g/cm^{-3} in CsCl compared to a density of 1.723 g/cm^{-3} for the nuclear DNA.[16,17] In the case of *Euglena,* ctDNA has been found to have a density of 1.685 g/cm^{-3} compared to 1.707 g/cm^{-3} for the nuclear DNA (nDNA). The large differences in the densities between the nDNA and ctDNA of *Euglena* and *Chlamydomonas* have been utilized to separate ctDNA from nDNA by preparative density gradient centrifu-

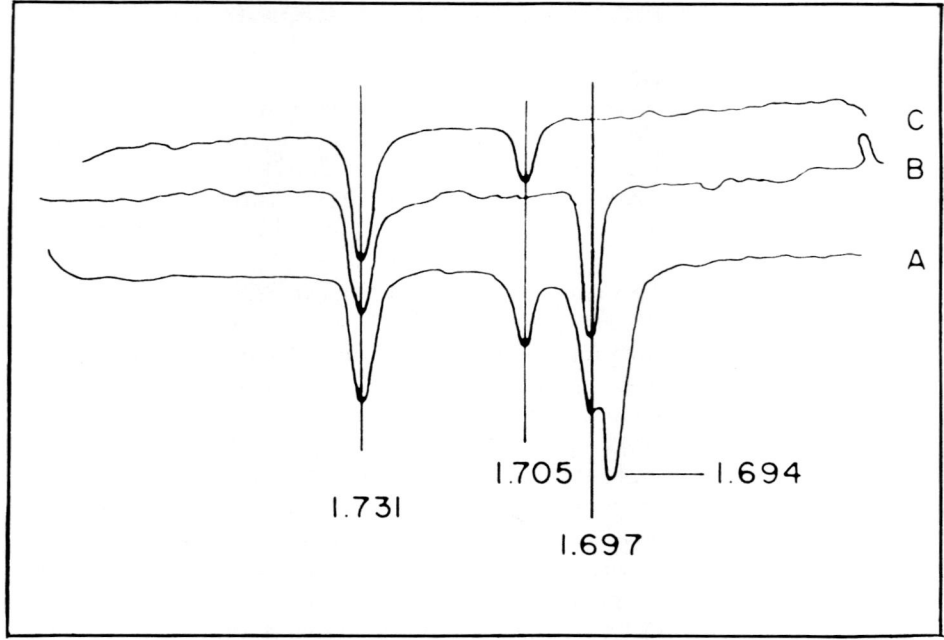

1.731   1.705   1.697   1.694

FIGURE 1.   Photoelectric scans of DNA centrifuged in CsCl of density 1.710 g/cm⁻³ for 18 hr at 28°
using a Beckman ® model E analytical ultracentrifuge equipped with photoelectric scanner. *Micrococcus
lysodeikticus* DNA of density 1.731 g/cm⁻³ was used as a marker. (A) DNA isolated from 12,000 × g
pellet containing chloroplasts and mitochondria. The picture shows a significant amount of nuclear contamination (DNA of density 1.694 g/cm⁻³). Although the difference in densities between nuclear and
chloroplast DNA is only 0.003 g/cm⁻³, the banding pattern clearly shows two components. (B) DNA
from chloroplast fraction after DNase treatment. (C) DNA from mitochondrial fraction after DNase
treatment.

gation. The buoyant density of the ctDNA of higher plants has been found to be very
close to that of nDNA. The density profile of the 12,000 × g pellet of a cell free homogenate of pea leaves (after removing the nuclear fraction by centrifuging at 100 × g
for 10′) without DNase treatment is shown in Figure 1. There are three DNA bands in
such a fraction. The major band at a density of 1.694 g/cm⁻³ represents the nDNA.
The shoulder of the nDNA at a density of about 1.697 g/cm⁻³ represents the ctDNA
(Figure 1A). The band at a density of 1.705 g/cm⁻³ arises from mitochondria. Thus,
the difference in densities between nDNA and ctDNA is only 0.003 g/cm⁻³. Kolodner
and Tewari,[15] and Tewari,[16] along with others, have examined the buoyant densities
of rDNAs and ctDNAs of a number of higher plants and shown them to differ from
each other by 0.003 to 0.004 g/cm⁻³. The data obtained from a number of such experiments are compiled in Table 1.
    Since the buoyant density of ctDNA from higher plants is very close to that of
nDNA, it is not always possible to ascertain the purity of ctDNA by analytical CsCl
gradient centrifugation. The most useful method for detecting the purity of ctDNA
from higher plants is to carry out an equilibrium CsCl gradient centrifugation on the
ctDNA which has been denatured and then renatured. For example, the ctDNAs from
pea, lettuce, spinach, beans, corn (Figure 2 (1a-e)), and oats (not shown) were found
to band at a buoyant density of 1.698 ± 0.001 gm/cm⁻³. On denaturation, the buoyant
densities of the different ctDNAs increased by 0.013 gm/cm⁻³ (Figure 2 (2a-e). Incubation of the denatured ctDNAs at 60°C for 4 hr at a concentration of 20 μg/mℓ in
0.15 *M* NaCl/0.015 *M* trisodium citrate, pH 7.0, resulted in about 90% renaturation,

as evidenced by their banding at a buoyant density of 1.699 ± 0.001 gm/cm^{-3} (Figure 2 (3a-e)). It is interesting to note that all of the different higher plant ctDNAs had practically the same buoyant density and this property was independent of whether the ctDNAs were native, denatured, or renatured. The buoyant densities of the nDNAs from pea, lettuce, spinach, and bean were 1.695 ± 0.001 gm/cm^{-3} and the nDNAs from corn and oats were 1.702 ± 0.001 gm/cm^{-3} (Table 1). These nDNAs increased in their buoyant densities by 0.013 gm/cm^{-3} on denaturation. Incubation of the denatured nDNAs at 60° C for 4 hr at 20 μg/mℓ in 0.15 M NaC1/0.015 M trisodium citrate, pH 7.0, only resulted in about 20% renaturation, which caused the denatured nDNA to decrease in density by 0.003 gm/cm^{-3}. Thus, if a denatured renatured ctDNA preparation is analyzed in CsC1 density gradient, the contaminated nDNA will band at a density of 1.708 g/cm^{-3} far removed from ctDNA which would band at a density of 1.699 g/cm^{-3}.

C. Thermal Denaturation of Chloroplast DNA

The thermal denaturation of the higher plant ctDNAs in 0.15 M NaCl/0.015 M trisodium citrate, pH 7.0, has been studied with T4 DNA included as a standard in all denaturation experiments. In Table 2, the T_m (transition midpoint), $\triangle$(ctT_m-T4T_m), h_{max} (maximal hyperchromicity), and the σ 2/3 (dispersion) of these ctDNAs are presented. Using T4 DNA as a standard, the ctDNAs were observed to have distinctly different T_m values. The most striking difference was between pea ctDNA, which had a T_m that was 1°C below the T_m of T4 DNA and corn ctDNA, whose T_m was 1.9° above the T_m of T4 DNA. These differences (see Table 2) in the T_m values of the

TABLE 1

Buoyant Densities of Higher Plant Chloroplast and Nuclear DNAs

Plant	Buoyant density (g cm^{-3})	
	Nuclear DNA	Chloroplast DNA
Pea	1.695	1.698
Spinach	1.694	1.697
Lettuce	1.694	1.697
Tobacco	1.695	1.697
Swiss chard	1.694	1.696
Broad and mung bean	1.695	1.697
Snapdragon	1.689	1.698
Onion	1.691	1.696
Wheat	1.702	1.697
Maize	1.702	1.697
Oat	1.701	1.697

TABLE 2

Thermal Denaturation Characteristics of DNA

DNA	T_m(°C)	T_m(ct)−T_m(T4)(°C)	h_{max}(%)	$\sigma2/3$(°C)
T4	83.8 ± 0.5	0	38	7.0
Pea	82.7 ± 0.7	-1 ± 0.05	36	9.3
Lettuce	85.1 ± 0.7	1.2 ± 0.04	37	8.3
Spinach	84.6 ± 0.5	0.7 ± 0.05	36	8.7
Maize	85.5 ± 0.7	1.9 ± 0.07	35	10

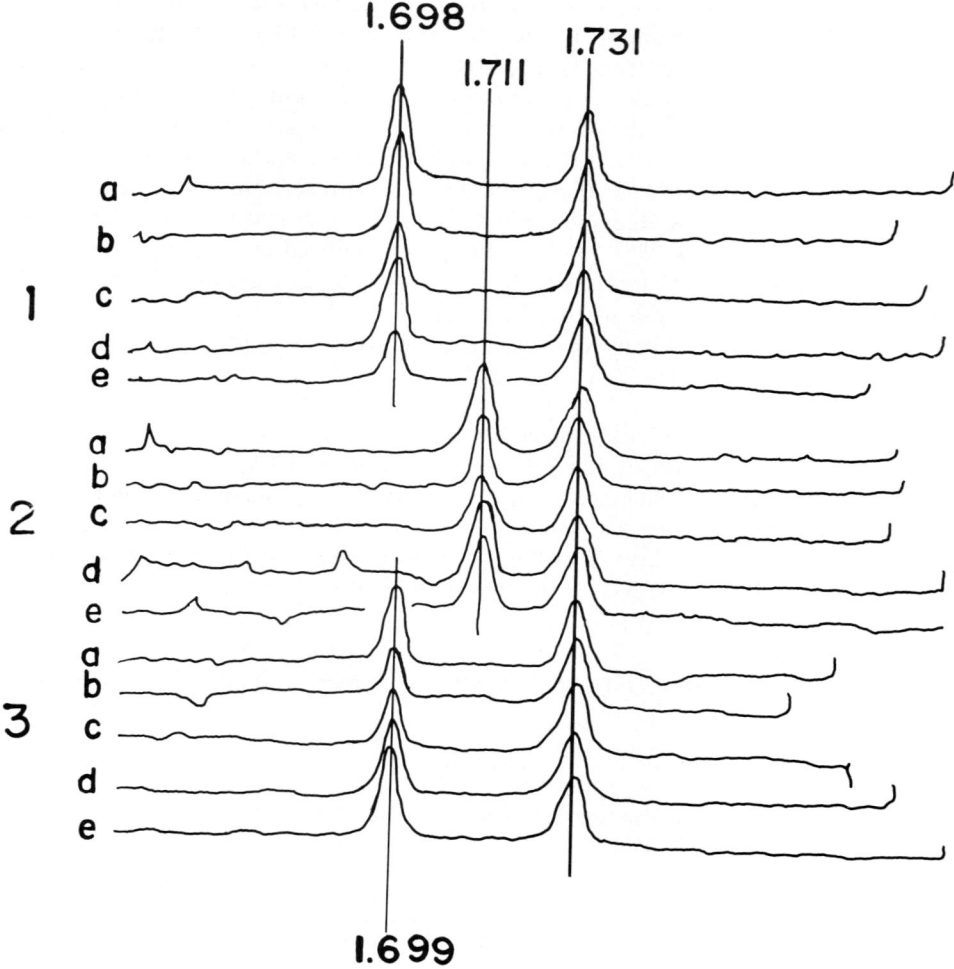

FIGURE 2.   Photoelectric scans of DNA centrifuged in CsCl of density 1.700 g/cm^{-3} for 18 hr at 18°C, 149,000 × g, in a Beckman® model E analytical ultracentrifuge. *Micrococus luteus* DNA of density 1.731 g/cm^{-3} has been used as a marker. (1) Native ctDNAs; (2) denatured ctDNAs; (3) renatured ct-DNAs. a, Pea; b, Lettuce; c, Spinach; d, Bean; e, Corn.

ctDNAs were independent of the rate of temperature rise used in these experiments; rates ranging from 1°C/min to 1°C/10 min were used.

In order to study the possible existence of heterogeneity in the melting patterns of the ctDNAs, their melting profiles were differentiated. The derivatives of the melting profiles are presented in Figure 3. The higher plant ctDNAs melted more broadly than T4 DNA. The pea, lettuce, and spinach ctDNAs melted as a single component, showing only one inflection point. The corn ctDNA melted much more broadly than the rest of the ctDNAs, which might suggest the presence of some heterogeneity. Similar experiments with the ctDNA from *Chlamydomonas* (Figure 4) showed it to melt in two components. The two DNA components were found to melt at 82.1 and 85.2°C, respectively. These values may be compared with the overall T_m of 83.5°C obtained from undifferentiated melting curves. The heterogeneity of the melting curve of *Euglena* ctDNA is the most pronounced, as seen in Figure 5. The *Euglena* ctDNA melted over a temperature range of 20°C and the melting profile clearly indicates the presence

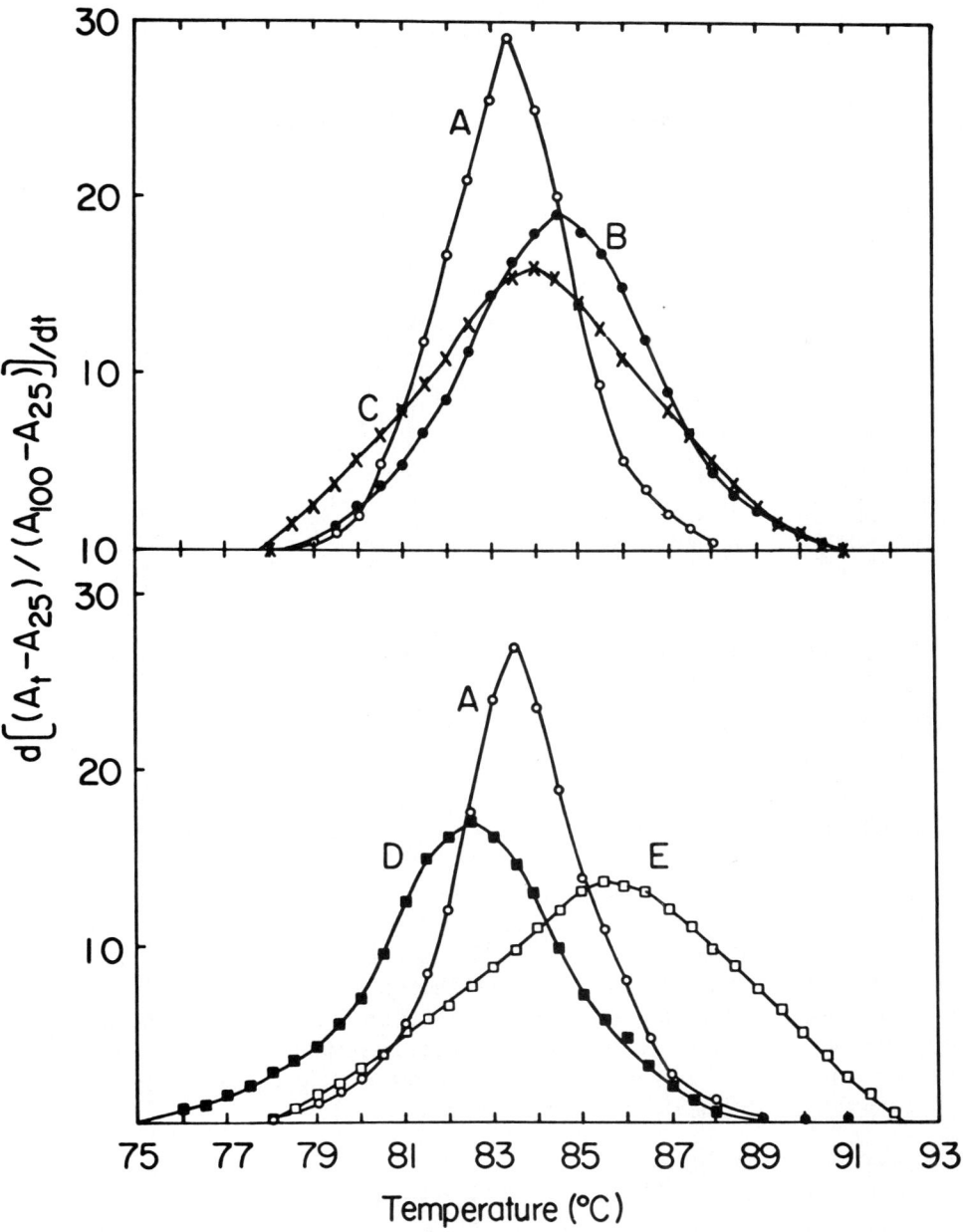

FIGURE 3. Derivative of the melting profiles of ctDNA. (A), T4 DNA; (B), Lettuce ctDNA; (C), Spinach ctDNA;(D), pea ctDNA; (E), Corn ctDNA. The meltings were carried out in $1 \times$ SSC using a rate of temperature rise of $1°C/10$ min. $d(A_T - A_{25})/(A_{100} - A_{25}/dt$ is the derivative of $(A_T - A_{25})/(A_{100} - A_{25})$ with respect to temperature. A_{25}, A_{260nm} of DNA at $25°C$; A_{100}, A_{260nm} of DNA at $100°C$; A_T, A_{260nm} of DNA at a given temperature.

of several components. These thermal denaturation experiments suggest that the base sequences of the different higher plants ctDNAs have diverged from each other even though the difference in their overall T_m does not amount to more than $2°C$. Furthermore, the higher plant ctDNAs and the algal ctDNA show drastic differences in their base composition. Thermal melting profiles of the higher plant ctDNAs do not show

CHLOROPLAST DNA OF *CHLAMYDOMONAS*

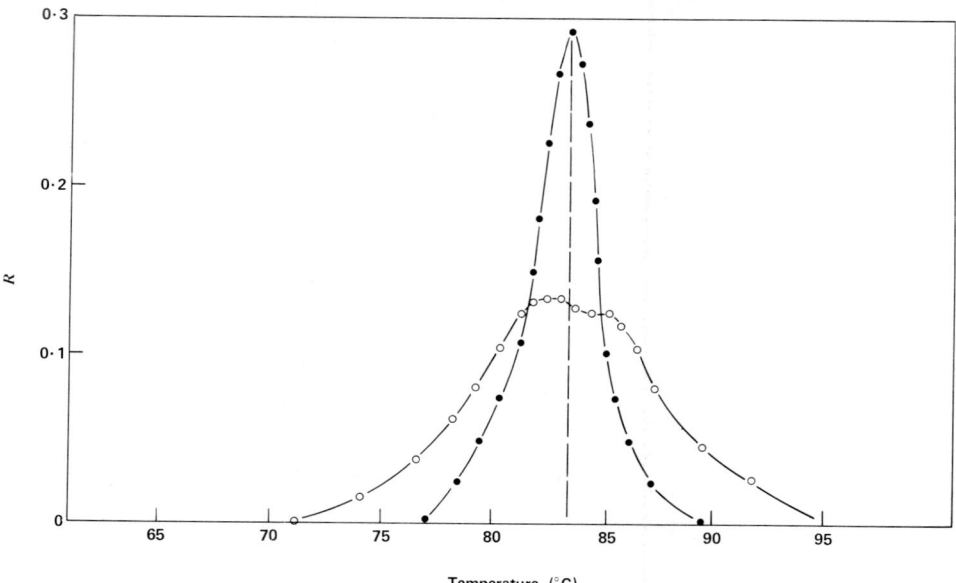

FIGURE 4.   Differential melting curves of native chloroplast (—O—O—) and T4 (—•—•—) DNA. Each point is the mean of values from three separate experiments. The ordinate represents the rate of melting of the DNA. R = $(A_{t2}—A_{t1})/(A_{100}—A_{20})/(t_2—t_1)$. A_{t2}, A_{t1}, A_{100}, A_{20} are the 260 nm absorbances (corrected for water expansion) at temperatures t_2, t_1, 100 and 20°C, respectively. The abscissa represents the mean temperature of the intervals and is equal to $(t_2—t_1)/2$. The denaturation conditions were 1 × SSC at a temperature rise of 10 min/°C. The DNAs were not sonicated and the single-stranded sizes were 1.2 × 10⁶ daltons and 1.8 × 10⁶ daltons for chloroplast and T4 DNA, respectively. (From Wells, R. and Sager, R. H., *J Mol. Biol.*, 158, 611, 1971. With permission.)

the presence of gross intramolecular heterogeneity, whereas the ctDNAs from *Euglena*,[18] *Chlamydomonas*,[19] and *Chlorella*[20] melt with a considerable amount of intramolecular heretogeneity. In the case of *Euglena* and *Chlorella* ctDNAs, their base sequence arrangement is so heterogeneous that satellite ctDNAs can be isolated by CsCl or Cs₂SO₄/Ag⁺ density gradient centrifugation. These differences in the arrangement of the base sequences between the ctDNAs from algae and higher plants further suggest that the sequence of the higher plant ctDNAs and the algae ctDNAs have undergone extensive changes during the course of evolution.

D. Structure of Chloroplast DNA

The circular structure of ctDNA was first demonstrated in *Euglena gracilis* by Manning and coworkers.[21] The structure of ctDNA in higher plants has been extens vely studied by Kolodner and Tewari.[15] The method which consistently yields large quantities of intact ctDNA is described below.

The chloroplast fraction from 1 kg of leaves is isolated, DNAase treated, and washed with EDTA as previously described. The resulting chloroplast pellet is suspended in 48 mℓ of Buffer C containing 0.05 *M* Tris/0.002 *M* EDTA, pH 8, 4°C. In the same buffer, 12mℓ of 10% sodium sarkosyl NL 97 and 0.03 mℓ of pronase (25 mg/mℓ, Calbiochem grade B, self digested 2 hr at 37°C) then are added and the suspension is incubated at 37°C for 45 min. Phenol (60 mℓ) saturated with 0.1 *M* Tris-OH was added and the suspension is shaken for 5 min at room temperature. The phases are separated by centrifugation at 12,000 × g for 5 min and the aqueous phase is removed. The aqueous phase is extracted two more times and the DNA is precipitated with two volumes of cold 95% ethanol. The precipitate can be stored at −20°C for up to 3 days. The precipitate is harvested by centrifugation at 12,000 × g for 15 min at 4°C and is rinsed once with 65% ethanol and recentrifuged. The pellet is gently suspended in Buffer C at room temperature; 30 mℓ of Buffer C is added for the DNA from 1 kg of pea leaves and 20 mℓ for the DNA from 1 kg of spinach and lettuce leaves. Each 5 mℓ of the DNA solution is adjusted to ϱ = 1.58 with CsCl; 0.15 mℓ of ethidium bromide is added and the solutions are centrifuged at 40,000 r/min for 36 to 44 hr in a Spinco® type 50 rotor. The lower DNA bands from 4 to 6 centrifuge tubes are collected into a cellulose nitrate centrifuge tube. Five mℓ of CsCl (ϱ = 1.57)/0.05 M Tris/0.01 M EDTA, pH 8, and 0.1 mℓ ethidium bromide (10 m/mℓ) are added. The solution is centrifuged for 24 to 36 hr at 33,000 r/min in a Spinco® SW 41 rotor. The DNA bands are collected in a glass tube and the ethidium bromide is removed by three batch extractions with Na⁺ Dowex AG 50W-X2 that had been extensively washed with 0.01 M EDTA, pH 8. The ethidium bromide is monitored by its fluorescence in long wavelength ultraviolet light. This ethidium bromide removal procedure preserves the ctDNA structure better than does extraction with isoamyl alcohol or dialysis against Dowex resin. The DNA solution is then dialyzed against three 500 mℓ changes of 0.1 M NaCl/0.05 M Tris/0.01 M EDTA/pH 8, over a 20 to 24 hr period. The centrifugation of DNA in CsCl-ethidium bromide density gradient results in a banding pattern shown in Figure 6. As much as 30% of the total ctDNA can be recovered in the lower band of the CsCl-ethidium bromide density gradients. In a large number of experiments with oats, corn, peas, lettuce, spinach, and beans, an average of 20% of the total ctDNA was recovered in the lower band, 70% of the ctDNA was recovered in the upper band, and 10% of the ctDNA was found in the area between the two bands. Centrifuging the ctDNA in a third CsCl/ethidium bromide density gradient did not significantly alter the banding pattern of the ctDNA, but did increase its stability to nicking during storage. From 100 g of pea leaves, 8 to 10 μg of lower band material was obtained, 4 to 6 μg of lower band DNA was obtained from 100 g of lettuce or spinach leaves, while approximately 1 μg of lower band DNA was obtained from 100 g of corn or oat leaves.

The ctDNA is mounted for electron microscopy using the following modification of the Kleinschmidt technique.[22] A typical 100 $\mu\ell$ spreading solution containing 1 M ammonium acetate pH 5.5/100 μg/mℓ cytochrome c/0.05 μg ctDNA/0.01 μg ϕX RF II monomer DNA is spread on a hypophase that consists of 0.3 M ammonium acetate, pH 5.5. After standing for 5 to 10 min, the protein monolayer is picked up on parlodion-coated 200 mesh Pellco-Cohen® handle grids, stained with uranyl acetate,[23] and then rotary shadowed with 0.8 cm of 8 mm 80% platinum per 20% palladium wire. The grids are examined in an electron microscope equipped with a 50 μm objective aperture. The molecules are photographed on Kodak® 4489 sheet film at a magnification of 4000 ×, and measured at a magnification of 120,000 ×. The magnifications are calibrated with a replica plating grid (Ladd®, 28,600 lines per inch).

When the lower band ctDNA from the second or third CsCl/ethidium bromide density gradient was examined in the electron microscope, more than 90% of the DNA molecules were present as supertwisted molecules (Figure 7B,C). Most of the remaining molecules were relaxed circles (Figure 7A). Treating this ctDNA with γ-rays, a treatment that produces single-strand breaks in DNA,[24] the supertwisted circular molecules were converted to relaxed open circular molecules. Approximately 50% of the supertwisted molecules were relaxed by treatment with 1000 rads of γ-rays and larger doses of

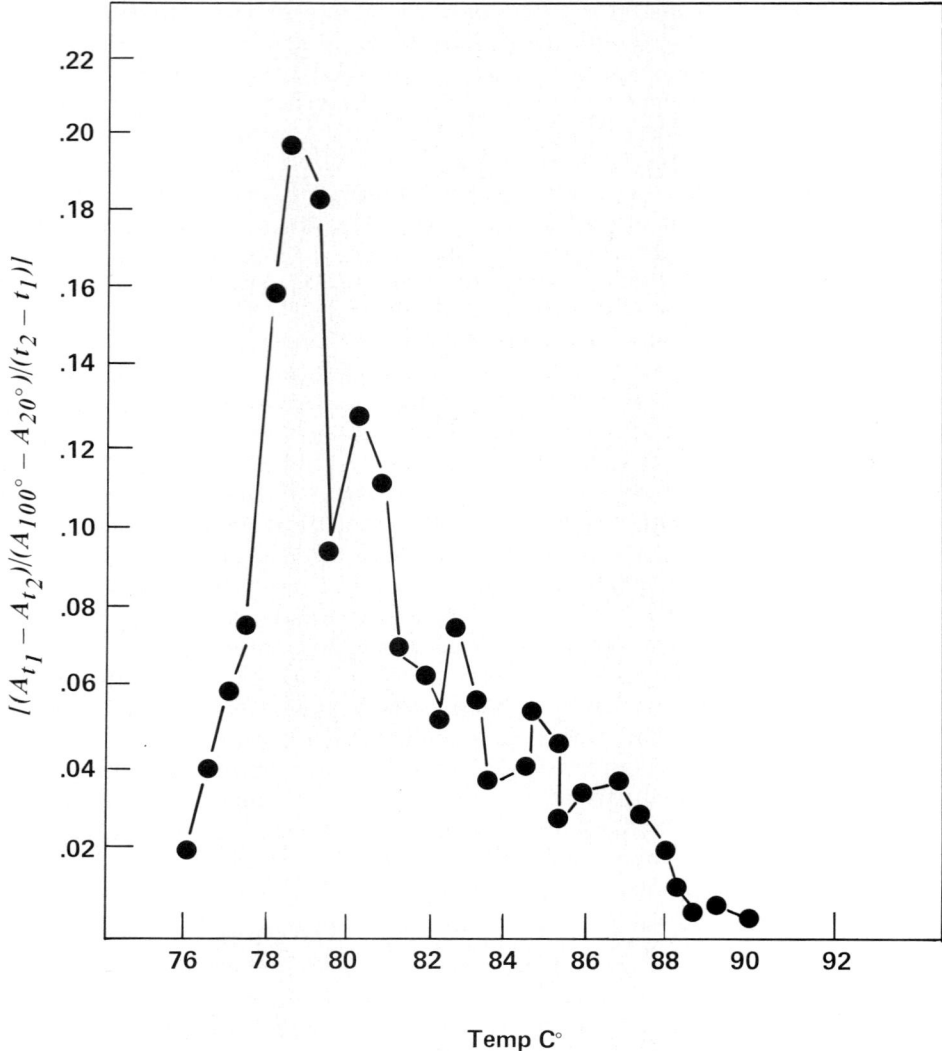

FIGURE 5.   Thermal transition of ctDNA. The thermal transition of ctDNA was measured in buffer containing 0.15 M NaCl and 0.015 M sodium citrate at pH 7.0. Measurements were performed in a Beckman® Acta III Spectrophotometer with automatic temperature programmer. The derivative of the thermal transition is shown: R vs. temperature where R equals $(A_{t1} - A_{t2})/(A_{100}° - A_{20}°/t_2 - t_1)$. Denaturation temperatures were equated into G + C mol percent. (From Slavik, N. S., and Hershberger, C. L., *FEBS Lett.*, 152, 171, 1975. With permission.)

γ-rays relaxed greater amounts of the supertwisted ctDNA. The ctDNA molecules that were found in the region of the gradient between the upper and lower bands mostly consisted of supertwisted circular DNA molecules. In an average preparation, 60% of the upper band ctDNA consisted of relaxed circular molecules while the rest of the DNA molecules were linear. No linear molecules longer than the circular molecules were observed. In general, 70% of the total ctDNA from any of the plants examined was recovered as supertwisted or relaxed circular DNA molecules. The best preparation of pea ctDNA contained 80% of the total DNA as circular DNA molecules.

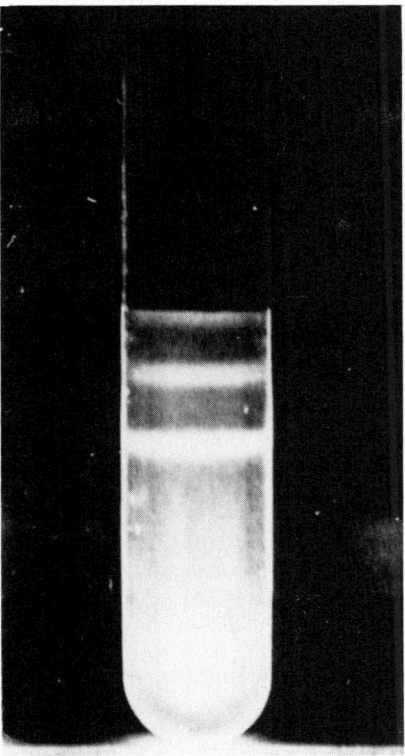

FIGURE 6.   Propidium diiodide — cesium chloride density gradient of pea ctDNA. Covalently closed circular pea ctDNA (25 μg) was nicked with 750 rads of gamma rays. The resulting DNA solution was adjusted to CsCl (ϱ = 1.57 g/cm^{-3})/0.01 M Tris/0.001 M EDTA/ 400 μg/mℓ propidium diiodide and was centrifuged for 36 hr at 32,000 rpm in a Spinco® SW41 rotor at 17°. The resulting density gradient was photographed under long wavelength (365 nm) light.

After nicking the supertwisted ctDNA molecules, there were three species of circular DNA molecules present: single-length circular molecules (Figure 7A), double-length circular molecules (Figure 8), and catenated dimers that appeared to consist of two topologically interlocked single-length circular molecules (Figure 9). The frequency of these species in the lower band ctDNA from pea, lettuce, spinach, corn, and oats is presented in Table 3. The frequency of circular and catenated dimers in the upper band of pea ctDNA is also presented in Table 3. It was calculated that 3.7% of the total circular pea ctDNA molecules were circular dimers, while 2.1% of the molecules are catenated dimers. These frequencies are very close to the frequencies of the circular and catenated dimers that were found in the lower band pea ctDNA. Therefore, only the lower band ctDNAs from spinach, lettuce, oats, and corn were analyzed for the frequency of circular and catenated dimers. All of the ctDNAs have approximately the same frequency of circular and catenated dimers.

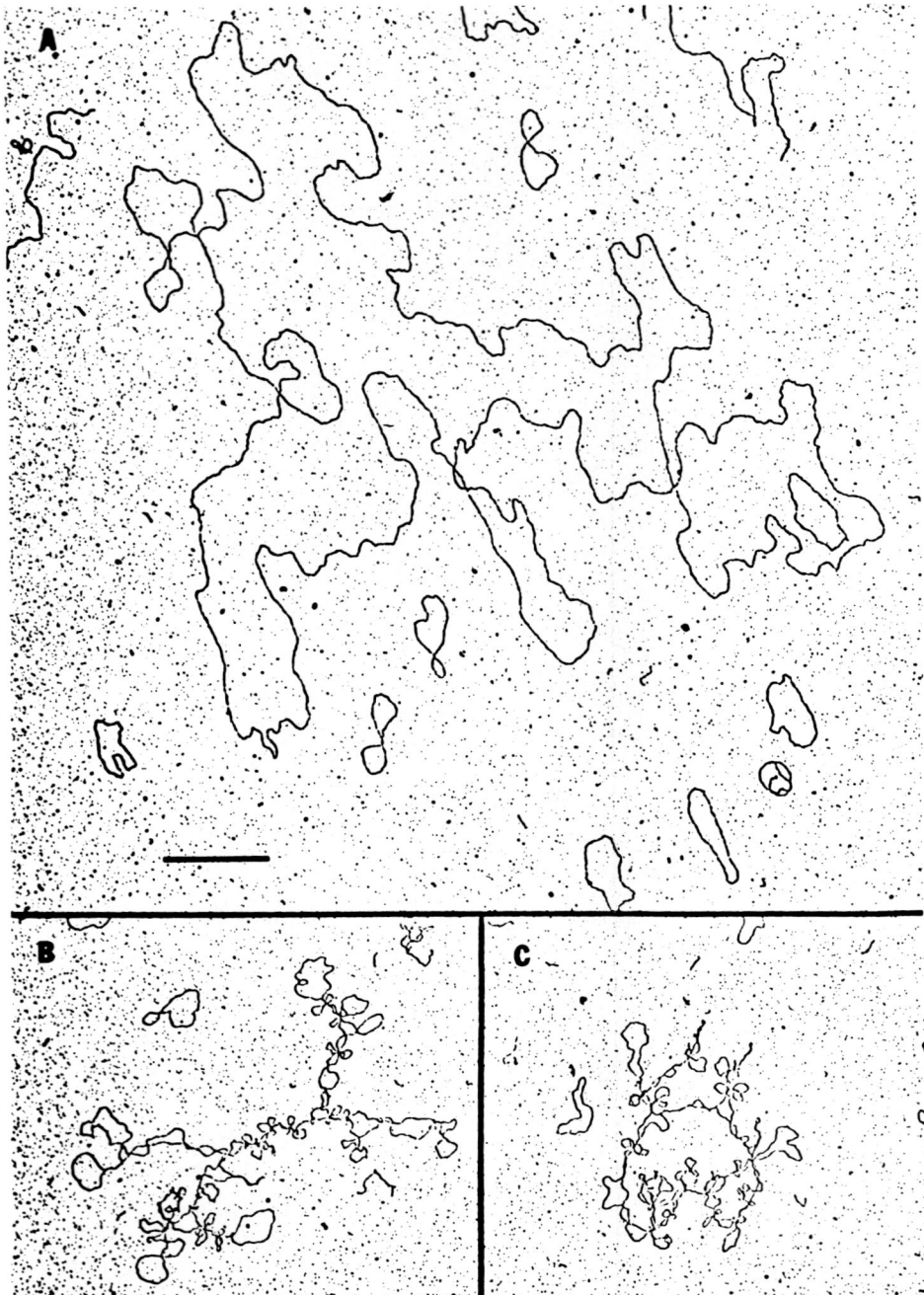

FIGURE 7.   Relaxed circular and supertwisted ctDNA molecules. (A), Relaxed circular lettuce ctDNA molecules. (B,C), Supertwisted pea and oat ctDNA molecules, respectively.

E. Molecular Size of the Circular Chloroplast DNA

The molecular sizes of the higher plant ctDNAs have been determined by electron microscopy. The closed circular ctDNA molecules were nicked with γ-rays and their size was determined by measuring their lengths relative to the length of ϕX RF II DNA,

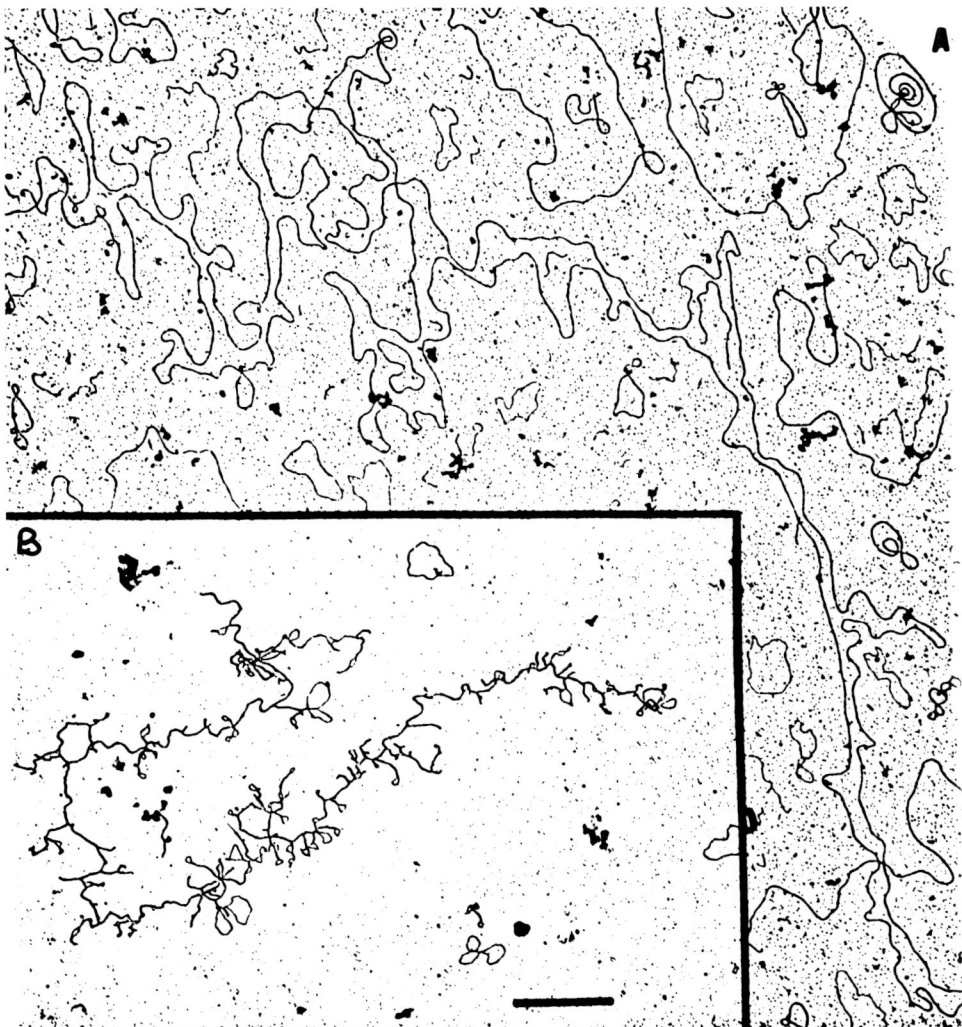

FIGURE 8.   Dimer length circular ctDNA molecules. (A), Relaxed double-length, circular pea ctDNA molecule; (B), Supertwisted, double-length, circular spinach ctDNA molecule. The small circular DNA molecules are φX RF II DNA. The bar indicates 1 μm.

which was used as an internal standard in all experiments. The distribution of the lengths of the circular pea, lettuce, spinach, corn, and oat ctDNA molecules (in φX units) is presented in Figure 10. The results show that each ctDNA contained both single-length and double-length circular molecules. Circular molecules of other sizes were not observed. The sizes of the single-length circular ctDNA molecules in φX units are presented in Table 4. The data show that there are significant differences between the sizes of all of the ctDNAs that were examined.

The commonly accepted method for determining the molecular weight of DNA by electron microscopy involves the use of an internal standard. The molecular weight of the unknown DNA is calculated from the ratio of its length to the length of the standard DNA that was present in the same spreading. This procedure assumes that the mass per unit length of two different DNAs is the same when the two DNAs are mounted for electron microscopy under identical conditions. This may not be a true

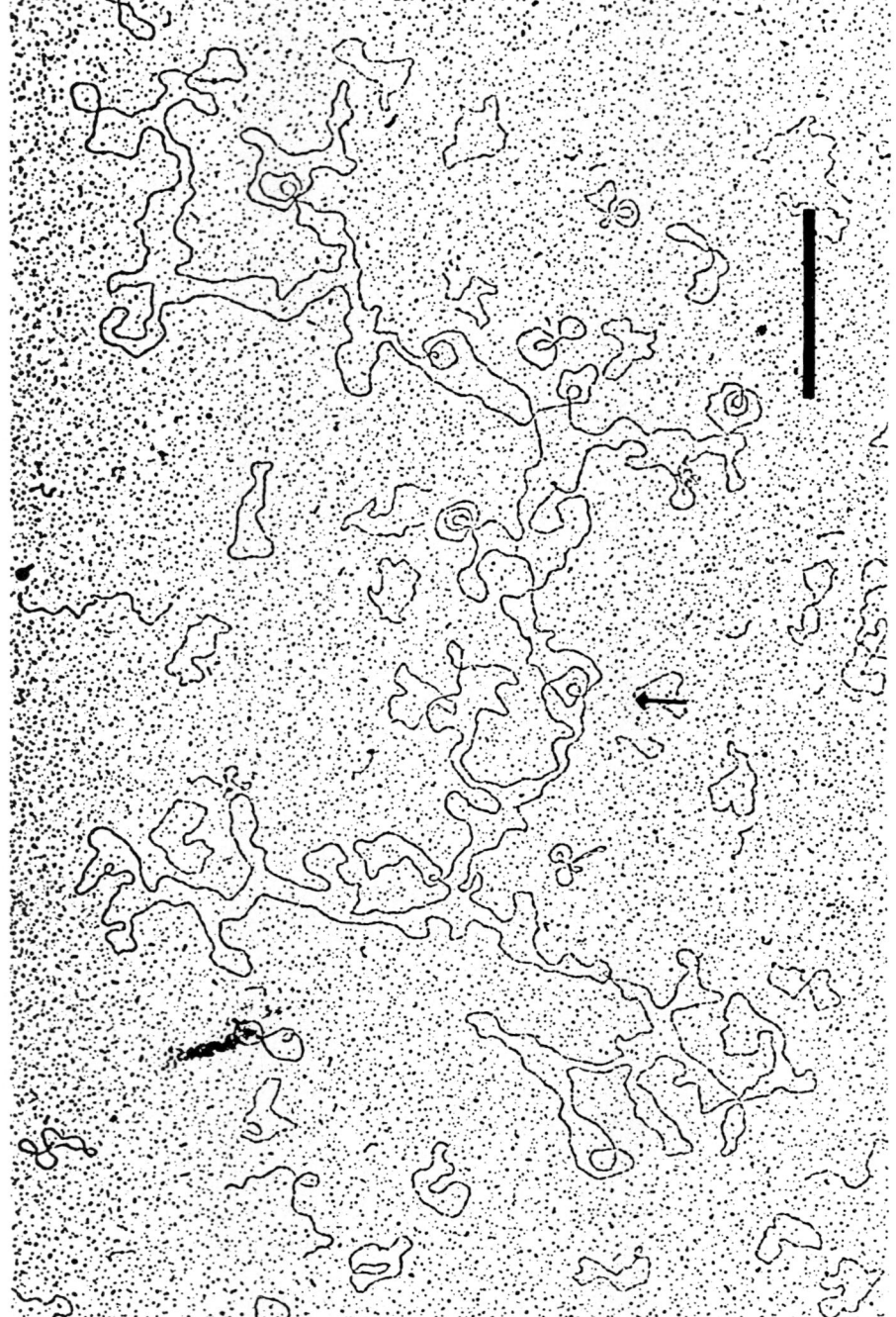

FIGURE 9.   Catenated dimer of pea ctDNA. The arrow marks the point of catenation between the two monomer units. The small circular DNA molecules are φX RF II DNA and the bar indicates 1 μm. This spreading only contained three circular ctDNA molecules per grid squares.

TABLE 3

The Frequency (N) of Complex Molecules in ctDNA

DNA	Monomers (%)	Circular dimers (%)	Catenated dimers (%)	N[a]
Pea (lower band)	94.2	4.1	1.7	520
Pea (middle band)	92.9	3.5	3.6	514
Pea (upper band)	94.4	3.6	2.0	506
Lettuce (lower band)	92.1	5.5	2.4	503
Spinach (lower band)	94.5	3.6	1.9	420
Maize (lower band)	95.8	2.9	1.3	409
Oats (lower band)	95.5	3.0	1.5	200

assumption. The analysis of the molecular weights of DNA by electron microscopy has been discussed at length by Kolodner and Tewari.[15] The molecular weights given in the Table 4 are calculated using the molecular weight of pea ctDNA that has been obtained by sedimentation analysis. The ctDNA from *Euglena* similarily has been found to be about 87×10^6 by electron microscopic measurements. The rigorous data on the molecular size of ctDNA from *Chlamydomonas* are not available. Behn and Herrmann[25] have reported the isolation of circular ctDNA from *Chlamydomonas reinhardi* with a molecular size of 134×10^6. This molecular size is in excellent agreement with the experiments of Rochaix[26] and Howell and coworkers,[27] who have determined the molecular size of ctDNA by summing up the molecular weights of the DNA fragments produced by endonucleases.

F. Physical Studies on the Chloroplast DNA

1. Sedimentation Properties of Chloroplast DNA

The pea form I ctDNA (supercoiled ctDNA), obtained from the lower band of a CsC1-ethidium bromide density gradient, was sedimented in neutral 3 M CsC1.[28] Only one component sedimenting at about 86 S was observed. When this DNA was irradiated with 1000 rads of γ-rays, a treatment known to produce single-strand breaks in DNA, 50% of the DNA was converted to a slower moving form sedimenting at about 55 S (Figure 11). Higher doses of γ-rays converted more of the faster moving form I DNA to the slower moving nicked circular form II DNA, but no other forms were produced. When the pea form I ctDNA was sedimented in 3M CsC1 containing 0.2 M NaOH, all of the DNA was present as the fast moving form IV, sedimenting at 240 S. The pea form I ctDNA was nicked with γ-rays (1000 rads) to produce single-strand breaks in 50% of the molecules and this DNA was sedimented through alkaline 3 M CsC1. The data showed that 50% of the DNA sedimented as form IV, while the rest of the DNA sedimented slowly at 53 S as single-stranded circular and linear molecules due to strand separation. All attempts to resolve single-stranded circles from single-stranded linears by alkaline sedimentation were unsuccessful even though our procedures were capable of resolving these components using φX RF I DNA that was nicked as described above. Similar results were obtained when these experiments were performed with lettuce and spinach ctDNA. The storage of form I ctDNA at −40°C for 3 months did not result in any detectable nicking as analyzed by sedimentation in neutral or alkaline 3 M CsCl.

2. Irreversible Denaturation of Form I Chloroplast DNA

Under alkaline conditions, form I ctDNA was denatured and converted to the com-

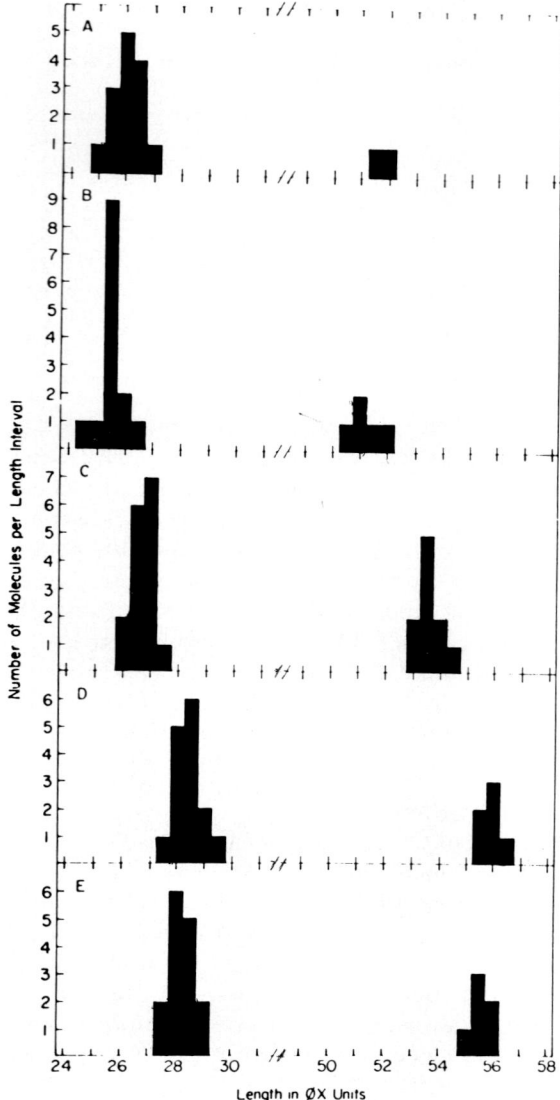

FIGURE 10. Length distribution of the circular ctDNA
molecules. (A), Oat ctDNA; (B), Corn ctDNA; (C), Pea
ctDNA; (D), Lettuce ctDNA; (E), Spinach ctDNA. The
lengths are expressed in units of ϕX length. The large classes
of circular DNA molecules are two times the size of the small
size class of circular DNA molecules.

pact, rapidly sedimenting form IV. This process was irreversible in the case of ϕX RF
DNA[20] and mitochondrial DNA[30] in the sense that form IV DNA is not converted to
form I DNA by neutralization, but requires special renaturation treatment. Pea form
I ctDNA was incubated for 1.5 min at pH 12.7 or pH 13 in 0.1 M NaCl/0.05 M Tris/
0.01 M EDTA/neutralized, and sedimented in neutral $3M$ CsCl. At pH 12.7, 56% of
the DNA was form I and 44% was form IV. At pH 13, 38% of the DNA was form I
while 62% of the DNA was form IV. The incubation of pea form I ctDNA at high

TABLE 4

Molecular Size of Chloroplast DNAs

Molecular size $\times 10^{-6}$

DNA	Electron microscopy	Renaturation kinetics
Pea	88.4 ± 0.3	86.9 ± 2.1
Lettuce	95.8 ± 0.3	90.9 ± 2.0
Spinach	94.3 ± 0.6	87.9 ± 2.7
Maize	84.7 ± 0.6	83.5 ± 2.8
Oats	85.7 ± 0.3	81.7 ± 3.0

pH for longer times did not change the proportions of forms I and IV. Similar results were obtained in experiments with lettuce ctDNA. When ϕX RF I DNA was denatured, neutralized, and sedimented under these conditions, all of the DNA was present as the irreversible denatured form[29] The mixture of forms I nd IV could result from the failure of some of the ctDNA molecules to reach a form IV configuration or from some of the form IV molecules being not stable on neutralization. To distinguish between these possibilities, the pea form I ctDNA was sedimented in the solvent made with H_2O-D_2O (density = 1.052) containing the identical salt and buffer concentrations that were used in the high pH incubations. In these experiments, all of the DNA sedimented as form IV. These results indicated that at pH 12.7 and 13, some of the form IV ctDNA molecules that were formed were not stable on neutralization.

3. Concentration Dependence of the Sedimentation Coefficient

In the experiments described above, except for the sedimentation of form IV in neutral 3 M CsC1, the DNA bands were found to be skewed toward the leading edge of the zone. Such a skewing of a DNA zone has been shown to result from the concentration dependence of the sedimentation coefficient. The sedimentation coefficients of each of the DNA forms (discussed above) of pea, lettuce, and spinach ctDNA were extrapolated to zero concentration by performing experiments at a number of different DNA concntrations. The results for pea ctDNA are presented in Figure 12. All of the DNA forms except form IV showed concentration dependence. The best values for the $S_{o,2O,\omega,Na^+}$ of the DNA forms studied, determined by a least squares fit of the data, are presented in Table 5. These sedimentation coefficients were found to be independent of the speed of centrifugation.

4. Superhelix Density of Chloroplast DNA

The superhelix density of pea, lettuce, and spinach form I ctDNA was determined by performing ethidium bromide sedimentation velocity titration. In Table 6, the free ethidium bromide concentration and the number of moles of dye bound per mole of nucleotide at the sedimentation minimum of pea, lettuce, and spinach form I ctDNA are presented. These are the experimental values that characterize the superhelix density of a molecule. From these values, one can determine the superhelix density of a DNA molecule if the angle by which each bound ethidium bromide molecule unwinds the DNA helix is known. This value is in dispute since Waring[31] have suggested it to be 12° while Wang[32] has recently redetermined it to be 26°. The value of the superhelix density (σ_o) for these ctDNAs, using both values for the angle of unwinding, is presented in Table 6. The superhelix densities of these three ctDNAs are almost the same and do not differ by more than 3%. This variation is much less than the variation in

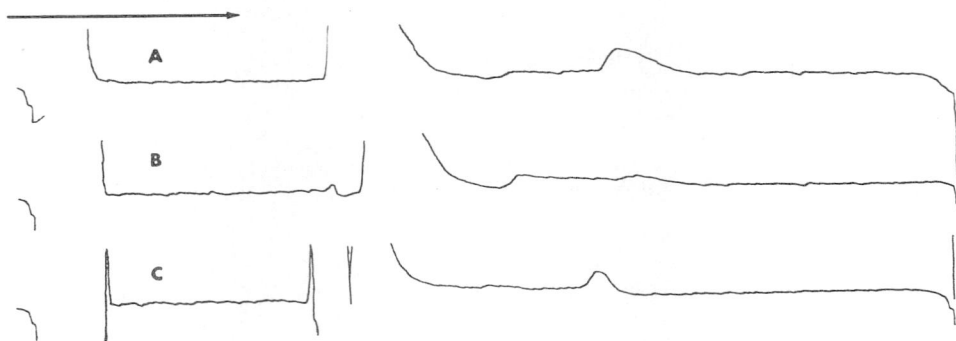

FIGURE 11.   Analytical ultracentrifuge scans of pea ctDNA. The samples were layered on neutral or alkaline 3 M CsCl at 20°. (A) Pea form I ctDNA on neutral 3 M CsCl; 24,000 rpm for 32 min, (B) Pea form I ctDNA that was nicked with 1000 rads of γ rays, on neutral 3 M CsCl; 24,000 rpm for 29 min, (C) Pea form I ctDNA on alkaline 3 M CsCl; 16,000 rpm for 20 min. The high absorption at the top of the scans is due to a high concentration of EDTA.

the superhelix densities that has been observed in the animal mitochondrial DNAs from normal and transformed cells.[28]

5. The Molecular Weight of Chloroplast DNA

The value for the molecular weight of pea chloroplast DNA as analyzed by buoyant sedimentation equilibrium was 89.1 (SD ±0.7) × 10⁶. This is in excellent agreement with the value for pea ctDNA of 88.4 × 10⁶ as determined by electron microscopy using φX RFII DNA and λDNA as internal standards.[33] Using the relationship between S and molecular weight and the value of So20,ω,Na⁺ for open circular lettuce and spinach ctDNAs of Table 5, their molecular weights have been calculated to be 98.2 (SD ± 1.5)×10⁶ and 97.2 (SD ± 1.5) 10⁶, respectively. These values are in good agreement with 96.7 (SD ± 0.5)×10⁶ for lettuce ctDNA and 95.1 (SD ± 1.0)×10⁶ for spinach ctDNA, which were calculated knowing that the lengths of lettuce and spinach ctDNAs are 1.085 (SD ± 0.006) and 1.067 (SD ± 0.011) times the length of pea ctDNA, respectively, as determined by electron microscopy.[15]

G. Renaturation Kinetics of Chloroplast DNA

The sequence complexities of the ctDNAs from higher plants were studied by examining their reassociation kinetics. The ctDNA was sheared into fragments having an average molecular weight of 1.4 × 10⁶ by passage through a 27 gauge needle. These DNA fragments were alkali denatured, neutralized, and then incubated at T_m − 25°C in a recording spectrophotometer. Either T4 DNA or pea ctDNA was included with each experiment as a standard. The kinetics of renaturation of lettuce, bean, and spinach ctDNA in 0.15 M NaCl/0.015 M trisodium citrate, pH 7.0, are plotted in Figure 13. The renaturation reaction of the ctDNAs showed second-order kinetics and the plot of their rates of reassociation showed the presence of a single class of molecules. The linearity of the plots (Figure 13) and the good extrapolation to 1.0 indicated the absence of any rapidly renaturing components within the limits of detection of the spectrophotometer. Renaturation kinetics experiments were also performed in 0.375 M NaCl/0.0375 M trisodium citrate, pH 7.0. They also gave second-order kinetics with no indication of rapidly renaturing sequences. Similar results were obtained in experiments with corn, oats, and pea ctDNA. The reassociation kinetics of the ctDNAs

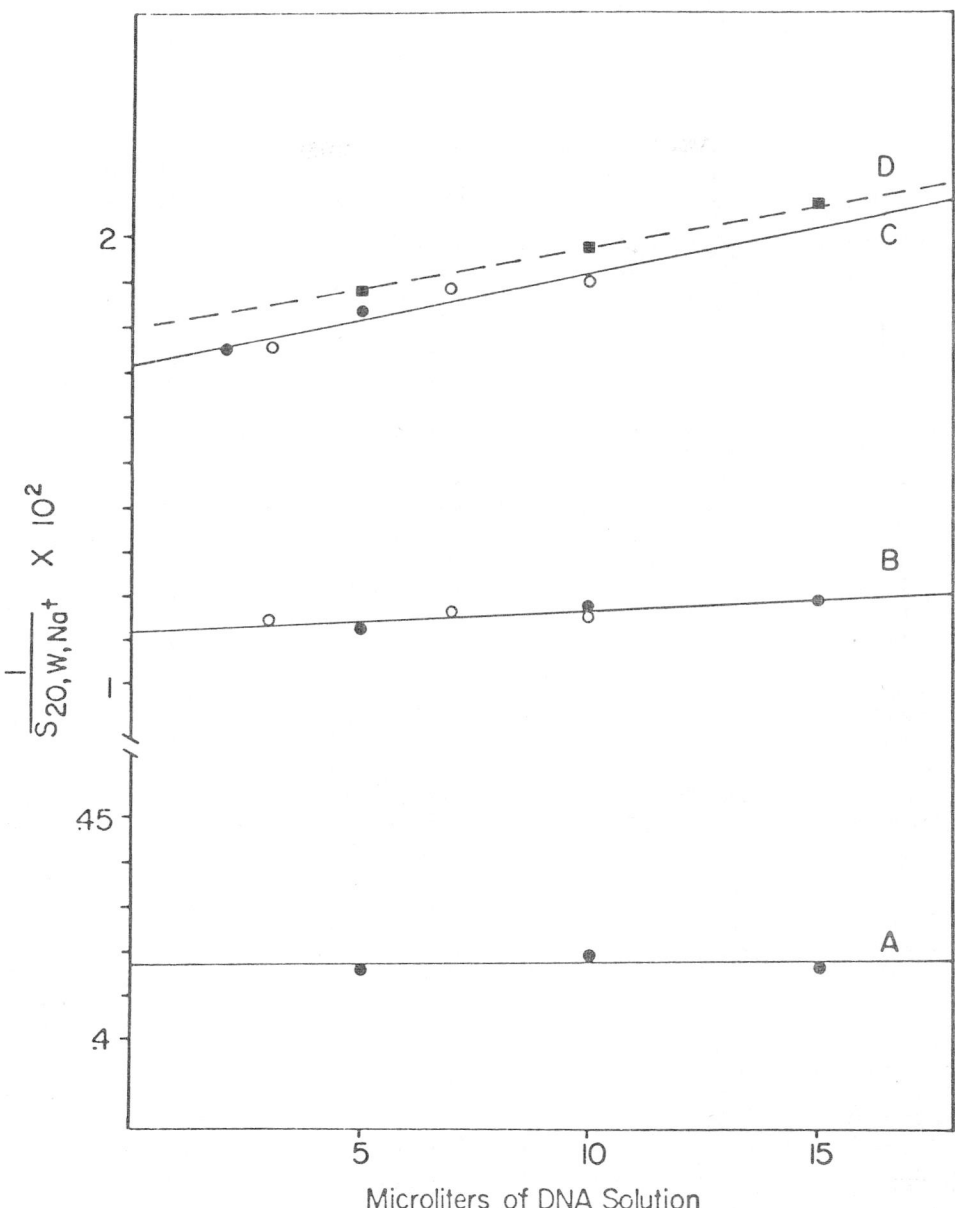

FIGURE 12.   The concentration dependence of the $s°_{20,w,Na}+$ of pea ctDNA. The DNA solution used in these experiments contained 50 μg/mℓ of pea ctDNA. (A) Form IV on alkaline 3 M CsCl; 16,000 rpm. (B) Form I on neutral 3 M CsCl; 24,000 rpm. (C) Form II on neutral 3 M CsCl; 24,000rpm. (D) Singly nicked form II on alkaline 3 M CsCl; 30,000 rpm.

were further studied using ctDNA that had been fragmented to a molecular weight of 1.5×10^5 by boiling in 0.2 M NaOH,[15] The ctDNA was incubated in 0.15 M NaCl/ 0.015 M trisodium citrate, pH 7.0, as described above. In this case, the ctDNAs and T4 DNA also reannnealed as a single component with no evidence of any repeating sequences. Similar results were obtained in experiments with pea, bean, lettuce, corn,

TABLE 5

The Values of $s^{\circ}{}_{20,w,Na}+$ of ctDNA

			Configuration		
DNA	Form I (neutral)	Form II (neutral)	Form II (alkaline)	Form IV (neutral)	Form IV (alkaline)
Pea	86.0 ± 0.4	58.3 ± 0.3	55.6 ± 0.08	146.8 ± 1.1	240 ± 0.7
Spinach	97.1 ± 0.6	60.6 ± 0.4	57.7 ± 0.5	—	264.0 ± 3.0
Lettuce	94.9 ± 0.3	60.8 ± 0.1	57.8 ± 0.5	170 ± 1.0	264.1 ± 1.5

TABLE 6

The Critical Dye Concentration and Superhelix Density of Various Closed Circular DNAS

DNA	Ethidium bromide concentration at the minimum ($\mu g/m\ell$)	The molar ratio of ethidium bromide bound per base pair at the minimum	Superhelix density $(-\sigma_o \times 10^2)$	
			12° Unwinding angle	26° Unwinding angle
Pea ct	5.2 ± 0.1	0.059	4.0 ± 0.1	8.7 ± 0.2
Spinach ct	5.4 ± 0.1	0.061	4.1 ± 0.1	8.9 ± 0.2
Lettuce ct	5.3 ± 0.2	0.060	4.0 ± 0.2	8.8 ± 0.4
ϕX RF I	—	0.055	3.7	8.0
SV 40	5.2 ± 0.3	0.059	3.9 ± 0.2	8.5 ± 0.4
Polyoma	4.2 ± 0.4	0.051	3.3 ± 0.3	7.2 ± 0.6
F-factor	4.5 ± 0.2	0.054	3.5 ± 0.2	7.2 ± 0.4
Escherchia coli 15 plasmid	—	0.058	3.9	8.5
PM 2	8.0 ± 0.5	0.080	5.3 ± 0.3	11.5 ± 0.6
HeLa mt	—	—	2.8 ± 0.4	6.1 ± 0.8
3T3 mt	—	—	2.0 ± 0.3	4.3 ± 0.6
Mouse L Cell mt	—	—	1.2 ± 0.3	2.6 ± 0.6
Rat mt	—	—	4.4	9.6

From Kolodner, R., Tewari, K. K., and Warner, R. C., *Biochim. Biophys. Acta*, 447, 144, 1976. With permission.

and spinach ctDNA that had been fragmented to a molecular weight of 1.5×10^5. In the above experiments, the reannealed DNAs were tested for nonspecific base pairing by thermally denaturing the reannealed ctDNAs. In all cases, the reannealed ctDNAs melted sharply with a T_m that was close to the T_m of the native DNA, indicating that specific base pairing had taken place during the renaturation reaction.

In Table 7 the $C_ot_{1/2}$ values for the higher plant ctDNAs and T4 DNA are presented. The C_ot ½ values for pea, spinach, lettuce, beans, corn, and oat ctDNA, determined under similar experimental conditions, were all very similar. For example, in 0.375 *M* NaCl/0.0375 *M* trisodium citrate, pH 7.0, the ctDNAs that were fragmented to a molecular weight of 1.4×10^6 had $C_ot_{1/2}$ values that ranged from 0.057 to 0.062; and in 1 × SSC, the $C_ot_{1/2}$ values ranged from 0.18 to 0.20. When the ctDNAs were fragmented by boiling in alkali, the $C_ot_{1/2}$ values were more variable. This effect probably was due to variable fragmentation in different experiments. When the $C_ot_{1/2}$ values of the alkali fragmented ctDNAs were normalized to a $C_ot_{1/2}$ value for T4 DNA of 0.76 (the average

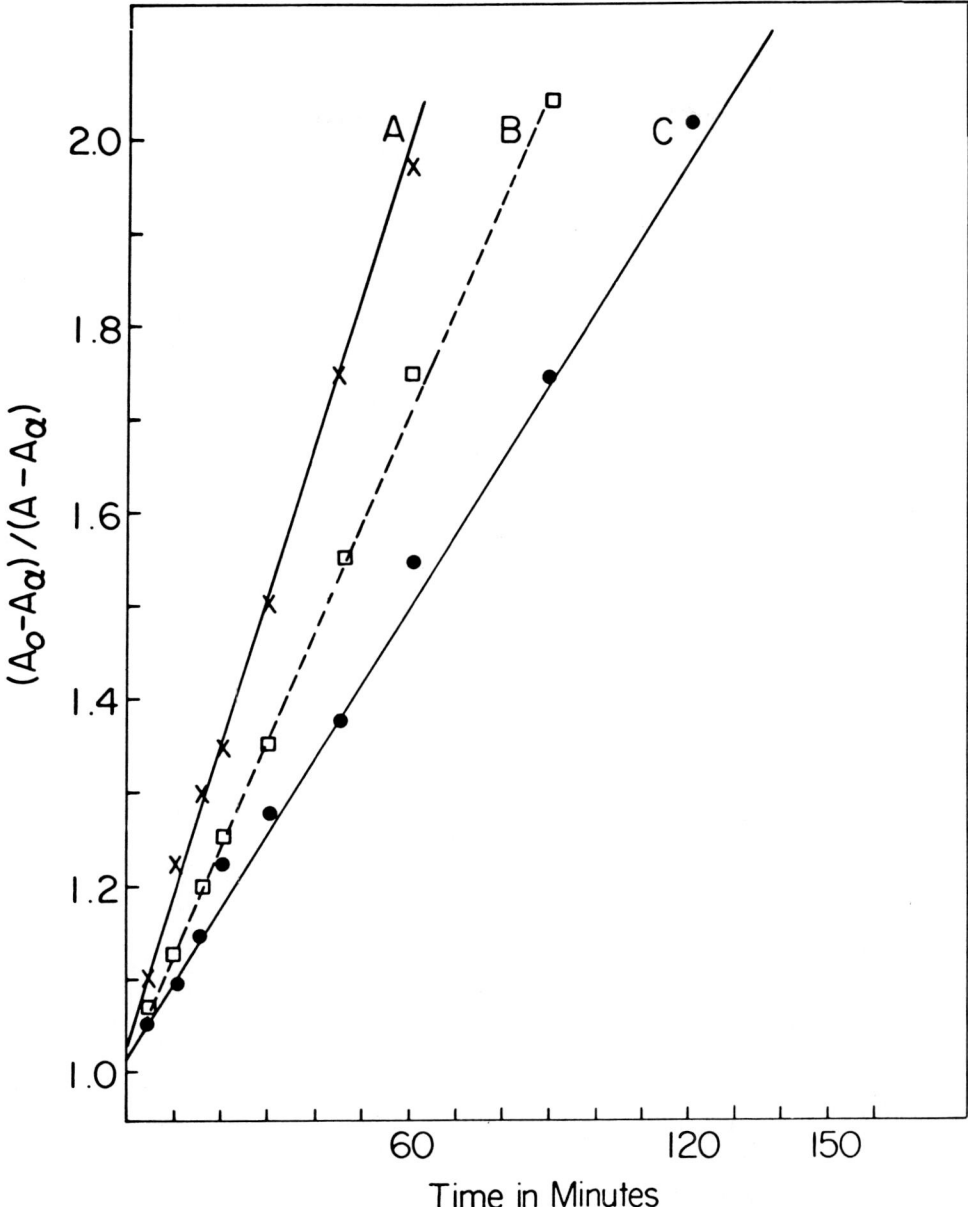

FIGURE 13. Second order rate plot of the renaturation of ctDNA. (A), Lettuce ctDNA, 20 μg/mℓ; (B), Bean ctDNA, 15 μg/mℓ; (C), Spinach ctDNA, 10 μg/mℓ. The DNA was fragmented by passing it through a 27 gauge needle, alkali denatured, and neutralized. The ctDNA was then allowed to renature $T_m-25°$ in 1 × SSC. A_o, $A_{260\ nm}$ of denatured DNA; $A\alpha$, $A_{260\ nm}$ of native DNA; A, $A_{260\ nm}$ of the sample at a given time.

experimental value), they were again similar and ranged from 0.57 to 0.65. This illustrates the necessity of including a standard DNA in all experiments in order to standardize the conditions among different experiments. The above data show that all of the ctDNAs that were examined are of similar size.

The molecular weight of the unique sequence of the ctDNAs from pea, lettuce, spin-

TABLE 7

Reassociation Rates of DNA

DNA	Fragmented size $(M, \times 10^{-5})$	Salt concentration $(\times 0.15\ M\mathrm{NaCl}/$ $0.015\ M$ trisodium citrate, pH, 7.0)	$C_ot\ \frac{1}{2}$[a]
T4	14	2.5	0.070
	14	1.0	0.24
	1.5	1.0	0.76
Pea ct	14	2.5	0.058
	14	1.0	0.19
	1.5	1.0	0.64
Lettuce ct	14	2.5	0.062
	14	1.0	0.20
	1.5	1.0	0.65
Spinach ct	14	2.5	0.059
	14	1.0	0.19
	1.5	1.0	0.65
Bean ct	14	2.5	0.060
	14	1.0	0.020
	1.5	1.0	0.63
Maize ct	14	2.5	0.056
	14	1.0	0.18
	1.5	1.0	0.62
Oat ct	14	2.5	0.057
	14	1.0	0.18
	1.5	1.0	0.57

[a] C_ot value at 50% denaturation. C_ot values were determined by the relation: $C_ot = (A_{260\ nm}$ of initial native DNA $\times$ time in min)/120.

ach, beans, corn, and oats were calculated using the ratio of the the ratio of the $C_ot_{1/2}$ values of the ctDNAs to T4 DNA or pea ctDNA. The molecular weight of T4 DNA was taken as 110×10^6 and the absolute molecular weight of pea ctDNA was 89.1×10^6. The values of the molecular weights of the higher plant ctDNAs are presented in Table 4. These values are the average values that were determined for each ctDNA under the different salt conditions and with different fragment sizes. The data suggest that corn and oat ctDNA are smaller than pea, spinach, bean, and lettuce ctDNAs.

The ctDNA from *Euglena gracilis* also was found to renature as a single component. After correcting for the low guanine and cytosine content of this ctDNA, a unique sequence length of about 90×10^6 daltons was obtained.[21] The renaturation kinetics of *Chlamydomonas* ctDNA have shown it to renature homogenously with a kinetic complexity (not the molecular weight) of 194×10^6 by Bastia and co-orkers.[34] However, Wells and Sager[18] have found it to renature with biphasic kinetics. There may be many explanations for these discrepancies.[25] The ctDNA from *Chlamydomonas* might be quite different from ctDNAs of higher plants and *Euglena*.

In higher plants, there is an excellent agreement between the molecular weights of the unique sequence of ctDNA determined by renaturation kinetics and the molecular weight of the circular ctDNA molecules determined by electron microscopy. This result suggests that the entire information content of higher plant ctDNA is coded by the sequence of a single circular ctDNA molecule. However, it may be pointed out that the spectrophotometric analysis cannot detect the presence of sequences that are only repeated once such as the chloroplast ribosomal RNA genes (see below); but this method of analysis shows that higher plant ctDNAs do not contain large numbers of highly reiterated sequences such as those found in the nuclear DNAs of eucaryotic organisms.[35]

H. Denaturation Mapping of the Chloroplast DNA

The studies on the molecular size of ctDNA by denaturation-renaturation experiments and electron microscopy indicated that the ctDNA consisted of homogeneous molecules with its entire information content present in the sequence of a circular DNA molecule. However, these techniques are only able to analyze the average properties of a very large number of molecules. Therefore, the genetic content of the pea ctDNA and the gross arrangement of its base sequences have been analyzed further by the denaturation mapping studies. The denaturation mapping technique of Inman[36] consistently is able to locate AT rich regions in individual molecules by mounting the molecules for electron microscopy under levels of partial denaturation.

Circular pea ctDNA was prepared for electron microscopic examination under partially denaturing conditions, using the isodenaturing technique of Davis and Hyman.[37] A typical spreading solution of 100 $\mu\ell$ contained 0.1 M Tris-OH/0.01 M Na$_2$EDTA (pH 8.5)/ 50 μg/mℓ of cytochrome c/0.02 μg of ϕX viral DNA/0.02 μg of ϕX RF II DNA/0.05 μg of circular pea ctDNA/and formamide. The covalently closed circular pea ctDNA used in these experiments was converted to a relaxed nicked circular form by irradiation with 1000 rads of γ-rays from a ^{137}Cs source. A typical hypophase contained 0.01 M Tris-OH/ 0.001 M EDTA (pH 8.5)/ and formamide. The formamide concentration in the hypophase was always 30% lower than the formamide concentration in the spreading solution. The DNA was prepared for electron microscopy as described before. Areas of the grids having good contrast were selected and systematically scanned. All measurable molecules showing partial denaturation were photographed. The partially denatured molecules and at least 100 of each double-stranded (dsDNA) and single-stranded (ssDNA) ϕX DNA standard molecules were measured at an enlargement of × 120,000. The lengths of the denatured and native regions of each Pea ctDNA molecule were expressed as ratios to the lengths of ssDNA and dsDNA ϕX DNA, respectively. The total length of each molecule then was normalized to a length of 26.7 × ϕX, the size of the pea ctDNA. These molecules then were used to construct denaturation maps.[39]

1. Denaturation Maps of the Circular Monomer

The circular ctDNA molecules were spread onto a hypophase of 45% formamide. At this concentration, 70% of all of the circular molecules were partially denatured, containing from one to seven denatured regions. The average amount of the denaturation was 2.5%. A typical molecule from this spreading is shown in Figure 14. All of the interpretable circular molecules containing two or more denatured regions were photographed, measured as described, and were arbitrarily linearized. These linearized molecules were arranged in a map, starting with the most denatured molecules first, so that the greatest number of denatured regions matched between the molecules. After the molecules were matched by this criterion, they were further matched with respect to the length of each individual denatured region. This denaturation map is given in Figure 15. In order to assess the accuracy of this method, denaturation maps also were produced from the same molecules using the least denatured molecules first, and arranging them as described above. A denaturation map was also constructed by picking the molecules randomly. All of the denaturation maps that were produced were identical. This map contained six major denatured regions, two minor denatured regions, and two places where only one molecule was denatured. A region corresponding to 34% of the pea ctDNA molecule contained no denatured regions. The circular pea

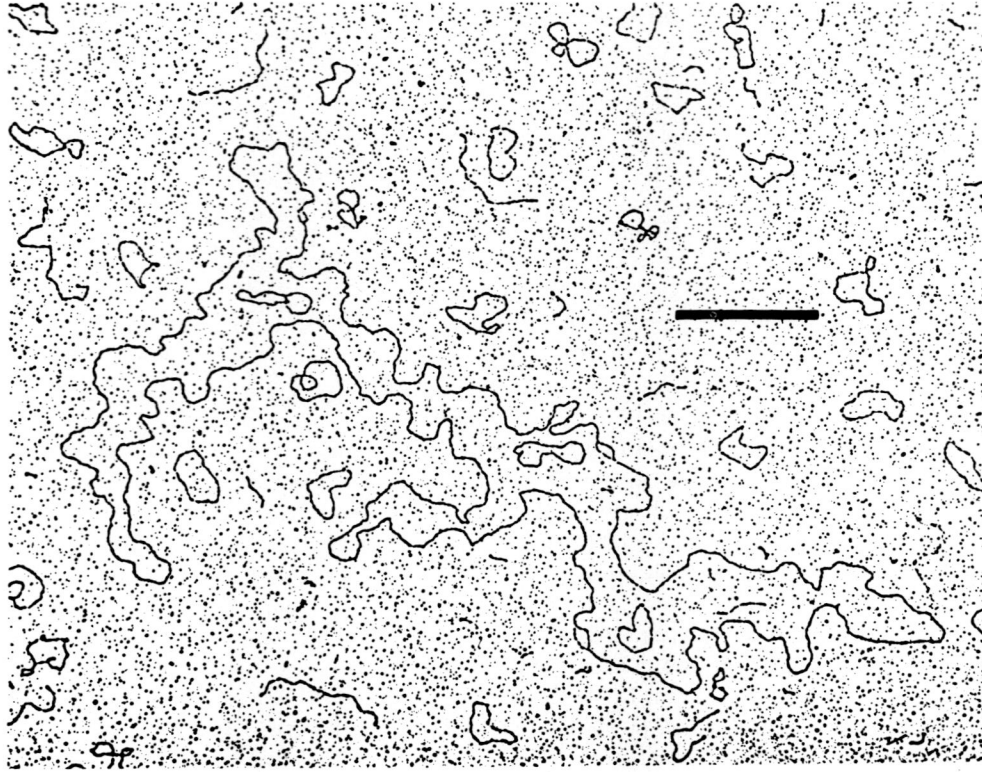

FIGURE 14.   A partially denatured pea ctDNA molecule showing three denatured regions. The hypo-phase contained 45% formamide in 1 × TE (10 m *M* Tris /1 m *M* EDTA) and the spreading solution contained 75% formamide in 10 × TE. The small circular molecules are SS and DS ɸ× DNA. The bar indicates 1 µm.

ctDNA molecules were spread into a hypophase of 47% formamide. All of the circular molecules were partially denatured, and contained from 11 to 33 denatured regions. The average amount of denaturation was 22%. A typical molecule from this spreading is presented in Figure 16. A denaturation map was constructed from these molecules by the methods described above. Because of a large number of denatured regions in these molecules, the map was constructed by using molecules showing intermediate levels of denaturation first. A map was also produced by randomly selecting the molecules. All of the maps were identical. This map contained 31 distinct denatured regions as well as six distinct regions showing no denaturation (Figure 17). It was possible to locate all of the denatured regions produced at the 2.5% level of denaturation. The circular pea ctDNA was spread onto a hypophase of 51% formamide. All of the circular molecules were partially denatured and contained from 18 to 43 denatured regions. The average amount of denaturation was 44%. A denaturation map was constructed using these molecules as described above (Figure 18). There were only two distinct undenatured regions which matched two undenatured regions in the map at 22% denaturation. The five denatured regions between these two undenatured regions were also present in the denaturation map at 22% denaturation.

2. Denaturation Maps of the Circular Dimers
In our preparations, 3% of the pea ctDNA molecules were present as circular di-

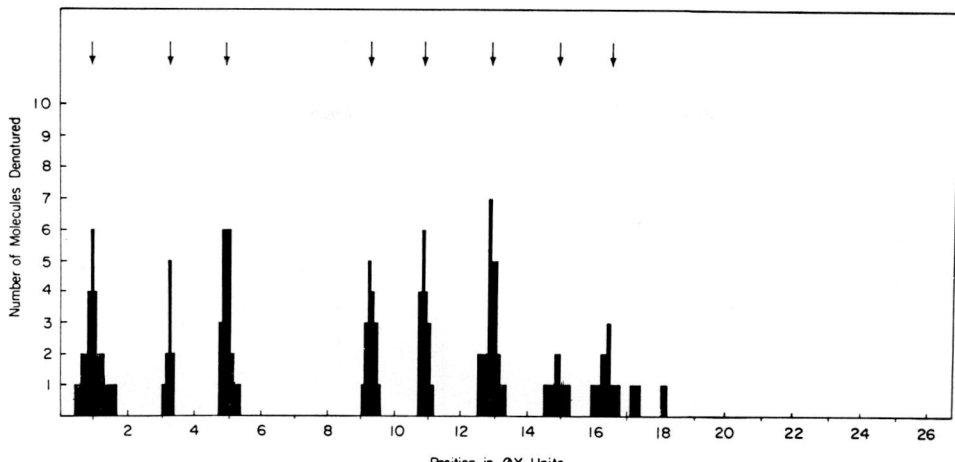

FIGURE 15.   The denaturation map of the pea ctDNA that was constructed with molecules from the spreading described in Figure 14. The average amount of denaturation was 2.5%. The map represents circular DNA molecules and has been linearized for display purposes. The ↓ designates the denatured regions that were found in more than one molecule. The six denatured regions on the left side of the figure are referred to as "major denatured" regions.

mers.[39] Our success in constructing denaturation maps of the circular monomers has enabled us to investigate the integration of the monomer units into the circular dimer. In a head to tail circular dimer, the two monomer units will be in tandem repeat. Therefore, in a partially denatured circular dimer, every denatured region will have a corresponding denatured region one monomer length away. In a head-to-head circular dimer, the two monomer units will be integrated in reverse repeat. Therefore, in a partially denatured circular dimer, it will be possible to locate two points of symmetry on the molecule. The corresponding denatured regions will then be equidistant to either side of these points. The structure of the circular dimer was examined by constructing individual denaturation maps of 10 circular dimers from three spreadings. The maximum number of matchings for each of the two types of integration was determined as described. There were more matched denatured regions when the data were analyzed by head to tail test than head to head test. Thus, the circular dimers are integrated in a head to tail fashion. This conclusion was confirmed by constructing a denaturation map of the circular dimer.

The denaturation mapping experiments described above further support the conclusion that large repeating sequences are not present in the denaturation maps of pea ctDNA. These results show that all of the pea ctDNA molecules are the same, and that the entire information content of pea ctDNA is coded for by the sequence of a circular DNA molecule having a molecular weight of about 90×10^6. However, it should be pointed out that this technique cannot detect small deletions, insertions, or point mutations in a molecule.

I. Presence of Covalently Linked Ribonucleotides in the Closed Circular Chloroplast DNA

1. Alkali Lability of Chloroplast DNA

Covalently closed circular pea ctDNA was centrifuged through 3 M CsC1/0.2M NaOH/0.01 M EDTA, and the centrifugation cell was scanned at 6 min intervals.[40] The DNA zone present in each scan is given in Figure 19. The observed decrease in

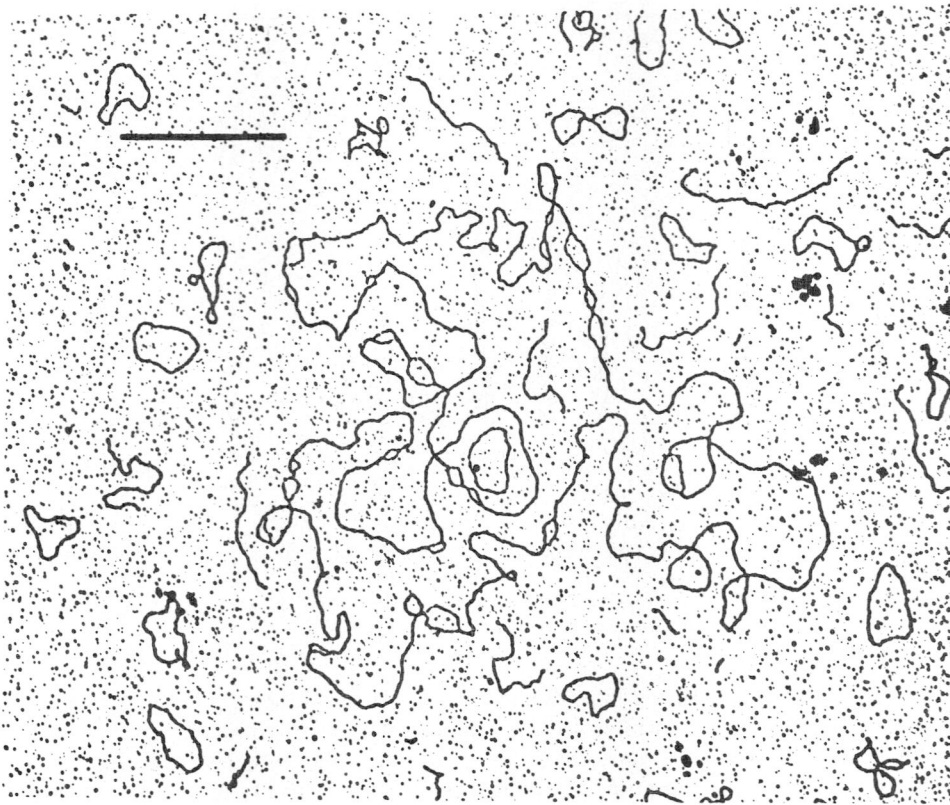

FIGURE 16.   A partially denatured pea ctDNA molecule showing 22 denatured regions. The hypophase contained 47% formamide in 1 × TE and the spreading solution contained 77% formamide in 10 × TE. The small circular molecules are SS and DS φ× DNA. The bar indicates 1 μm.

the size zones could not be accounted for by the effect of radial dilution. By plotting the log of the percent of the DNA remaining against time for spinach ctDNA, the half-life of spinach form I ctDNA was 10 ± 1 min (Figure 20). Similar alkaline sedimentation experiments were performed with both φX × 174 RFI and G4 RFI monomers and dimers. There was no detectable loss of either of these DNAs during alkaline sedimentation. Similarly, there was no loss of DNA when closed circular ctDNA was centrifuged through neutral sedimentation solvent. Thus, the disappearance of closed circular spinach ctDNA from the form IV zone during sedimentation through the alkaline sedimentation solvent was due to a specific effect of alkali on the DNA. The loss of spinach ctDNA from the form IV peak was due to alkali-induced single strand breaks which converted the rapidly sedimenting form IV configuration (264S) to the more slowly sedimenting denatured single stranded form of spinach ctDNA (57.7S). The closed circular ctDNAs from pea and lettuce plants were also analyzed by the above methods. Under identical conditions, pea form IV ctDNA disappeared with a half-life of 10 ± 1 min, while lettuce form IV ctDNA disappeared with a half-life of 15 ± 1 min. The kinetics of degradation of pea, lettuce, and spinach form I ctDNAs were first order and it was possible to observe the degradation of each ctDNA for a minimum of four half-lives with the kinetic data being first order, showing only one component, in all cases.

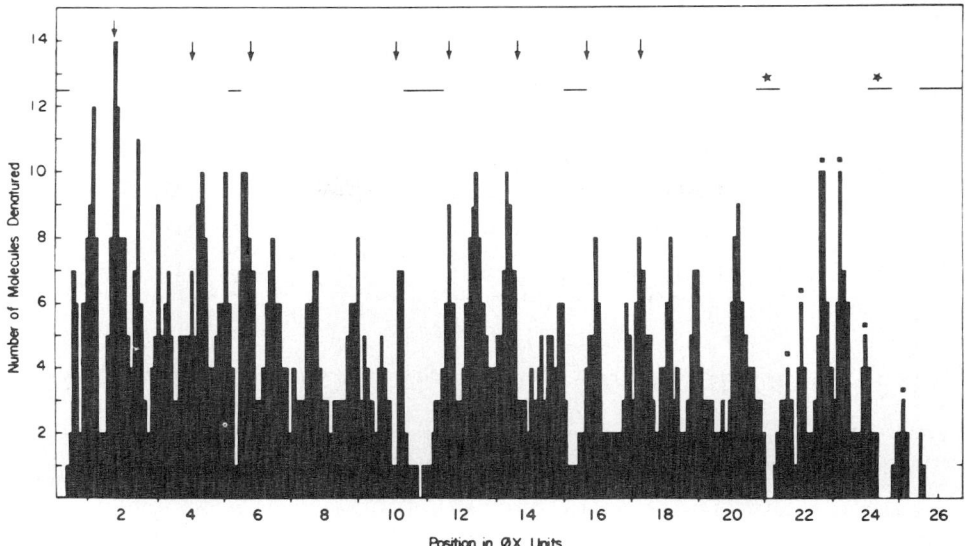

FIGURE 17.   The denaturation map of pea ctDNA constructed with the molecules from the spreading described in Figure 16. The average amount of denaturation was 22%. The ↓ designates the positions of the denatured regions that were present at the 2.5% level of denaturation (Figure 15.) The — indicates the positions of regions that consistently remain undenatured. The   marks two undenatured regions that appear in Figure 18 and the □ indicates six denatured regions that are also present in Figure 18.

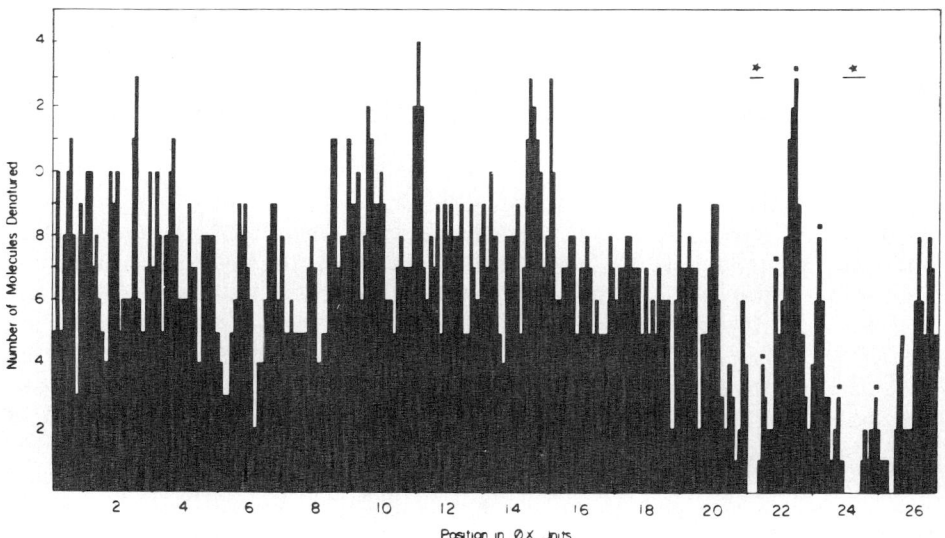

FIGURE 18.   The denaturation map of pea ctDNA constructed from molecules having an average denaturation of 44%. The   indicates the two undenatured regions present in the 22% denaturation map (Figure 17). The □ indicates six denatured regions that are also present in the denaturation map of Figure 10.

2. Ribonuclease Lability of Chloroplast DNA

The alkali sensitivity of ctDNA could be due to the presence of covalently inserted

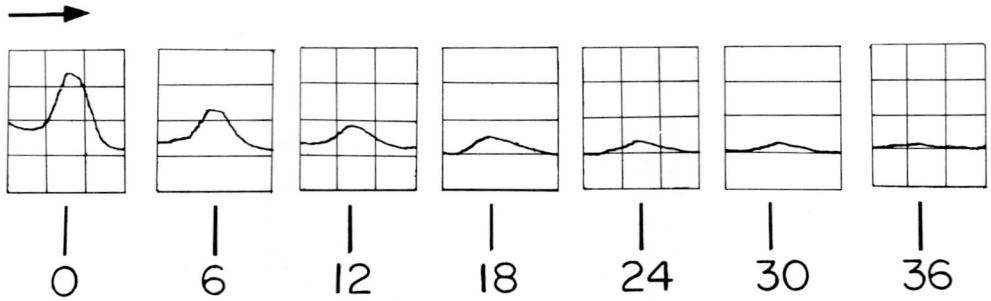

Time Interval From The First Scan In Minutes

FIGURE 19. The alkali breakdown of pea ctDNA. Pea form I ctDNA was centrifuged in the alkaline sedimentation solvent at 16,000 rpm. The form IV peak from scans taken at min intervals during this experiment is shown to illustrate the loss of DNA from the form IV zone. No other zones were present during this experiment, or in other experiments of this type. The zero time peak presented here was scanned at 4 min after layering and the 36-min peak represents 4% of the applied DNA. The experiment presented here represents a period of 4.2 half lives. Sedimentation is from left to right.

ribonucleotides. This was tested by incubating ctDNA with RNases A and T1 under conditions where these enzymes would digest RNA in an RNA-DNA duplex as well as double stranded RNA. The pea form I ctDNA was incubated with a mixture of RNase A and RNase T1 for increasing lengths of time. The covalently closed circular pea ctDNA (89.1S) was successively converted to the open circular form (58.3S). A 3-hr incubation with RNase A and T1 quantitatively converted pea, lettuce, and spinach form I ctDNA to the open circular form, but did not produce any unit length linear (50S) or smaller molecules. G4 RFI monomers and dimers were also incubated with RNase A and T1 under identical conditions. These molecules were not nicked in these experiments.

3. Number of Ribonucleotides in Chloroplast DNA

To determine the number of ribonucleotides present in the ctDNA, the rate of alkaline hydrolysis of ctDNA was compared to the rate of alkaline hydrolysis of *E. coli* [^{32}P] RNA under identical conditions. The kinetic data of hydrolysis of the [^{32}P] RNA were first order and the RNA had a half-life of 180 min. This half-life represents the rate of breakage of a single RNA to RNA phosphodiester bond. The rate of nicking pea and spinach ctDNA was 18 times as fast as the rate of breaking a single RNA to RNA phosphodiester bond. Therefore, pea and spinach form I ctDNA nick at rates that would be expected if they each contained a maximum of 18 ± 2 ribonucleotides. Similarly, the lettuce form I ctDNA was nicked 12 times as fast as an RNA phosphodiester bond, which corresponds to the presence of 12 ± 2 ribonucleotides in the lettuce ctDNA. The ctDNAs we examined were nicked at rates that are 100 to 150 times faster than the rates of nicking of viral and *E. coli* DNAs, which do not contain covalently inserted ribonucleotides.

4. Structure of Nicked Chloroplast DNA

The ctDNAs could contain the ribonucleotides located at one or several sites in either one or both strands of the DNA. To investigate this, nicked ctDNA was studied by sedimentation analysis. Each of the three form I ctDNAs was incubated for 3 hr in 0.2 *M* NaOH at 20°C, and was then sedimented in alkaline 3 *M* CsCl. In each case,

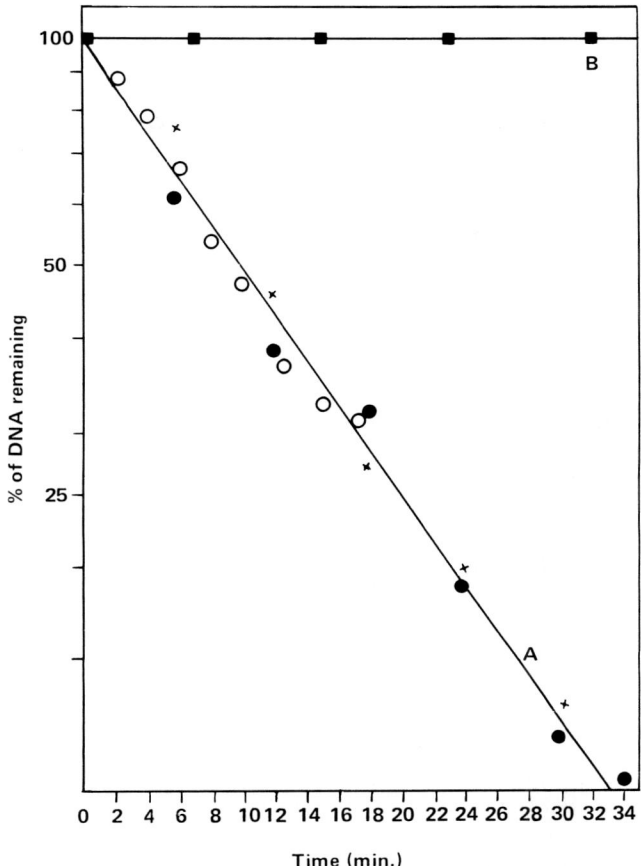

FIGURE 20. The kinetics of the alkali nicking of ctDNA. Spinach ctDNA Form I was sedimented in the alkaline sedimentation solvent at 16,000 rpm and the size of the zones present at various times was determined. The log of the percent of the DNA remaining at various times is plotted. (A), ctDNA; (B), G4 RF I DNA.

20 to 40% of the DNA sedimented at the position of intact single strands, while the rest of the DNA sedimented more slowly as a broad zone. This indicated that the alkaline labile sites could be located at multiple positions in both strands. When the form I ctDNAs were incubated for 3 hr with RNases A and T1, a similar sedimentation pattern resulted. Because centrifugation techniques would require large quantities of ctDNA to determine if RNase treatment of alkaline hydrolysis of form I ctDNA produces specific size classes of fragments, this problem was studied by electron microscopy. Pea form I ctDNA was incubated with 0.2 M NaOH at 20°C for 16 hr (96 halflives) and was mounted for electron microscopy by the formamide technique to visualize single stranded DNA. The length distribution of the fragment sizes produced by this treatment is presented in Figure 21. A large number of fragment size classes were observed. The largest fragment was 12.4 φX units long. The length of intact pea ctDNA is 26.7 φX units. These results indicated that the alkali labile sites in pea ctDNA were located in both strands of the DNA since unit length single stranded circular molecules were not observed.

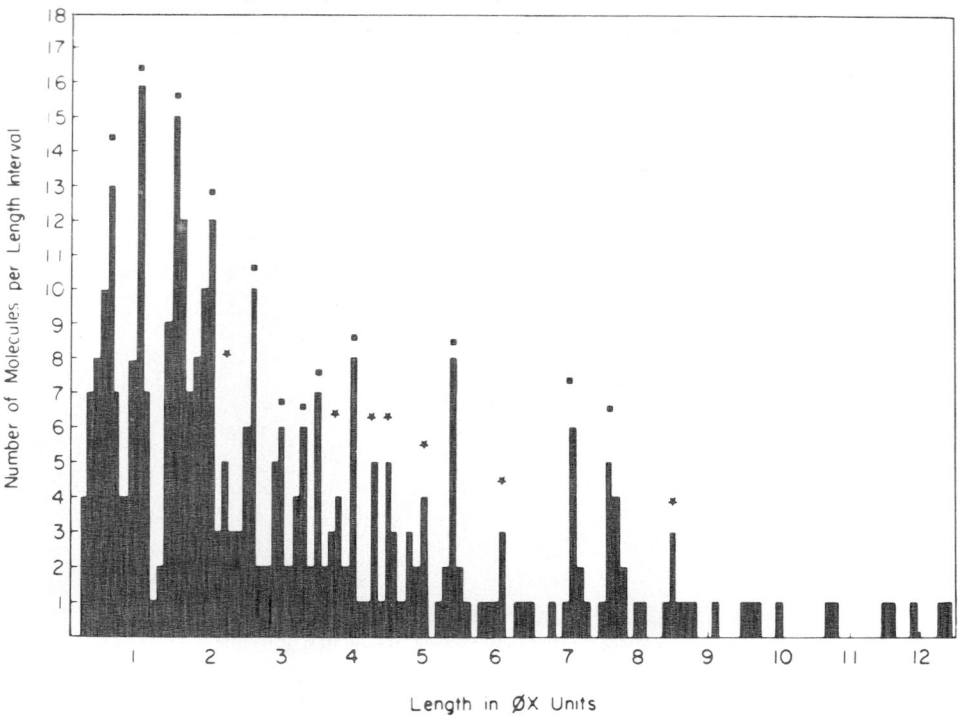

FIGURE 21.   The length distribution of fragments produced by the alkali hydrolysis of pea form I ctDNA. Pea form I ctDNA (10 μg/mℓ) was incubated at 20°C in 0.2 M NaOH, 0.04 M EDTA for 16 hr (≃96 half lives) and was neutralized with 1.8 M Tris HCl, 0.2 M Tris. This DNA was spread with single-stranded φ× 174 DNA by the formamide technique; the spreading solution contained 50% formamide and the hypophase contained 20% formamide. Fields were selected and photographed randomly, and all of the molecules on a negative were measured. The ■ indicates the size classes that match the fragment lengths predicted by the map that is presented in Figure 22. The ✱ indicates the size classes that match fragment lengths resulting from incomplete digestion of pea form I ctDNA, as predicted by the map presented in Figure 22.

5. Map of the Nicks in Pea Chloroplast DNA

The previous experiments indicated that pea ctDNA might contain covalently inserted ribonucleotides located at specific sites. In order to investigate this further, the pea form I ctDNA was incubated in 0.2 M NaOH for 3 hr (18 half-lives), which will nick more than 99% of the molecules, but will not digest all of the alkali labile sites. These fragments then were partially reannealed to produce molecules that generally had single stranded tails and internal duplex sections. This procedure will generate a number of overlapping molecules from which a map containing the positions of the nicks relative to each other can be constructed. This map should be circular with a monomer repeat length of 26.7 φX units (the length of pea ctDNA), if the alkali labile sites are at specific sites. A map of the relative positions of the alkali labile sites in pea ctDNA is presented in Figure 22. At distinct sites, 19 single strand breaks were located. All of the molecules were consistent with this map. It should be noted that at sites 6 and 16, two single-strand breaks were mapped on opposite strands at the same position. When two nicks mapping on opposite strands at the same site were found on a hybrid molecule, one end of the molecule appeared to be fully duplex. The repeat length of the map was 26.7 φX units.

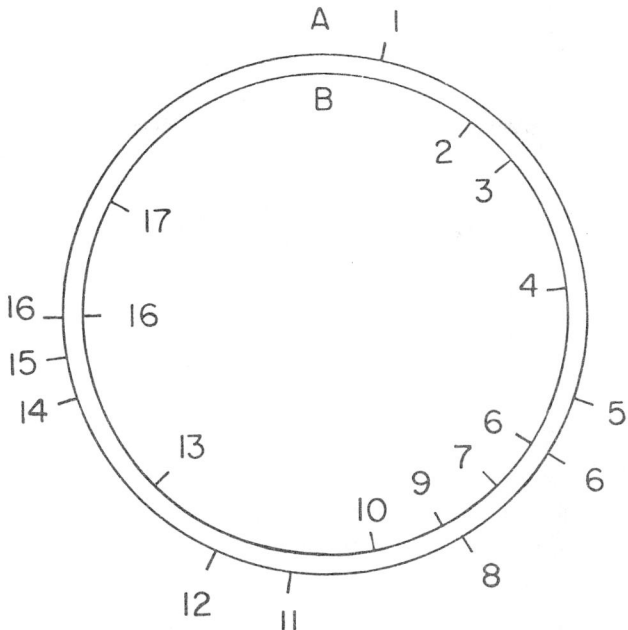

FIGURE 22.   The map of the positions of the alkali labile sites in
pea ctDNA.

6. Strand Location of the Alkali Labile Sites of Pea Chloroplast DNA

If the map presented in Figure 22 accurately represented the locations of the alkaline labile sites, it should be possible to make unambiguous strand assignments for each of the nicks. The two nicks that define a single-strand tail of a reannealed molecule are located on opposite strands. Using these criteria, it was possible to make a list of the nicks that were located on opposite strands from each other. The nicks at positions 1, 5, 8, 11, 12, 14, and 15 (Figure 22, Strand A) were located on one strand, while the nicks at positions 2, 3, 4, 7, 9, 10, 13, and 17 (Figure 22, Strand B) were located on the other strand. This method located nicks on both strands at positions 6 and 16, which agrees with the previous finding that there was one nick located on each strand at these two positions.

The significance of the individually inserted ribonucleotides in form I ctDNA is not understood at this time. The ribonucleotides could arise from a nonstringent DNA polymerase, but in that case, they would not be located at specific positions in the ctDNA molecule. It is possible that the ribonucleotides are remnants of RNA primers which now generally are believed to initiate DNA replication. In pea ctDNA, replication is initiated at two sites located on opposite strands of the DNA molecule. Replication is bidirectional and at least 50% of the ctDNA is synthesized bidirectionally. If the ribonucleotides resulted from incomplete excision of the primers that are used for the initiation of pea ctDNA replication, we should expect to observe only two alkali labile sites. If they were remnants of primers for Okazaki fragments, we would expect to observe on the order of 40 to 50 alkali labile sites. In addition, they would be located more uniformly than the alkali sites we have observed. It should be pointed out that all systems in which RNA primers for DNA replication have been studied, the primer is completely excised in the mature DNA except under abnormal conditions. We consider it possible that these ribonucleotides have some function in ctDNA. These sites

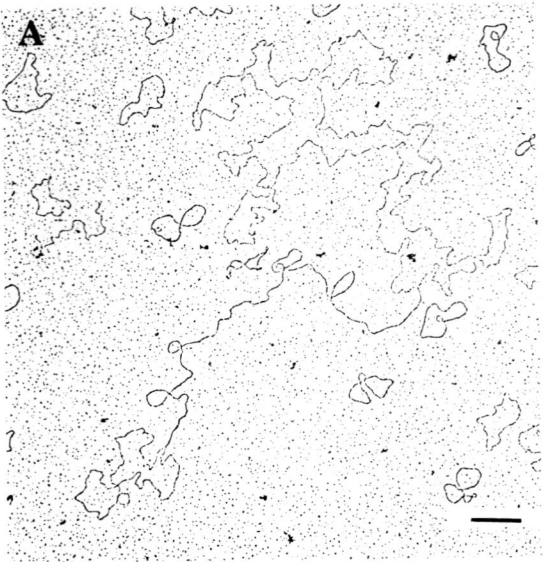

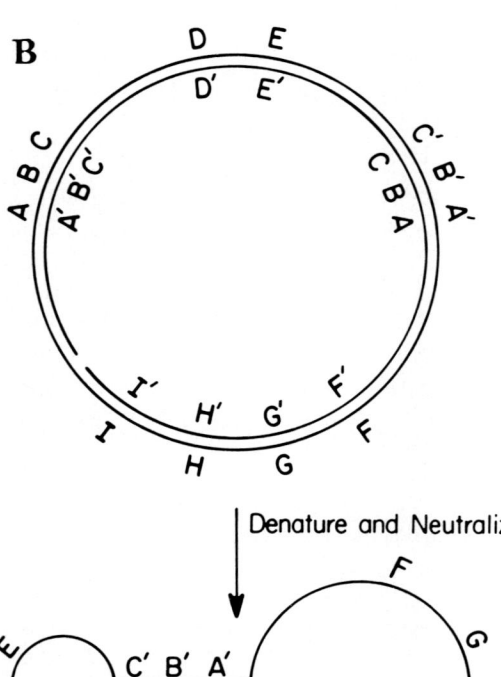

FIGURE 23.   Self renatured circular ctDNA molecule. (A)
Electron micrograph of a self-renatured spinach ctDNA
molecule. The small circles are single-stranded and double-
stranded φ× DNA. The bar indicates 1.0 μm. (B) Illustration
of the formation of a self-renaturated molecule from a circular molecule containing an inverted repeat.

TABLE 8

Length Measurements on Self-renatured ctDNA Molecules

	Small loop (ϕX units)[a]	Duplex region (ϕX units)	Large loop (ϕX units)	N
	N			
Lettuce	3.64(S.D. ± 0.13)	4.56(S.D. ± 0.08)	16.21(S.D. ± 0.48)	21
Spinach	3.44(S.D. ± 0.15)	4.53(S.D. ± 0.15)	16.0 (S.D. ± 0.84)	18
Maize	2.34(S.D. ± 0.20)	4.18(S.D. ± 0.06)	14.9 (S.D. ± 0.62)	18

[a] The data are presented as the ratio of a given length to the length of single-stranded or double-stranded ϕX DNA which were used as internal standards.

could be involved in transcription, recombination, or some other process that requires specific recognition sites. It will require further experimentation to test these possibilities.

J. Inverted Repeats in the Chloroplast DNA of Higher Plants

The structural relationship between the ctDNAs from different species has been further studied by denaturing the nicked circular DNA in alkali followed by neutralization.[40a] When preparations of nicked circular lettuce, spinach, and corn ctDNA molecules were denatured and examined in the electron microscope, DNA molecules that contained one duplex region were observed. A typical example of a partially-duplex spinach ctDNA molecule is presented in Figure 23. About 40% of the observed partially-duplex ctDNA molecules had a central duplex region, a small single-stranded loop on one end of the duplex region and a large single-stranded loop on the other end of the duplex region. The remaining partially-duplex ctDNA molecules contained a single-stranded loop at one end of the duplex region and had two single-stranded branches at the other end of the duplex region. As many as 70 to 90% of the denatured ctDNA molecules were found in the form of partially-duplex ctDNA molecules (400 DNA molecules from each DNA preparation were examined). The remaining DNA molecules were single-stranded linears which were less than the length of ctDNA.

A summary of length measurements of partially-duplex lettuce, spinach, and corn ctDNA molecules is presented in Table 8. The duplex segment of lettuce and spinach ctDNA was 24.4 kilobases (kb) long and amounted to approximately 16% of the native length of these two ctDNAs. The corn duplex segment was 22.5 kb long, smaller than the lettuce and spinach duplex segments, but still amounting to approximately 16% of the native length of corn ctDNA because the molecular size of corn ctDNA is smaller than lettuce and spinach ctDNAs. The small single-stranded loop in partially-duplex spinach ctDNA appeared to be slightly smaller (1 kb) than the corresponding loop in lettuce ctDNA, while the large single-stranded loops of these two ctDNAs were quite similar in size. Both the small and large single-stranded loops in partially-duplex corn ctDNA molecules were significantly smaller than the corresponding structures in lettuce and spinach ctDNA. These results were consistent with corn, lettuce and spinach ctDNA molecules containing a sequence that is repeated once in reverse polarity and can participate in intramolecular renaturation. The structure and formation of a partially-duplex ctDNA molecule from a circular ctDNA molecule containing a sequence repeated in reverse is illustrated in Figure 23B.

Denatured pea ctDNA molecules did not form any self-renatured DNA molecules. Four different pea ctDNA preparations were studied as described above and no self-renatured molecules like those observed in corn, spinach, and lettuce ctDNA were

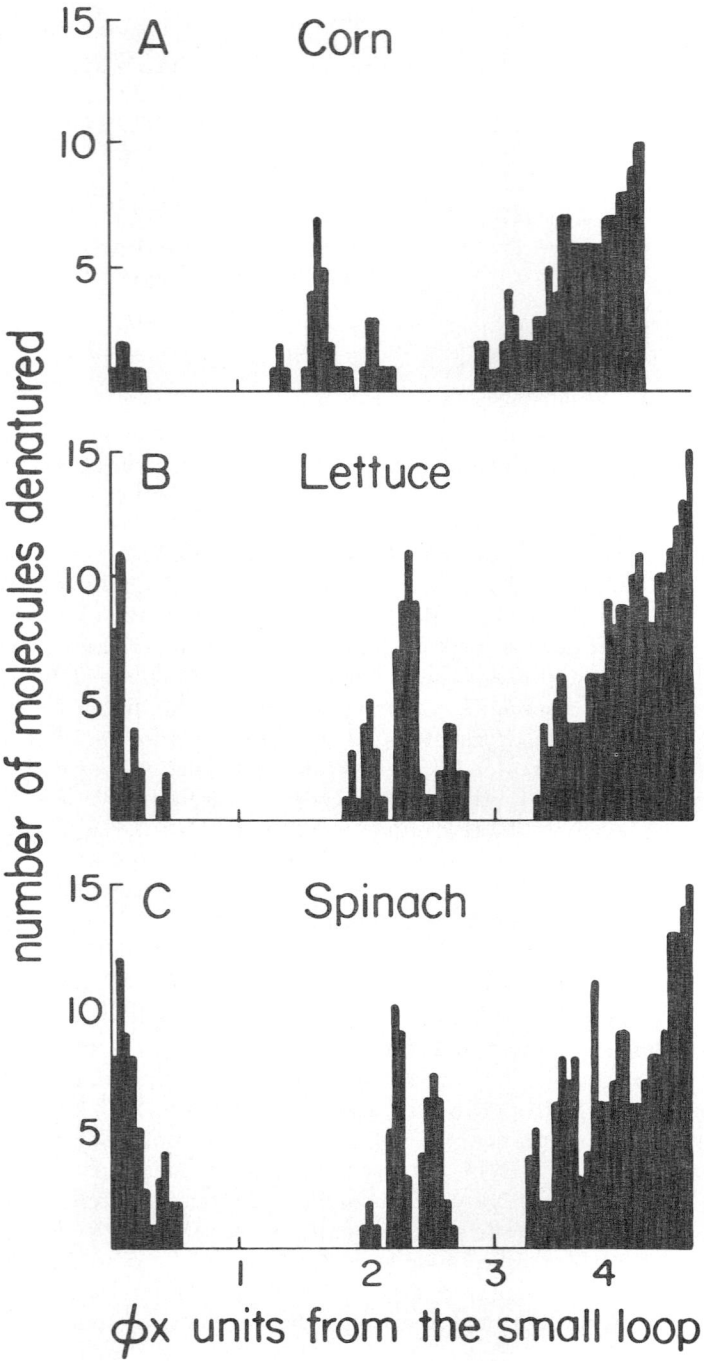

FIGURE 24.   Denaturation maps of the duplex segment of self-renaturated ctDNA molecules. (a) Corn , (b) Lettuce, (c) Spinach. The denaturation maps are oriented so that the small loop of the self—renatured molecules is on the left side of the figure and the large loop is on the right side. All lengths were standardized by using φ× DNA as the internal standard.

found (<0.05%). The pea ctDNA preparations used for these experiments were highly intact, with approximately 40% of the single-stranded ctDNA molecules being unit length single-stranded circles while most of the rest of the single-stranded DNA molecules were unit length linear molecules. Incubating the denatured ctDNA molecules under less stringent renaturing conditions did not result in any intramolecular renaturation. Mixing experiments showed that preparations of pea ctDNA would not inhibit self-renaturation by denatured spinach and lettuce ctDNA molecules. The pea ctDNA molecules did not self-renature during the mixing experiments and were observed as linear and circular single-stranded DNA molecules the size of pea ctDNA.

1. Denaturation Mapping of the Inverted Repeat

When self-renatured ctDNA molecules were mounted for electron microscopy, from a spreading solution containing 78% formamide, onto a hypophase containing 48% formamide, the duplex region of these DNA molecules was found to be partially denatured. The extent of denaturation ranged from 12% for corn ctDNA to 17% for spinach ctDNA. Denaturation maps of the duplex region of self-renatured corn, lettuce, and spinach ctDNA molecules were constructed from partially denatured DNA molecules as previously described and are presented in Figure 24. The three denaturation maps show a striking degree of similarity. At the large loop side, they all have a highly denatured region that is 1.3 ϕX units long. Further in from the large loop, all three denaturation maps have a native region that is 0.55 to 0.65 ϕX units long followed by three denatured regions covering a distance of 0.75 to 0.95 ϕX units. To the small loop side of the three denatured regions, spinach and lettuce ctDNA have a native region 1.4 ϕX units long followed by two small denatured regions at the small loop end of the duplex segment. Corn ctDNA is somewhat different in this region. It has a smaller native region that is 0.95 ϕX unit long followed by a single denatured region at the small loop end of the duplex segment. It is not clear if the denatured region at the small loop end of the corn duplex segment is the same as the corresponding denatured region in the spinach and lettuce duplex. However, the shortening of the native region in this area could fully account for the difference in length between the corn duplex segment and the duplex segments of lettuce and spinach ctDNA. With the exception of the differences at the small loop side of corn ctDNA, the sequences of the inverted repeats in these three ctDNAs appear to be highly related.

2. The Structure of Circular Dimers

Circular dimers constitute as much as 3 to 4 % of the circular ctDNA molecules from higher plants. A circular dimer can conceivably consist of two monomers joined in either a tandem repeat (head-to-tail circular dimer) or in an inverted repeat (head-to-head circular dimer). These two arrangements can be distinguished after denaturation of a relaxed circular dimer since head-to-tail circular dimers should yield dimer length single-stranded circular and linear molecules while a head-to-head circular dimer should renature to form a monomer length double-stranded linear molecule. This approach was used to examine the structure of pea, lettuce, and spinach ctDNA circular dimers.

After denaturation and neutralization of pea ctDNA, 1% of the single-stranded molecules were either dimer length linears or circles (16 circles and 23 linears were observed). No monomer length duplex molecule was found (<.0.025%). These results were consistent with most pea ctDNA circular dimers being the head-to-tail conformation and confirmed the previous results obtained by denaturation mapping of pea ctDNA circular dimers.

When relaxed circular lettuce and spinach ctDNA molecules were denatured and neutralized, two new types of dimer length molecules were found. A typical electron micrograph of the first type of dimer is presented in Figure 25. The molecule has a small, equal length, single-stranded loop on each end. Each loop is attached to a separate duplex segment of equal length and the two duplex segments are connected to each other by two equal length single-stranded segments. The duplex segments are equal in length to the inverted repeat discussed above. The small single-stranded loops are the size of the small spacer between the inverted sequences of the monomer, and each internal single-stranded segment of the dimer is the same length as the large spacer between inverted sequences of the monomer. This molecule is consistent with its being formed from a head-to-tail circular dimer. Its formation is illustrated in Figure 25B. An electron micrograph of the second type of circular dimer is presented in Figure 26. The molecule consists of two equal length single-stranded loops joined by a long duplex segment. The length of the single-stranded loops is equal to the length of the small spacer between the inverted sequences of the monomer and the length of the duplex segment is equal to the sum of the lengths of two of the inverted sequences and one large spacer. This molecule is consistent with its being formed from a head-to-head circular dimer. Its formation is illustrated in Figure 26B. Approximately 1.5% of the self-renatured lettuce and spinach ctDNA molecules were circular dimers (2000 ctDNA molecules from each plant were examined) and the head-to-head dimers were approximately 4 to 5 times as prevalent as the head-to-tail dimers (24 out of 30 dimers in spinach ctDNA and 24 out of 29 dimers in lettuce ctDNA).

The results presented above are consistent with the idea that lettuce, spinach, and corn ctDNAs each contain a sequence amounting to approximately 16% of their genome length repeated once in reverse polarity. Denaturation mapping studies suggest that the repeated sequence in the ctDNAs is highly related. The greater similarity between the repeated sequences of lettuce and spinach ctDNA than between either of these two ctDNAs and corn ctDNA is in line with the established evolutionary divergence of these plants. It is not obvious why these ctDNAs have such a large sequence repeated in reverse. However, this organization does have some advantages over two sequences repeated in tandem. An intramolecular recombination event between two tandemly repeated sequences could lead to the excision of the segment located between the two repeated sequences while an intramolecular recombination event between two inverted sequences could, at worst, reverse the polarity of segment between the two sequences, but would not result in the physical loss of any genetic material. Intramolecular recombination events between the two inverted sequences also would tend to prevent their divergence from each other. In any event, the inverted conformation probably is not strictly required for ctDNA function since pea ctDNA does not have this sequence repeated in reverse. It will be interesting to see if pea ctDNA has a similar sequence that possibly is repeated once in a tandem repeat.

A sequence that is 16% of the length of ctDNA is a large proportion of the ctDNA to be repeated a second time and suggests that the genes encoded by this sequence are important for the function of the ctDNA. Studies on the structure of corn ctDNA suggest that one copy of the two sets of chloroplast specific rRNA genes is located on each of the corn ctDNA inverted sequences (see below). In view of the striking similarity between the denaturation maps of the corn, spinach, and lettuce ctDNA inverted sequences, it is probable that the inverted DNA sequence also codes for the chloroplast rRNA genes in spinach and lettuce ctDNA. In any event, a minimum of 12 to 15% of each inverted sequence would be required to code for one set of chloroplast rRNA genes (not including spacers), which probably leaves room for other genes on the inverted sequences.

The studies presented here also have provided information on the structure of circular dimers of ctDNA. The finding that most, if not all, pea ctDNA circular dimers are in a head-to-tail configuration is consistent with the denaturation maps described

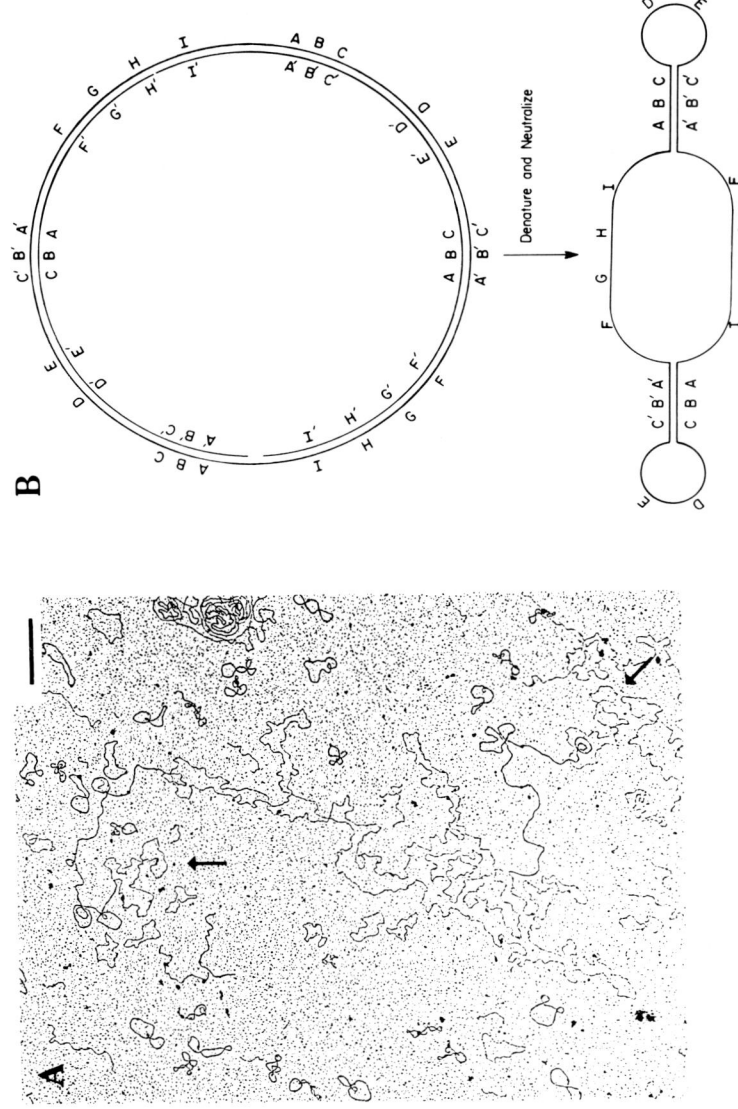

FIGURE 25. Self-renatured head to tail circular dimer. (A) Electron micrograph of a self-renatured ctDNA head to tail circular dimer. The arrows point to the single-stranded loops at the end of the molecule. The small circles are single-stranded and double-stranded φ DNA. The bar indicates 1 μm. (B) Illustration of the formation of a self-renatured molecule from head to head circular dimer containing two sets of inverted sequences.

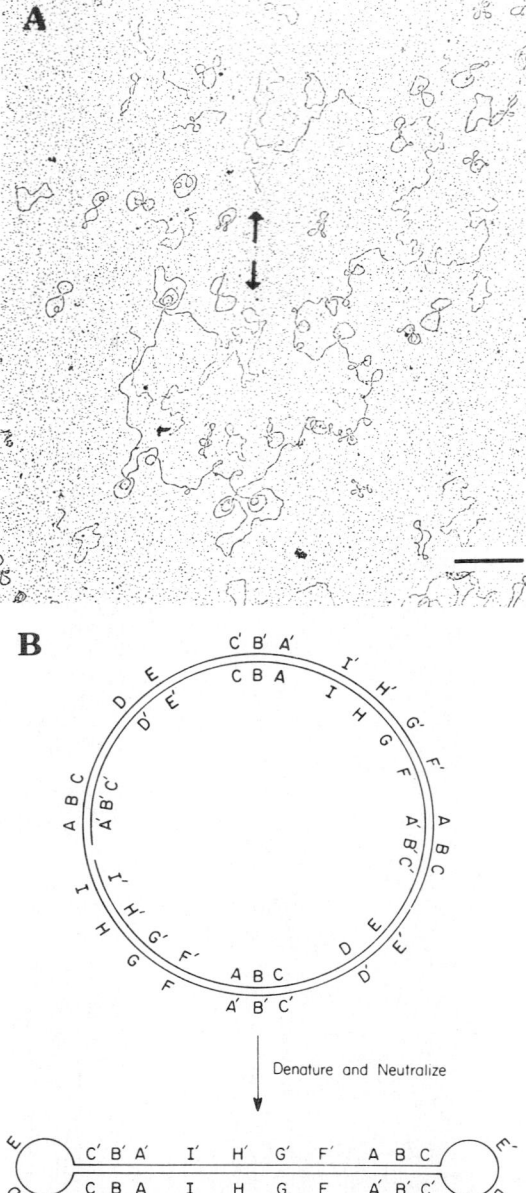

FIGURE 26.   Self-renatured head to head circular dimer. (A) Electron micrograph of a self-renatured
ctDNA head to head circular dimer. The arrows point
to the single-stranded loops at the end of the molecule. The small circles are single-stranded and double-
stranded φx DNA. The bar indicates 1 μm. (B) Illustration of the formation of a self-renatured molecule from
head to head circular dimer containing two sets of inverted sequences.

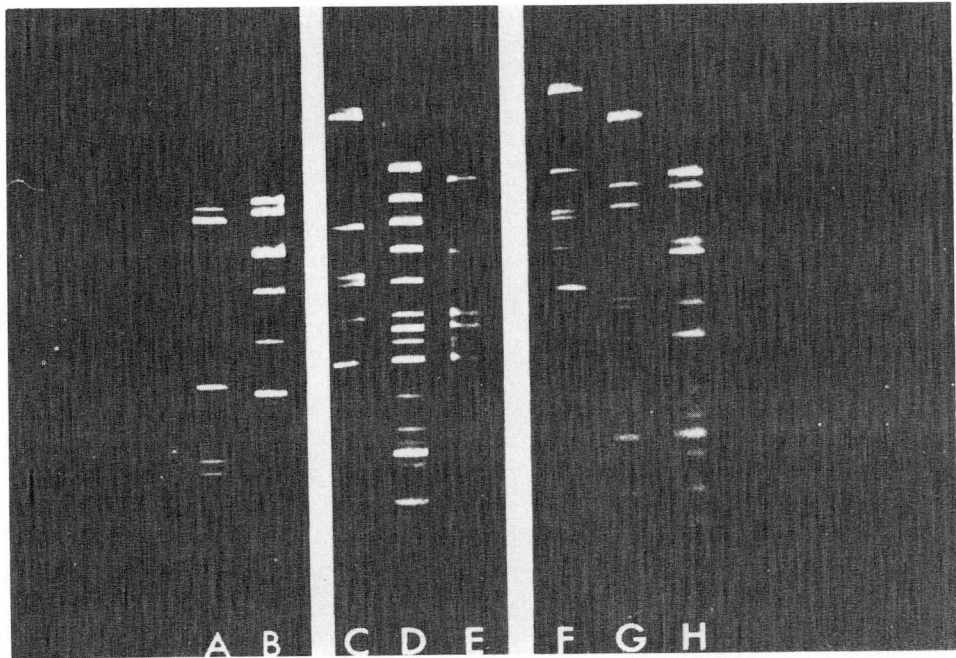

FIGURE 27.   Fractions of chloroplast-DNA fragments by agarose gel electrophoresis. (A) Lambda phage DNA digested with EcoRl, 0.75% agarose gel. (B) Chloroplast DNA digested with Sal I, 0.75% agarose gel. (C) Lambda phage DNA digested with Hae III, 0.85% agarose gel. (D) Chloroplast DNA digested with Bam I, 0.85% agarose gel. (E) Chloroplast DNA digested with both Bam I and Sal I, 0.85% agarose gel. (F) DNA as in (C), 1.0% agarose gel. (G) Chloroplast DNA digested with EcoRl, 1.0% agarose gel. (H) Chloroplast DNA digested with EcoR1 and Sal I, 1.0% agarose gel. (From Bedbrook, J. R., and Bogorad, H., Proc. Natl. Acad. Sci., U.S.A., 73, 4309, 1976. With permission.

before and is similar to the results obtained with circular dimers of φX RF DNA and mitochondrial DNA. The finding of circular dimers of lettuce and spinach ctDNA that are in a head-to-head configuration is unusual. The formation of head-to-head circular dimers appears to be associated with the presence of the inverted repeat in ctDNA. This could most easily be explained if circular dimers of ctDNA were formed by a recombination event between two circular monomers. A head-to-head circular dimer would be formed by a recombination event between two circular monomers at their inverted sequences, in which one monomer was inserted into the other monomer in reverse polarity.

K. Endonuclease Recognition Sites in Chloroplast DNA

The discovery of restriction endonucleases (the enzymes which recognize specific sequences in the DNA molecules) has brought a new era to the molecular biology of chloroplast DNA. It is now possible to produce physical maps of ctDNAs with the help of restriction endonucleases. The availability of physical maps of ctDNA helps us to map the genes in this organelle DNA. This was only possible before by genetic studies. However, the genetic studies have not been able to contribute significantly to our understanding of the genetic information in ctDNA. This was mainly due to the complex genetic analysis required to understand the mutation, segregation and recombination of ctDNA.

The corn ctDNA has been elegantly studied for restriction endonuclease sites by

TABLE 9

Size of Chloroplast DNA fragments Produced by SAL I restriction Endonuclease

Fragment Size $\times 10^{-6}$

Maize	Spinach	Pea
16.5	29	32
13.9	13.9	15
10.6	13.0	10.5
10.6	8.6	8.5
8.4	6.8	6.5
5.9	6.8	6.2
4.4	5.3	6.2
4.4	3.8	1.5
4.4	3.3	
4.2	2.4	
4.0	0.45	

Bedbrook and Bogorad.[41] The gel electrophoresis patterns of corn ctDNA fragments produced by terminal digestion with Sal I, Bam I, and EcoR1 are given in Figure 27. The sizes of the DNA fragments produced by Sal I digestion of corn ctDNA with their molar ratios are given in Table 9. Pea ctDNA has been similarly studied by restriction endonucleases and the DNA fragments produced by digestion with Sal I, and Sma and Sal are shown in Figure 28. The molecular sizes of the DNA fragments produced by digestion of pea ctDNA with their molar ratios is shown in Table 9 along with the DNA fragments obtained from spinach ctDNA.[42] The data clearly show that the ctDNAs from corn, pea, and spinach have different endonuclease sites for the enzyme Sal I. The number of DNA fragments produced by this enzyme and the molecular sizes of the DNA fragments are entirely different in ctDNA for each of these higher plants. These data again confirm the earlier conclusion that the ctDNA from higher plants differ in their base sequences.

The physical map of ctDNA can be produced using a number of restriction endonucleases. For example, when corn ctDNA is digested with Bam I and Sal I together, the pattern of DNA bands obtained is shown in Figure 27E. Similar experiments by digesting corn ctDNA with both R1 and Sal I is shown in Figure 27H . By comparing Figure 27D and E and Figure 27G and H, it was determined that several Bam I fragments and RI fragments are further fragmented by Sal I. Figure 27 demonstrated that Bam I fragments 1, 2, 3, 4, 5, and 12 contain recognition sites for Sal I. Sal I digestion of isolated Bam I fragments demonstrated that fragment 7 and 13′ were further digested by Sal I. Figure 27G and H show that fragments a, c, d and g are fragmented by Sal I. These fragments containing Sal I sites were isolated and copy RNA was transcribed (using *Escherichia coli* polymerase and labeled nucleotide triphosphates) from isolated DNA fragments. The copy RNA was then hybridized with the Sal I restriction fragments of the ctDNA by the Southern technique.[43] The hybridization to two or more Sal I fragments would indicate that those fragments are linked in ctDNA. This technique of DNA sequence homology between an overlapping fragment and the fragments being overlapped was the major method used to arrive at the physical map of corn ctDNA by Bedbrook and Bogorad[44] (Figure 29). The ability to construct the physical map of ctDNA supports the view that the majority of the circular DNA molecules

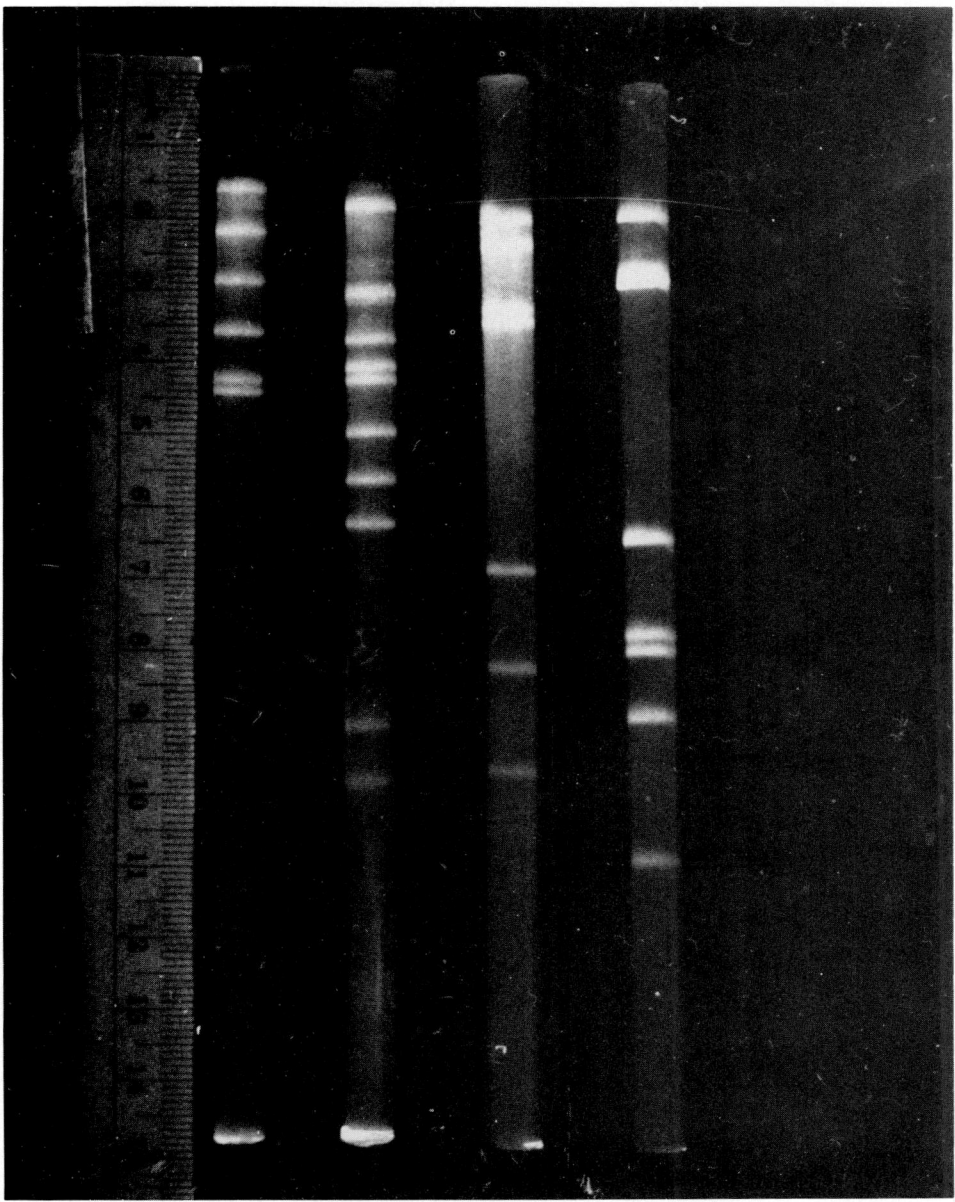

FIGURE 28.   Fractionation of pea ctDNA fragments by agarose gel electrophoresis. From left to right;
pea ctDNA digested with Sal I, pea ctDNA digested with Sal I and Sma, pea ctDNA digested with Sma,
and λDNA digested with EcoR1 and intact λDNA.

isolated from corn chloroplasts represent a single homogeneous species. The restriction
map of corn ctDNA also shows that 15% of the ctDNA genome is repeated. The two
copies of this sequence are in inverted orientation with respect to one another and are
separated by a nonhomologous sequence representing approximately 10% of the genome. These data are in agreement with those obtained by electron microscopic studies
and described earlier.[45]

Similar experiments have been carried out with ctDNA of *Euglena gracilis*. The frag-

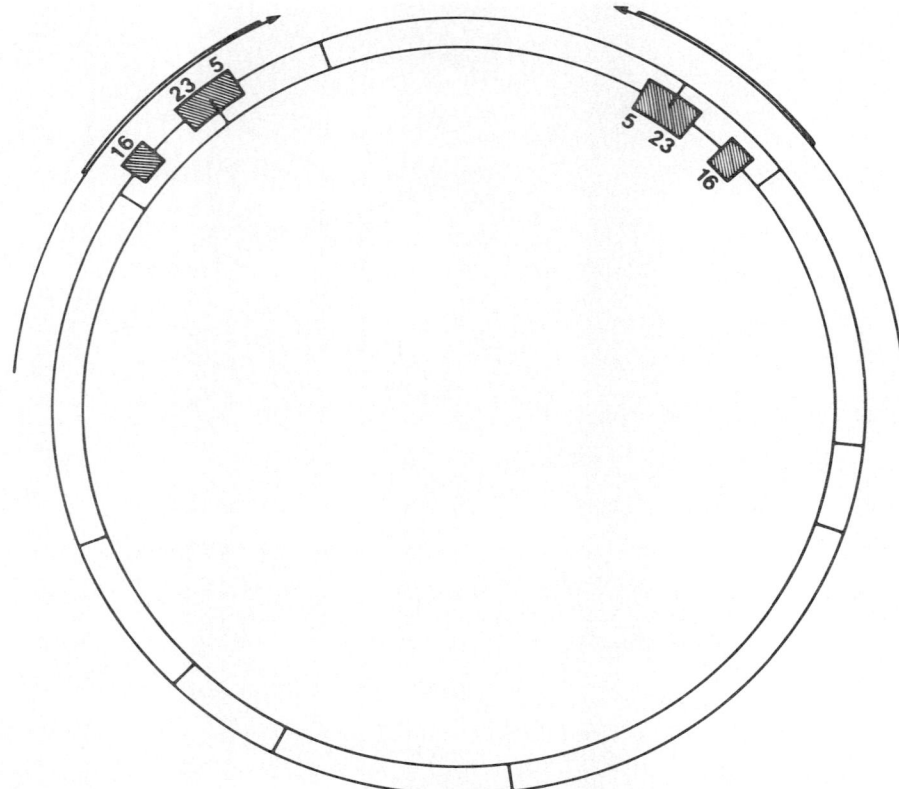

FIGURE 29.   The location of the two sets of rRNA genes on chloroplast DNA relative to recognition sites for the restriction endonuclease Sal I and the inverted repeat sequence. The two
concentric circles represent the two DNA strands of corn chloroplast DNA. The bars on the circles
represent the known location of recognition sequences for the restriction endonuclease Sal I. The
two arms of the inverted repeat are represented by arrows. The location of Eco Rl fragments
within the inverted repeat is indicated by a double thickness line. The position and strand of DNA
coding the maize chloroplast rRNAs are illustrated by cross-hatched boxes. (From Bedbrook, J.
R., Kolodner, R., and Bogorad, L., *Cell*, 11, 739, 1977. © by The MIT Press. With permission.)

ments produced by the enzyme Hae III are given in Table 10. This enzyme produces
about 51 to 52 fragments which added up to a molecular size of 87×10^6. The molecular
weight of *Euglena* ctDNA obtained by the sum of the molecular weights of the restriction fragments also was very close to that obtained by electron microscopy and renaturation kinetics.

The restriction endonuclease digestion of ctDNA from *Chlamydomonas* with
EcoR1, Bam I, Hind III, and Sal I has been studied.[26] In digestions with all these
enzymes, there were DNA fragments in nonstoichiometric amounts unlike the results
obtained with ctDNA from *Euglena* and higher plants. EcoR1 digestion generated at
least 35 bands; and if one sums up the molecular weights of the fragments, taking into
account that some of them comigrate in the gel, one obtains a molecular weight of
$120 \pm 5 \times 10^6$. Thus, it is not known whether there is some heterogeneity in the base
sequences of ctDNA from *Chlamydomonas*.

III. GENETIC INFORMATION IN CHLOROPLAST DNA

The genetic information contained in the ctDNA largely has been studied by molec-

TABLE 10

Molecular Weights of endo R HaeIII DNA Fragments from *Euglena* Chloroplast DNA

Band number	$10^{-6} \times M$, of DNA	Fragment copies	$10^{-6} \times M$, of DNA per band
1	5.7	3	17.1
2	5.3	2	10.6
3	4.3	1	4.3
4	3.9	1	3.9
5	3.7	2	7.4
6	3.2	2	6.4
7	2.4	3	7.2
8	2.2	1	2.2
9	2.1	1	2.1
10	1.9	1	1.9
11	1.8	1-2	1.8-3.6
12	1.6	1	1.6
13	1.4	1	1.4
14	1.2	1	1.2
15	1.1	1	1.1
16	0.88	1	0.88
17	0.82	1	0.82
18	0.81	1	0.81
19	0.80	3	2.40
20	0.78	1	0.78
21	0.73	2	1.46
22	0.72	1	0.72
23	0.69	1	0.69
24	0.68	1	0.68
25	0.64	2	1.28
26	0.59	1	0.59
27	0.51	1	0.51
28	0.50	1	0.50
29	0.38	1	0.38
30	0.36	1	0.36
31	0.29	3	0.87
32	0.28	1	0.28
33	0.20	2.	0.40
34	0.19	2	0.38
35	0.16	1	0.16
36	0.15	1	0.15

Note: The bands resolved in six gels are ranged according to molecular weights. The numerical values in the third column indicate the relative staining intensities of the bands.

From Kopecka, H., Crouse, E. J., and Stutz, E., *Eur. J. Biochem.*, 72, 525, 1977. With permission.

ular DNA-RNA hybridizations. Such studies have shown that ctDNA codes for ribosomal RNA (rRNA), transfer RNAs (tRNA), and messenger RNAs (mRNA).

A. Ribosomal RNA Genes in Chloroplast DNA

Chloroplasts have been found to contain 70 S ribosomes as compared to 80 S ribosomes found in the cytoplasm (Figure 30). The 70 S ribosomes from chloroplasts can be prepared by adding Lubrol® to a final concentration of 0.25% to the chloroplast fraction obtained by differential centrifugation. The lysed chloroplasts are centrifuged

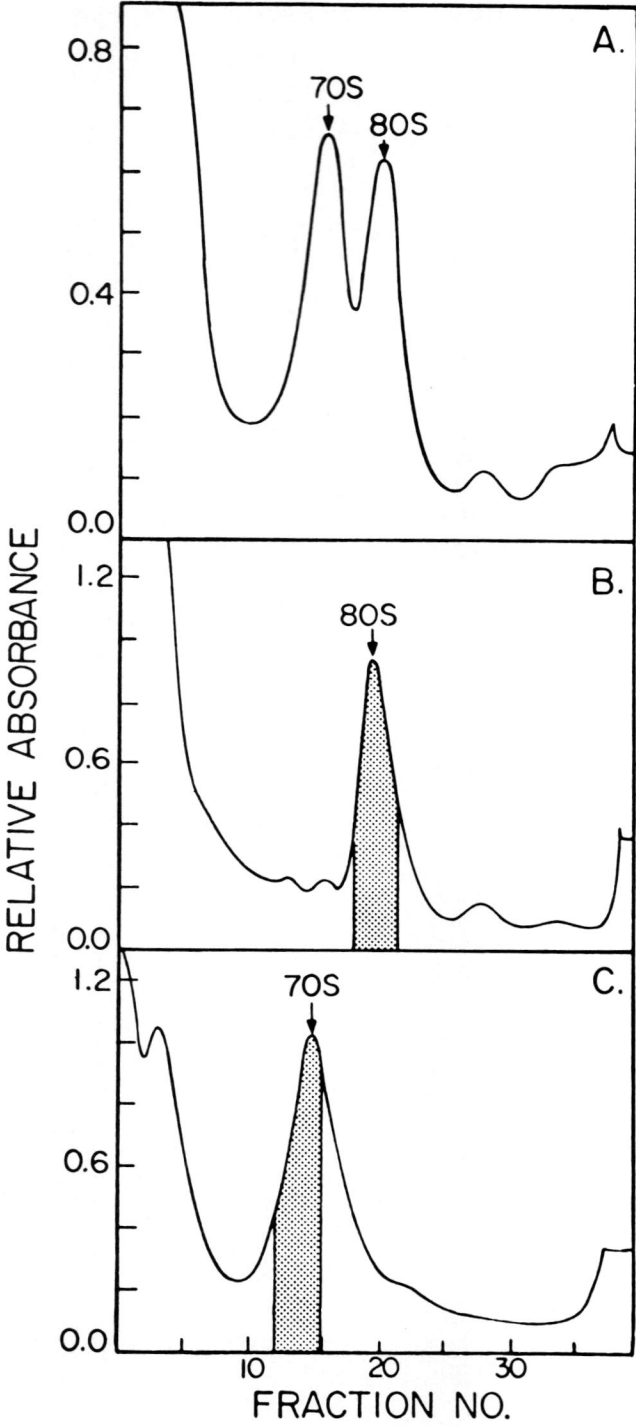

FIGURE 30.   Sedimentation of ribosomes on sucrose gradient. Ribosomes were centrifuged in a 15 to 30% linear sucrose gradient at 24,000 rpm for 5 hr in SW27 rotor. (A) Total ribosomes of leaf; (B) Purified cytoplasmic ribosomes; (C) Purified chloroplast ribosomes.

at 17,000 × g for 30 min and the top three fourths of the supernatant solution above the loosely packed pellet is collected. Aliquots containing 60 to 80 A_{260} units are then loaded on 10 to 30% linear sucrose gradients. After 4 hr of centrifugation at 27,000 rpm in a SW-27 rotor, the gradients are fractionated for 70 S ribosomes. The sedimentation pattern usually obtained is shown in Figure 30C. The RNA from the pooled fractions of ribosomes is extracted by adding 30 mℓ of buffer containing 0.05 M Tris·HC1/25 mMMgC1$_2$/25 mMKC1/2% Triton X-100 (pH 7.5) and 30 mℓ of water-saturated phenol. After three such extractions, two volumes of 95% ethanol are added to the final aqueous phase and the precipitated RNA is suspended in buffer containing 2 mM Mg(OAc)$_2$/0.1 mM EDTA/0.05 M Tris·HC1 (pH 8.0). The same procedure is used for the isolation of ct-rRNA from other plants. The RNA obtained by this extraction procedure yields pure undegraded rRNA as shown in Figure 31.

The number of rRNA genes in ctDNA was determined by hybridizing pure ctDNA with the in vivo labeled rRNA. The purified ctDNA had a hyperchromicity of 36%. The concentration of DNA in different preparations was calculated on the basis of their hyperchromicity. Such a procedure eliminates variations in the DNA concentrations resulting from spurious absorption or nonspecific color formation with diphenylamine. Covalently closed circular molecules were obtained by the equilibrium density-gradient centrifugation of ctDNA in CsC1-ethidium bromide gradients. The hybridizations were carried out as follows: ctDNA was denatured by adding 1 part of 1 M NaOH to 8 parts of DNA in 1 × SSC. After 15 min at room temperature, the solution was neutralized by the addition of 1 part of 2 MNaH$_2$PO$_4$. Aliquots containing 1.0 to 10.0 μg of DNA in 3.0 mℓ volume were passed through prewashed nitrocellulose filters (S and S, B6). The filters were then washed with 45 mℓ of 6 X SSC, dried in a vaccum desiccator overnight at room temperature, heated for 6 hr at 75°C in a vacuum oven and used within 24 hr for hybridization. Closed vials containing both a pea ctDNA filter and a control filter containing an equal amount of calf thymus DNA were incubated for 16 hr at 65°C in 2.0 mℓ of 4 × SSC. After the incubation, each filter was treated with 10 μg/mℓ of ribonuclease in a 5.0 mℓ volume of 2 × SSC for 2 hr at room temperature. Each side of the filter was then washed with 60 mℓ of 2 × SSC, dried, and counted in a Beckman LS-230 scintillation counter.

1. Number of Ribosomal RNA Genes in Chloroplast DNA

Purified total ctrRNA was hybridized with 1.0 and 2.5 μg of pea ctDNA. From the saturation-hybridization curve, the maximum hybridization of 4.2% was obtained. The shape of the saturation curve showed that there were no heterogeneous populations in the ct-rRNA preparations. In different experiments involving varying concentrations of RNA and DNA, 4.07% ± 0.35% of the ctDNA was found to be complementary to rRNA. DNA-rRNA hybridization was also carried out in solution and the hybrids fractionated by hydroxyapatite chromatography. The maximum hybridization obtained was 4.0% at the end of 3 hr. In this experiment, the size of the fragmented DNA was 2.8×10^6 daltons and expected self annealing in 2 hr at this size and concentration would be about 10%. Therefore, the amount of hybridization obtained with solution hybridization was very close to that obtained by filter hybridization.

The amount of hybridization obtained in the above experiments would be affected if ct-rRNA preparations were contaminated with mRNAs, tRNAs, or cytoplasmic rRNAs. It has been found that ctDNA does not contain base sequences complementary to rRNA of the cytoplasm. Furthermore, cytoplasmic rRNA was not found to compete with ct-rRNA for hybridization with ctDNA. As analyzed by acrylamide gel electrophoresis, ct-rRNA preparations were found to be essentially free of mRNA or tRNA. The specificity of ctDNA and ct-rRNA hybrids also has been confirmed in competition experiments involving cold ct-rRNA. The thermal stability characteristics of the DNA-

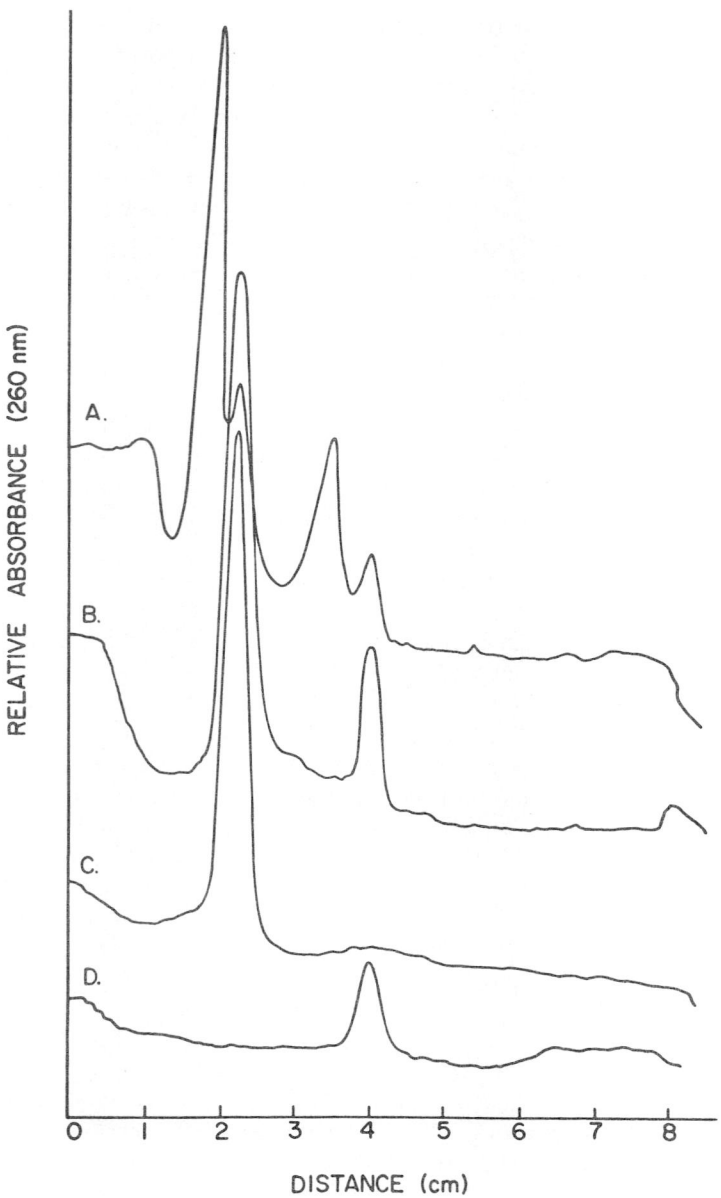

FIGURE 31.   Acrylamide gel electrophoresis of fractionated and unfractionated plant rRNAs. (A) Pea total leaf rRNA; (B) Purified pea ctRNA; (C) Pea
23S rRNA; (D) Pea 16S rRNA. Electrophoresis was carried out in MgB-buffer
(9) for 5 hr at 5 ma per tube using 2.6% acrylamide gels (5% *bis*-acrylamide).

rRNA hybrids for the presence of mismatched base sequences was studied. The hybrid
was relatively stable up to 70%C, after which it began to melt with a sharp T_m (temperature of half dissociation) of about $82 \pm 2°C$.

2. Hybridization with the Subunits of Chloroplast-Ribosomal RNA

The fractions corresponding to 23S and 16S regions were collected after ^{32}P-labeled

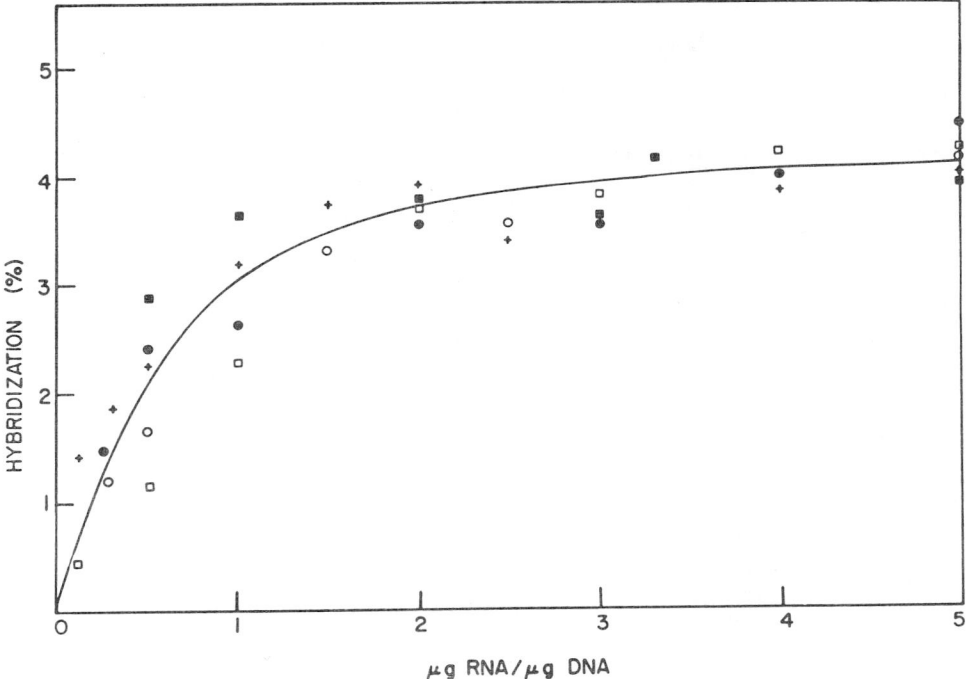

FIGURE 32. Hybridization of ^{32}P-labeled pea ct-rRNA with ctDNA from different plants. Each filter contained 1.0 μg ctDNA which were prepared as described (9). (□) peas; (○) bean; (●) lettuce; (+) spinach; (■) corn. The specific activities (cpm/μg) of pea ctrRNA preperations used in the experiments with pea, spinach, bean, lettuce, and corn ctDNAs were 13,560, 11,400, 11,400, 10,800, and 7540, respectively. The percent hybridization on the ordinate is based upon the weight of double-stranded ctDNA.

total ct-rRNA was fractionated in 15 to 30% linear sucrose gradients. The purity of these rRNAs was confirmed by acrylamide gel electrophoresis and is shown in Figure 32. Both 23S and 16S rRNA fractions were found to be essentially free of each other and relatively intact. No cross-hybridization between 23S and 16S rRNA was found.

The data of saturation hybridization using purified rRNA subunits are presented in Table 11. At saturation, the purified 23S rRNA was found to hybridize with 2.5% of the ctDNA. At saturation, 1.88% of ctDNA was found to have sequences complementary to the 16S rRNA subunit. Different experiments carried out with various DNA and subunit-rRNA concentrations showed that 2.35% ± 0.27% and 1.45% ± 0.27% of ctDNA was complementary to 23S and 16S rRNA, respectively. The total amount of hybridization of rRNA to ctDNA was calculated to be in the range of 3.80% by adding the values obtained from the hybridization of the purified 23S and 16S rRNA. These values are in reasonable agreement with the level of hybridization obtained when the total rRNA was employed in the hybridization experiments. The data, therefore, suggest that 23S and 16S rRNA have distinct genes on the ctDNA.

The hybridization level between ctDNA and ct-rRNA from pea leaves has been found to be 4.07%. The molecular weight of pea ctDNA has been reported to be 90×10^6. Therefore, an amount of ctDNA equivalent to a size of 3.6×10^6 contains base sequences similar to the ct-rRNA. Assuming a molecular weight of 1.7×10^6 for ct-

TABLE 11

Hybridization of Chloroplast DNA with Ribosomal RNA subunits

Experiment number	Hybridization with 23S RNA (%)	Hybridization with 16S RNA (%)
1	2.41	1.57
2	2.37	1.62
3	2.66	1.18
4	2.06	1.57
5	1.92	1.21
6	2.62	1.14
7	2.41	1.86

rRNA, the hybridization data show that there are two genes for the rRNA in pea ctDNA.

3. Number of Ribosomal RNA Genes in Chloroplast DNAs of Other Higher Plants

The detailed hybridization studies using pea ctDNA and ct-rRNA from pea leaves have shown that there are two rRNA genes in the ctDNA. Such studies have been extended to other higher plants.[46] The ct-rRNA from pea leaves ^{32}P-labeled in vivo was hybridized with 1.0 μg of ct-DNA from spinach, lettuce, bean, and corn. The saturation curves presented in Figure 32 show that the level of hybridization increases with increasing RNA to DNA ratio (R/D) and reaches a maximum when this ratio is 5:0. It may be noted that most of the hybridization has taken place when R/D is only 2:0. For example, spinach ctDNA has hybridized to the extent of 3.18% at the R/D of 1:0. This hybridization increases to 3.90% when R/D is 2:0. Further increase of R/D results in a maximum hybridization of 4:00%. Even when R/D is increased to 20:0, the level of hybridization remains at about 4.1%. These experiments indicate that all the ct-rRNA sites on the ct-DNA are saturated and the ct-rRNA preparations are not heterogeneous. In a control experiment, ct-rRNA from pea was hybridized with pea ctDNA. The rate of hybridization of this homologous system follows closely that of the heterologous system involving spinach ctDNA. Hybridization reaches a maximum of 4.22% when R/D is 5:0. Under similar conditions, ctDNAs from bean and lettuce hybridize to 4.14 and 4.46%, respectively. Experiments with ctDNA from corn leaves, a monocotyledonous plant, mimic the saturation curve. Thus, the data show that spinach, lettuce, bean, and corn ctDNAs hybridize between 3.96 and 4.46% with ^{32}P-labeled pea ct-rRNA. This range of hybridization with the different ctDNAs is of the same order as that obtained in hybridization with pea ctDNA. The molecular weights of the ctDNAs from pea, spinach, bean, corn, and oats have been found to be 89 $\times$ 10^6, 97 $\times$ 10^6, 85 $\times$ 10^6, and 86 $\times$ 10^6, respectively. Thus, the hybridization obtained with pea ct-rRNA indicates that the amount of single-stranded ctDNA complementary to ct-rRNA ranges from 3.5 to 4.1 $\times$ 10^6 daltons. This level of complementarity shows that there are two gene equivalents of ct-rRNA in the ctDNAs of higher plants, assuming a molecular weight of 1.7 $\times$ 10^6 for total ct-rRNA.

4. Specificity of Heterologous Chloroplast-Ribosomal RNA-DNA Hybridization

In the above experiments, the number of ct-rRNA gene equivalents in the ctDNAs of higher plants was studied using ^{32}P-labeled pea ct-rRNA. In order to find out the specificity of such heterologous hybridizations, competition experiments and thermal

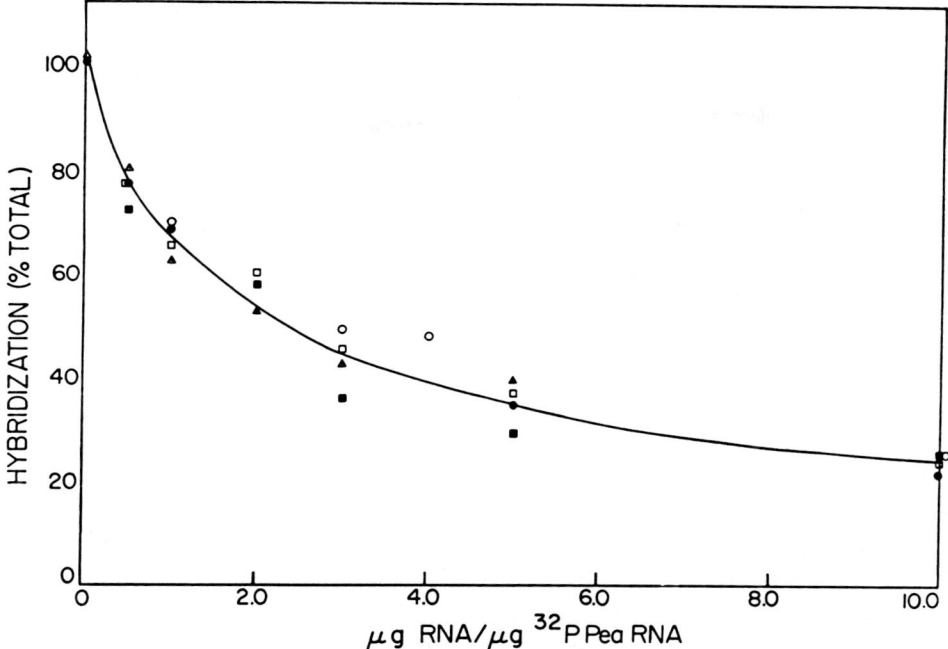

FIGURE 33. Competition hybridization experiments. ctDNAs from different plants were hybridized with ^{32}P-labeled pea ct-rRNA in the presence of increasing amounts of cold ct-rRNA from plants. All the data reported are an average of three experiments. (●) Hybridization with 5.0 μg bean ctDNA in the presence of unlabeled ct-rRNA from lettuce. (▲ Hybridization with 1.0 μg corn ctDNA in the presence of increasing amounts of ct-rRNA from spinach. (○) Hybridization of 1.5 μg ctDNA in the presence of increasing amounts of ct-rRNA from oats. (■) Hybridization of 1.5 μg pea ctDNA with 3.75 μg ^{32}P-labeled pea ct-rRNA in the presence of increasing amounts of ct-rRNA from corn. (□) Hybridization of 2.5 μg pea ctDNA with ^{32}P-labeled pea ct-rRNA in the presence of increasing amounts of rRNA from peas.

stability analysis of the DNA-rRNA hybrids was carried out. In the competition hybridization experiments of Figure 33, it is clearly seen that heterologous rRNA is equally effective in competition compared to homologous rRNA. These competition experiments were confirmed using homologous and heterologous 23S and 16S rRNA.

The competition experiments demonstrate that base sequences of ct-rRNA from the various plants are similar to the base sequence of pea ct-rRNA. The possible differences between ct-rRNAs has been further analyzed by studying the thermal stability of pea-ct-rRNA-ctDNA hybrids. A thermal stability profile of pea ctDNA with ^{32}P-labeled pea ct-rRNA is shown in Figure 34. This plot indicates that the hybrid formed between pea ctDNA and pea ct-rRNA is quite stable up to 70°C, after which it begins to melt with a T_m (temperature of half dissociation) of about 85.5 ± 1.0°C. The thermal stability curve does not reveal significant heterogeneity in the hybrids and the high T_m value demonstrates their stability. The melting curve of corn DNA-rRNA hybrid has a T_m of 85.5 ± 1.0°C. The lettuce DNA-rRNA hybrid has a T_m of 83.5 ± 0.5°C. Spinach and bean DNA-rRNA hybrids have T_ms of 80.5 ± 1.0°C and 81.0°C, respectively. From the plot of thermal stability presented in Figure 34, the T_m of spinach is the lowest at 81.0°C and corn and pea are the highest at 86.0°C. The maximum difference between the lowest and the highest T_m could result from no more than 3% difference in the base sequences of rRNA. It should be pointed out that the maximum difference may only be experimental variation because the hybrid of corn ctDNA, which

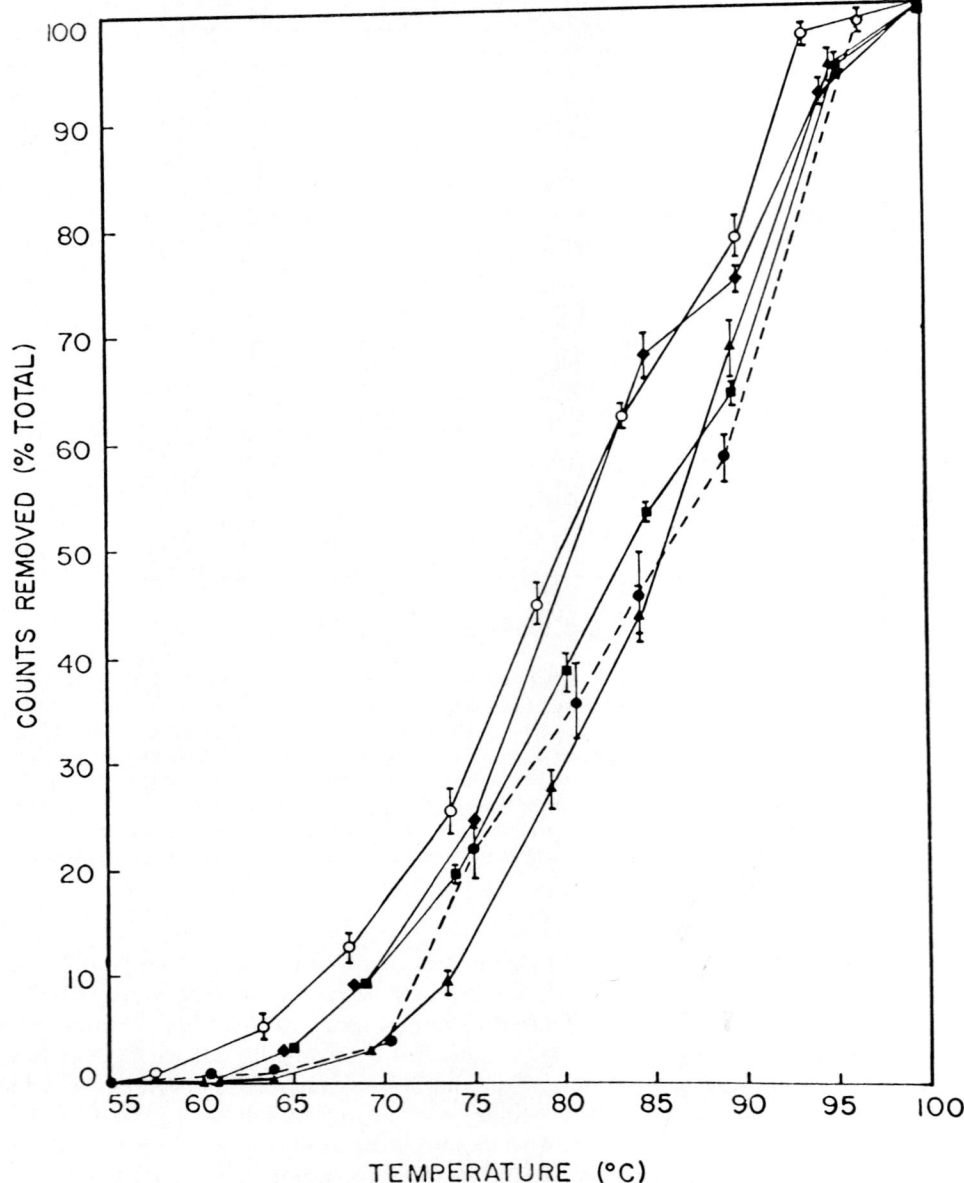

FIGURE 34.   Thermal stability of DNA-rRNA hybrids. The hybridization was carried out with an R/ D of 3.0. After the hybridization, the filters were incubated at the indicated temperature for 10 min. in 3mℓ af 2 × SSC. The released RNA was precipitated with trichloroacetic acid after addition of 100 μg of carrier DNA, washed, and dried, and its radioactivity was measured. ▲ Pea ctDNA, hybridization was 4.84% using 5.0 μg of ctDNA and ct-rRNA of specific activity of 5800 cpm/μg; ■ Lettuce ctDNA, hybridization was 4.65% using 1.5 μg of ctDNA and ct-rRNA of specific activity of 12,850; ○ Spinach ctDNA, hybridization was 4.43% using 5.0 μg of ctDNA and ct-rRNA of specific activity of 6680; ● Corn ctDNA, hybridization was 5.05% using 4.0 μg of ctDNA and ct-rRNA of specific activity of 8200; ◆ Bean ctDNA hybridization was 4.45% using 7.5 μg of ctDNA and ct-rRNA of specific activity 5360. The bars in the figure represent the variation in four different experiments.

is from a monocotyledonous plant, has the same T_m as that of pea ctDNA, which is from a dicotyledonous plant. On the other hand, the ctDNA from bean, which belongs to the same family as pea, gives a lower T_m. However, minor differences in the base sequences might go undetected because of the limitation of chemical hybridization studies.

The above experiments have shown that ctDNA in higher plants contains two gene equivalents for the ct-rRNA. The nucleotide sequence of these genes appears to have been exceptionally invariant during evolution and divergence of higher plants.

The detailed studies on the rRNA genes in the ctDNA of *Euglena* and *Chlamydomonas* have not been carried out, but the available evidence suggests that they contain at least two rRNA genes.[13]

5. Mapping of Ribosomal RNA Genes in Corn Chloroplast DNA

Bedbrook and coworkers[44] have mapped the rRNA genes in the restriction endonuclease map of the corn ctDNA (Figure 29). Corn ct-rRNA has been found to exist in two identical units. Their studies have shown that each unit contains one sequence for the 16, 23, and 5S rRNAs in the order given. The 16 and 23S sequences in each unit are separated by a 2100 base pair (bp) spacer. The DNA sequence for 5S RNA is closely linked to that for the 23S RNA. Within the above unit, the three RNAs are transcribed from a single DNA strand. The two rDNA units have an inverted orientation and are part of 22,500 bp sequences, which are repeated with inverted orientation as described before.

B. Transfer RNA Genes in Chloroplast DNA

Chloroplasts have been found to contain tRNAs and aminoacyl-tRNA synthetases.[16] Weil and coworkers[47] have extensively characterized the properties of the cytoplasmic and chloroplastic aminoacyl-tRNA synthetases and tRNAs. (See Chapter by Weil, Section II.) These studies do not indicate whether the chloroplast enzyme is coded by the ctDNA and synthesized in the organelle, or if it is coded by a nuclear gene, made on cytoplasmic ribosomes, and imported into the chloroplasts. Some of these questions have been investigated by Hecker and coworkers[48] in *Euglena*. Cells grown in the presence of streptomycin (which completely blocks the translation of proteins on chloroplast ribosomes) were found to contain in ct-aminoacyl-tRNA synthetases of phenylalanyl- and valyl-tRNAs. When cells were grown in the presence of cycloheximide (which blocks translation of proteins on cytoplasmic ribosomes), there was no demonstratable activity of these enzymes. In addition, an aplastidic mutant of *Euglena* B, W₃BUL (which contains no detectable ctDNA or chloroplast structure), was found to contain low levels of detectable ct-phenylalanyl-tRNA synthetase. These two lines of evidence indicate that the localization of synthetases in *Euglena* chloroplasts reflects intracellular compartmentalization of nuclear coded, cytoplasmically translated proteins rather than either compartmentalization of a nuclear-derived messenger RNA in the organelle for translation, or a purely chloroplastic origin for the synthetases.

The presence of ct-tRNA genes in ctDNA was first demonstrated by Tewari and Wildman (see Tewari[16]) in tobacco leaves. The aminoacyl-tRNA synthetases were isolated from purified tobacco chloroplasts and any endogenous tRNAs were removed by passing through DEAE-cellulose. This enzyme preparation was able to charge radioactive amino acids to the tRNAs obtained from chloroplasts. The level of hybridization of tobacco ctDNA with in vivo labeled ct-tRNAs was found to range from 0.4 to 0.7%. This level of hybridization was found to reflect about 20 to 30 ct-tRNA genes in the ctDNA. Haff and Bogorad[49] have recently shown that about 0.6 to 0.75% of corn ctDNA contained sequences complementary to tRNAs, which again accounts for 20 to 26 genes for tRNAs. The total tRNA from green and etiolated maize leaves was labeled with a mixture of 21 labeled amino acids using a mixture of aminoacyl-tRNA synthetases. The amounts of the different sets of isoaccepting plastid tRNAs for which genes occur in maize plants were determined by amino acid analysis of those amino-acyl-tRNAs that hybridized specifically to ctDNA. The data indicated that at least 17 distinct aminoacyl-tRNAs hybridized to ctDNA. The tRNAs for the amino acids cysteine, methionine, and glutamine were not detected. Weil and co-workers[47] have studied the hybridization of bean ct-^{3}H-leucyl-tRNA and ^{3}H-phenylalanyl-tRNAs with ctDNA. Both aminoacyl-tRNAs were found to hybridize with ctDNA. These experiments, performed with total ct-tRNA, were confirmed using the fractionated chloroplast specific tRNA species; namely, three isoacceptors in the case of leucyl-tRNA and two in the case of phenylalanyl-tRNA. Chloroplast specific leucyl-tRNA$_1$, leucyl-tRNA$_2$, leucyl-tRNA$_3$, phenylalanyl-tRNA$_1$, and phenylalanyl-tRNA$_2$ each hybridized with ctDNA. In order to determine whether the chloroplast specific isoacceptors are coded for by the same or different genes, experiments were performed to check if the hybridizations of isoacceptors were additive. All three combinations of two chloroplast-specific tRNAs showed no additivity. The combination of the two chloroplast phenylalanyl-tRNAs did not show any additivity either, whereas control experiments in which one chloroplast leucyl-tRNA and one chloroplast phenylalanyl-tRNA were mixed showed the expected expected additivity. These results suggest that the three chloroplast leucyl-tRNAs are coded by the same gene(s) and that this is also the case for the two chloroplast phenylalanyl-tRNAs.

Williams and co-workers[50] have carried out hybridization studies between ^{3}H-leucyl-tRNAs of bean and the ct- and nDNAs of bean, tobacco, and corn. Their results indicated that the bean leucyl-tRNAs hybridized to bean ctDNA. Hybridization of the bean leucyl-tRNAs was also observed to take place with the bean nuclear DNA, but at a much lower level. When the bean leucyl-tRNAs were hybridized with tobacco and corn ctDNAs, the level of hybridization was greatly reduced compared to the level observed between bean ctDNA and bean ^{3}H leucyl-tRNA. Under nonstringent hybridization conditions corn tRNAs could partially compete with the bean leucyl-tRNAs for the base sequences of bean ctDNA, but when stringent hybridization conditions were used, no competition between the corn tRNA and the bean leucyl-tRNAs took place. This result shows that there is partial sequence homology between bean and corn leucyl-tRNAs. Williams and co-workers[50] have also shown that all of the isoaccepting species of the bean leucyl-tRNAs, as separated by reversed phase chromatography, could hybridize with both the nDNA and ctDNA of bean in the same ratio as observed with unfractionated tRNAs. Their data are unique in finding complementary sequences for all cellular leucyl-tRNAs in the bean ctDNA.

Using in vitro ^{125}I-labeled ct-tRNAs, Schwartzbach and co-workers[51] found that 0.74% of the base sequences of *Euglena* ctDNA were complementary to the ct-tRNAs. This level of hybridization would indicate that the ctDNA coded for about 26 ct-tRNA genes. When ^{125}I-labeled phenylalanyl-tRNA and ct-aspartyl-tRNA were hybridized to the *Euglena* ctDNA, it was found that the data were consistent with the ctDNA containing one gene for each of these ct-tRNAs. If total ct-tRNA was hybridized to nuclear and ctDNA that had been fractionated on a CsC1 equilibrium buoyant density gradient, it was found that the hybridiztion took place with the ctDNA banding at a density of 1.685 g/cm^{-3}. The rRNA was found to hybridize with the DNA banding at a density of 1.707 g/cm^{-3}. When ^{125}I-ct-phenylalanyl-tRNA and ct-aspartyl-tRNA were tested as above, essentially the same pattern of hybridization was observed. Finally, the [125]I-*Euglena*-ct-tRNAs were competed by the cold *Euglena* ct-tRNAs, but not by the tRNAs from a blue-green algae or the *Escherchia coli* tRNAs.

Chang and co-workers[52] recently determined the primary sequence of phenylalanyl-tRNA from the chloroplasts of *Euglena gracilis*. The ct-phenylalanyl-tRNA was found to be 76 nucleotides long. The ct-phenylalanyl-tRNA was found to resemble more closely procaryotic phenylalanyl-tRNA than it resembled those from the cytoplasm of eucaryotes, both in the nature of its modified nucleotides and in the nature of its sequences. There were eight positions in the phenylalanyl-tRNA molecules where nucleotides were invariant in procaryotes, but differed from invariant nucleotides in eucaryotes. At five of these positions, chloroplastic phenylalanyl-tRNA is similar to procaryotes. Weil and co-workers[47] have determined the base composition of bean phenylalanyl-tRNA and found it to contain modified bases similar to those found in procaryotic phenylalanyl-tRNA.

Further studies on the hybridization of pea ctDNA with pea ct-tRNAs has been recently carried out by Meeker and Tewari[53] The ct-tRNAs were obtained as follows. In a typical experiment, 10- to 14-day-old pea leaves were homogenized in 3 l of cold (4°C) STM Buffer (0.3 M Sucrose/50 mM Tris-OH (pH 7.6)/5 mM MgCl$_2$/0.1 mM SHEtOH) in two 500-g batches, using two 5-sec bursts in a gallon-size Waring® blender. All steps were carried out at 4°C unless otherwise stated. The homogenate was filtered through four layers of cheesecloth and four layers of Miracloth® and centrifuged at 1500 × g for 15 min in a Sorvall® centrifuge. The chloroplast pellet (containing nuclei and some mitochondria) was gently suspended in 225 ml of STM and recentrifuged at 1500 × go for 15 min. The procedure was repeated one more time. The washed chloroplast pellet was then resuspended with 125 ml TM (50 mM/Tris (pH 7.6)/5 mMMgCl$_2$) and Lubrol® (Lubrol WX, ICI) added to a final concentration of 0.2%. After incubating for 15 min on ice, the lysed chloroplasts were centrifuged for 20 min at 35,000 × g. The pellet (containing nuclei and chloroplast membranes) was discarded and the supernatant was further centrifuged for 5 hr at 80,000 × g in a Beckman® L3-30 preparative ultracentrifuge. The top 2/3 of the 80,000 × g supernatant was then carefully pipetted off using a sterile glass syringe and to the supernatant a half volume of extraction buffer (50 mMTris, (pH 7.6)/25 mMKCl /25 mMMgCl$_2$/ 2% SDS) was added. The solution was incubated for 5 min at room temperature and RNA was obtained by phenol extraction procedure. Typically, this isolation procedure yielded 1 to 2 mg of pea ct-4S RNA per kilogram of pea leaves.

The tRNAs obtained as above were radioiodinated in the presence of $2.5 \times 10^{-5} M$ T1Cl$_3$, 7×10^{-5} MKI, and 1 mC of ^{125}I for each 20 μg of tRNAs. The iodinated RNA was heated with alkaline acetate buffer, precipitated with alcohol, and purified through a hydroxyapatite column. The iodinated RNA was further fractionated on a Sephadex® G-25 column. The saturation hybridization data with the ^{125}I tRNA and pea ctDNA are shown in Figure 35. The data show that about $1.2 \pm 0.2\%$ of ctDNA contain sequences complementary to ctDNA. Since the molecular weight of pea ctDNA has been found to be about 90×10^6 and tRNAs average about 25,000 daltons, it can be calculated that pea ctDNA contain 43 ± 10 tRNA genes. These hybridization data are a pretty reliable estimate of the number of tRNA genes in ctDNA because tRNAs used in these experiments are not contaminated with rRNAs. The use of in vitro iodination of tRNAs has also circumvented the possible contamination of tRNAs by RNAs that are rapidly labeled in vivo. The hybridization between ctDNA and tRNAs has been found to result from specific base pairing between the two components as reflected by the Tm of 81°C between DNA-tRNA hybrids. The T$_m$ of ctDNA has been found to be 85°C. Thus, the T$_m$ of DNA-RNA hybrids is only 4°C lower than the T$_m$ of DNA-DNA hybrids, an observation which has been reported for many such systems. The data have also shown that the base sequences of ct-tRNAs are unique since the tRNAs from cytoplasm of pea roots, *Escherchia coli,* yeast, and calf (liver) did not compete for the complementary sequences in ctDNA (Figure 36). In contrast, when the competition was carried out using unlabeled ct-tRNAs with ^{125}I-ct-tRNAs, expected levels of competition was observed. The uniqueness of the base sequences of ctDNA

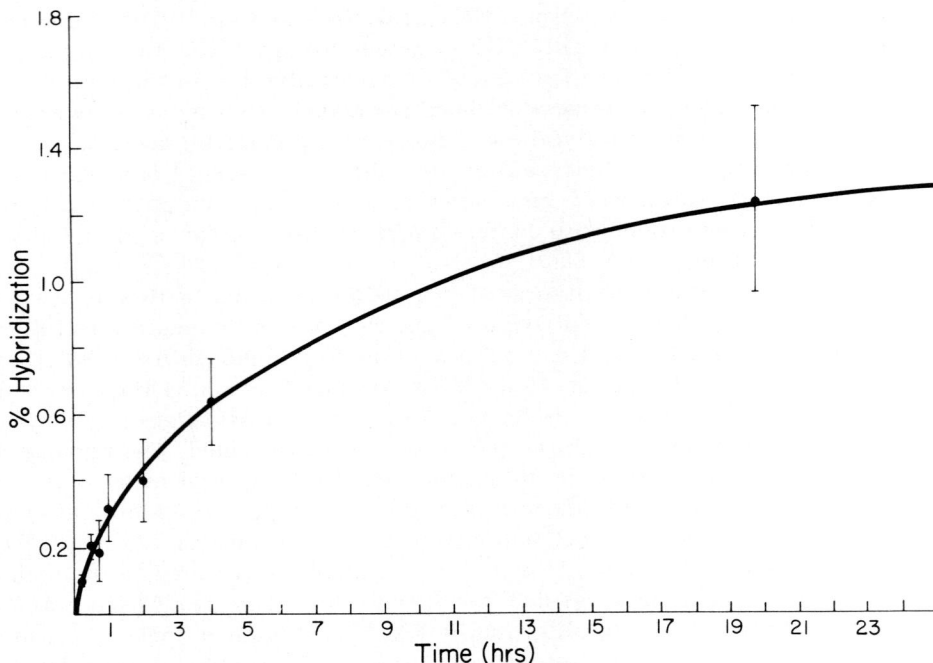

FIGURE 35.   Hybridization of ctDNA with in vitro ^{125}I-labeled tRNA. 1 μg of ctDNA and 2 μg of labeled tRNAs were used in the experiment.

was also evident from the data where the saturation hybridization between ctDNA and cell-tRNAs give the same level of hybridization (1.2%) as obtained using tRNAs from chloroplasts. These data indicated that cytoplasmic tRNA did not hybridize with the ctDNA.

The hybridization of ct-tRNAs with an excess of ctDNA has shown that only about 20 to 30% of the ct-tRNAs are coded by ctDNA. These data demonstrate that cytoplasmic tRNAs are present in chloroplasts. The consistent presence of cyt-tRNA in chloroplasts has also been observed by Weil and co-workers.[47] It is not quite clear whether the presence of cytoplasmic tRNAs in chloroplasts represents the actual in vivo state of the cell or is an artifact of the isolation procedures. However, the importation of cytoplasmic tRNAs into organelles has been reported. The possibility of export of tRNAs from organelles also has not been eliminated. The observation that 20% of the cell-tRNAs are coded by ctDNA is extremely significant although the data presented here do not offer any explanation for their role in protein synthesis of the cell. However, one has to bear in mind that ctDNA has a molecular size of only 90×10^6 compared to a molecular size of 3×10^{12} for nDNA. The studies of Meeker and Tewari[53] have identified the presence of at least 17 aminoacyl-tRNA synthetases and their tRNAs in chloroplasts. The individuality labeled aminoacyl tRNA has been shown to hybridize with ctDNA. The saturation hybridization curves have been generated for each of the labeled aminoacyl-tRNAs. An example of such saturation experiments is shown in Figure 37. These hybridizations have been found to result from specific base pairing because the T_m of ctDNA-aminoacyl-tRNA hybrids was found to be very close to the T_m of ctDNA (Figure 38). In addition, the amount of hybridization obtained was directly proportional to the amount of ctDNA. From the hybridization of labeled aminoacyl-tRNA to ctDNA, it is not possible to calculate the number of genes for each of the tRNAs because the amounts of specific aminoacyl-tRNA in total ct-tRNAs are not known and it is not possible to know whether all of the specific

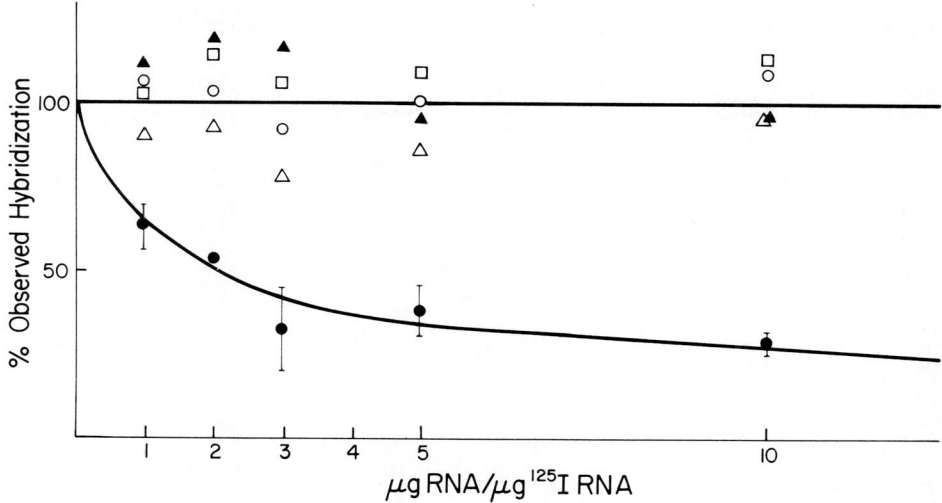

FIGURE 36. Competition hybridization experiments. Pea ctDNA was hybridized with 2 μg of ^{125}/ct-tRNAs (●———●), pea cyt. tRNAs (o———o), *Escherchia coli* tRNAs (▲———▲), yeast tRNAs (△———△), and calf liver tRNAs (□ ———■).

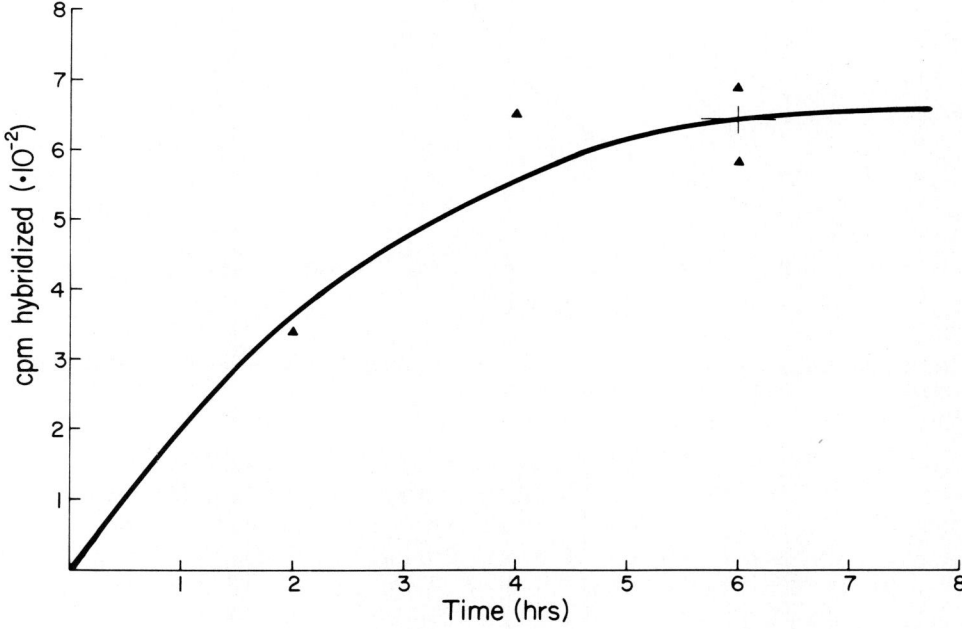

FIGURE 37. Saturation hybridization of ctDNA with increasing concentrations of ^{3}H-tyrosyl tRNA.

aminoacyl-tRNAs have been acylated. However, such experiments identify the presence of at least one aminoacyl-tRNA gene in the ctDNA.

The results from such hybridizations using individual aminoacyl-tRNA indicate that about 17 aminoacyl-tRNA genes are present in ctDNA. All of these aminoacyl-tRNA hybridize to a significant amount of ctDNA as shown in Table 12. The amount of hybridization reflects the minimal amount because of the nature of the experiments. In experiments with pea leaves, the tRNAs of cysteine and glutamine have been found

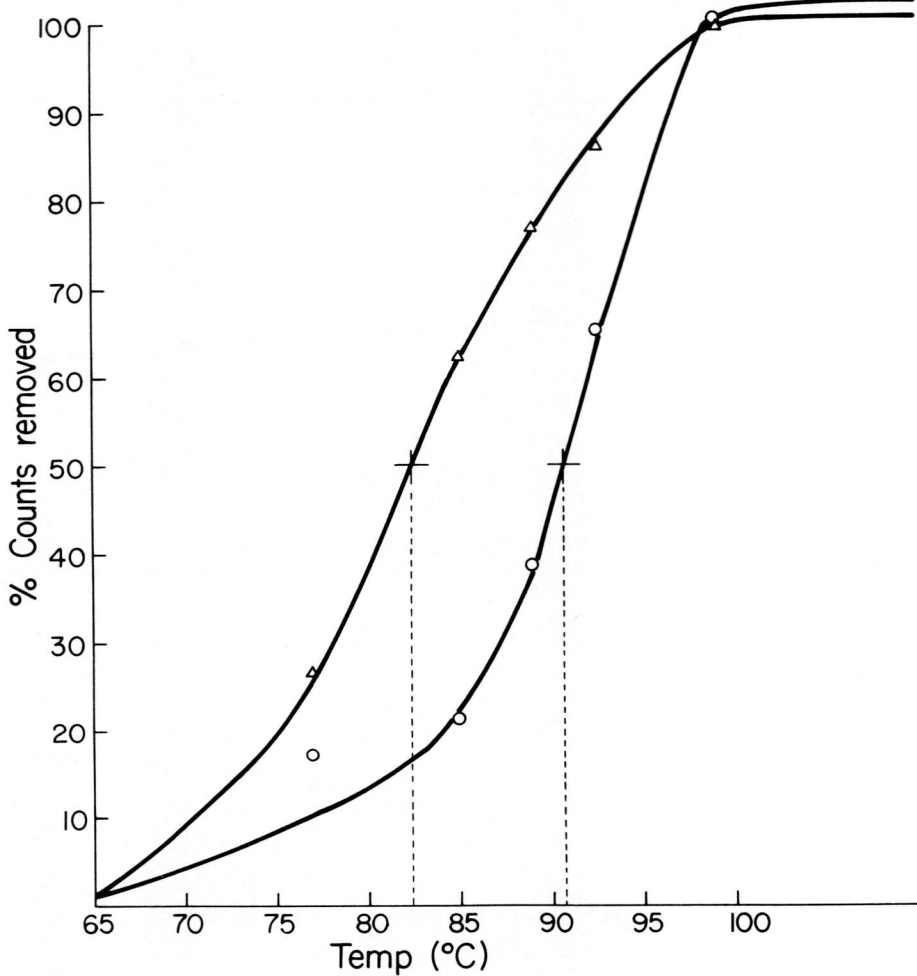

FIGURE 38.   Thermal denaturation of ³H-lysyl- and ³H-isoleucyl-ct-tRNA hybridized to 20 μg of pea ctDNA.

to hybridize very poorly with ctDNA so that their presence in ctDNA cannot be detected with certainty. However, the methionyl-tRNA gene has been identified in pea ctDNA, which has not been shown to be present in corn ctDNA. On the other hand, glutamyl-tRNA has not been detected in pea, although it has been found to be present in corn ctDNA. The inability to localize three tRNA genes in ctDNA of corn and pea probably results from inadequate charging of the tRNAs rather than the lack of their genes in ctDNA. A number of isoaccepting species of tRNAs have been reported in chloroplasts. In experiments using three isoaccepting species of leucyl-tRNAs and two isoaccepting species of phenylalanyl-tRNA, the hybridization data obtained by Weil and co-workers[47] have shown that the isoaccepting species of leucyl-tRNA are coded by one gene. The same was found to be true for the isoaccepting species of phenylalanyl-tRNA. The multiplicity of tRNA genes must be tightly restricted because we have found that there are only about 40 genes of tRNAs in ctDNA.

The hybridization of labeled tRNAs with Sal I restriction endonuclease have shown that tRNA genes are distriuted throughout the genome. The exact locations of tRNA genes in ctDNA have yet not been determined.

TABLE 12

Hybridization of Aminoacyl-tRNA Synthetases with
Pea ctDNA

Amino acid	Specific activity of aminoacyl tRNAs cpm/μg	Counts bound to 100 μg of ctDNA
Ala	102	130
Arg	800	650
Asn	1,595	660
Asp	1,463	1,600
Cy-SH	25	16
	33	21
Glu	30	14
	15	12
	14	15
Gln	43	45
	61	12
Gly	480	290
His	132	210
Ile	1,185	595
Leu	420	520
Lys	503	440
Met	2,964	1,090
Phe	323	303
Pro	182	138
Ser	592	220
Thr	297	120
Trp	120	107
Tyr	1,225	688
Val	444	315

1. Divergence of Transfer RNA Genes in Chloroplast DNA of Higher Plants

The hybridization experiments between the spinach ctDNA and spinach ct-tRNA ct-tRNAs have shown that there are 42 tRNA genes in spinach ctDNA.[54] The number of tRNA genes in corn ctDNA under similar experimental conditions have been found to be 27, in agreement with the values reported by Haff and Bogorad.[49] The heterologous hybridizations between ctDNA and tRNAs have shown that the genes for tRNAs have underrone much more divergence than that observed for rRNA genes. The ctDNA from pea and spinach share about half the base sequences of tRNAs with each other. The ctDNA from pea and spinach, however, share only about one third of the base sequences of tRNAs with corn ctDNA. The amount of cross hybridization between tRNAs and ctDNA have been critically analyzed by various combinations of ctDNA and tRNAs, thermal stability of DNA-tRNA hybrids, and competition hybridization experiments.

C. Messenger RNA Transcripts of Chloroplast DNA
I. Complete Transcription of Chloroplast DNA

It is apparent from the above data that only about 5×10^6 daltons of ctDNA are utilized in the coding of rRNA and tRNAs. The remaining 45×10^6 daltons of ctDNA are available for the formation of mRNA transcripts. Are all of the base sequences of ctDNA transcribed in chloroplasts? The saturation hybridization with the in vivo labeled RNA and ctDNA has shown that about 50 to 60% of the base sequences of ctDNA are transcribed. The in vivo labeled RNA was fractionated on a poly U Sepharose® column or on poly U cellulose filters. About 0.5% of the total RNA was found to bind with the poly U in the presence of 25% formamide/0.7 M NaC1/50 mM Tris/10 mM EDTA, pH 7.5 (Figure 39). The bound RNA (presumably containing poly A) was eluted from the column by passing 90% formamide-10 mM EDTA-10 mM KPO$_4$, pH 7.5. Nonpoly A and poly A containing RNAs were found to hybridize with 50% and 15 to 20% of the ctDNA, respectively (Figure 40). Those experiments indicate the presence of both poly A- and nonpoly A- RNA transcripts of ctDNA in chloroplasts. The presence of poly A in the mRNA transcripts of ctDNA was confirmed as follows. Fifty μg of ctDNA was hybridized with 100 mg of ^{32}P labeled total RNA from chloroplasts. The DNA-RNA hybrids were incubated with 90% formamide at 50°C and the ctDNA specific RNA was precipitated with alcohol. This RNA was then incubated with RNase and RNase T1. The undigested RNA was analyzed in 10% acrylamide gel for their molecular sizes. The data showed that the poly A tails in the mRNAs of ctDNA range from 40 to 150 adenosine residues. The presence of poly A in chloroplast RNAs has also been shown by Haff and Bogorad.[55] They have reported that about 6% of the total polyadenylated RNA from chloroplasts hybridized with ctDNA. When the polyadenylated RNA was isolated from RNase treated chloroplasts, it was found that 65% of the RNA was hybridized to corn ctDNA. The size of the poly A tracts of chloroplast specific RNAs were analyzed and were found to be about 45 nucleotides.

The ctDNA specific RNA was analyzed in 2.5% acrylamide gels. The data are plotted in Figure 41. It is interesting to note that the mRNA transcripts of ctDNA have specific size classes and are found to range in sizes from 6 to 30S.

The transcription of ctDNA has been studied in *Euglena gracilis* by Chelm and Hallick[56] using nick translated ctDNA for the hybridization with unlabeled cell RNA. The total cell RNA was isolated at 0, 4, 8, 12, 24, 48, and 72 hr after exposing the dark grown cells to light. The results of these studies showed that dark-adapted cells, containing no detectable chloroplast structure, contained ctRNA transcripts equivalent to 17% of the ctDNA. The results also demonstrated that there was an initial decrease of the level of ctRNA transcript on exposure to light; and, that after 72 hr of chloroplast development (i.e., when the development is complete), the level of transcription was equivalent to 23% of the chloroplast genome. Rawson and Boerma[57] explored the question of how much of the ctDNA of *Euglena* is being transcribed into RNA by separating the strands of the ctDNA in alkaline CsC1 equilibrium buoyant density gradients, labeling the separated DNA strands with ^{125}I, and hybridizing the DNA with unlabeled total cell RNA. By this procedure, it was observed that 9.5% of the heavy ctDNA strand and 43% of the light ctDNA strand was complementary to the total cell RNA.

2. Specific Messenger RNA Transcripts of Chloroplast DNA

The specific mRNA of the large subunit of ribulose biphosphate carboxylase enzyme has been identified in *Chlamydomonas* and mapped in corn.[27,58] The partially purified mRNA from *Chlamydomonas* was hybridized to the Eco R1 fragments of ctDNA. The data showed that the labeled mRNA hybridized to essentially only one ctDNA fragment of 3.9×10^6 molecular weight. This fragment was also found to hybridize with 23 and 16 S rRNA subunits. However, these RNA subunits were not found to compete with the mRNA. Coen and co-workers[58] have used an in vitro linked transcription-translation system to show that the large subunit of ribulose-1,5, biphosphate carboxylase is encoded in ctDNA. A Bam H1-generated ctDNA sequence cloned in *Escherichia coli* was shown to direct the in vitro synthesis of this protein identified as the large subunit by its size, serological properties, and limited proteolytic digestion products.

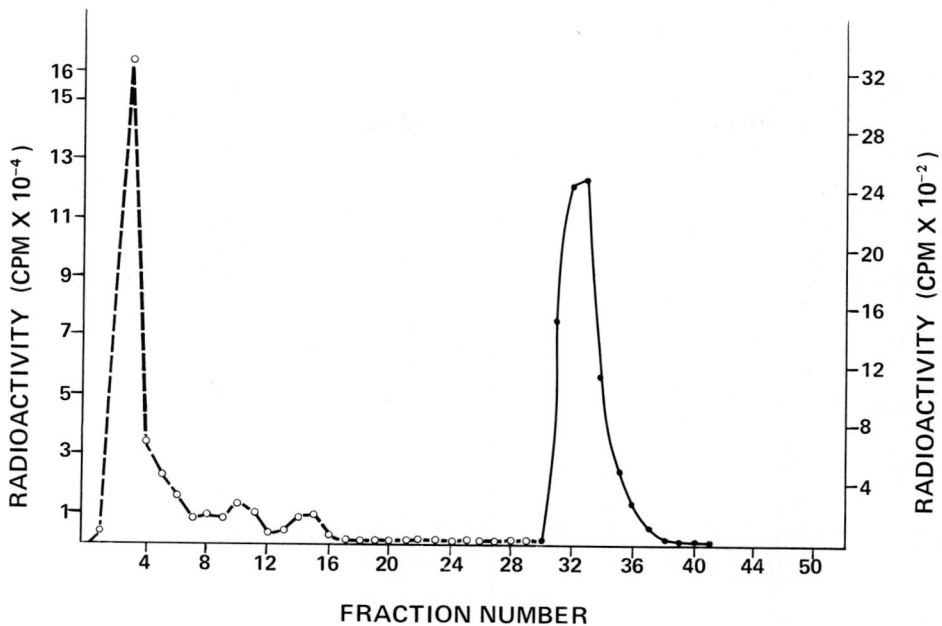

FIGURE 39. Fractionation of in vitro labeled ct-RNAs on a poly U sepharose ® column.

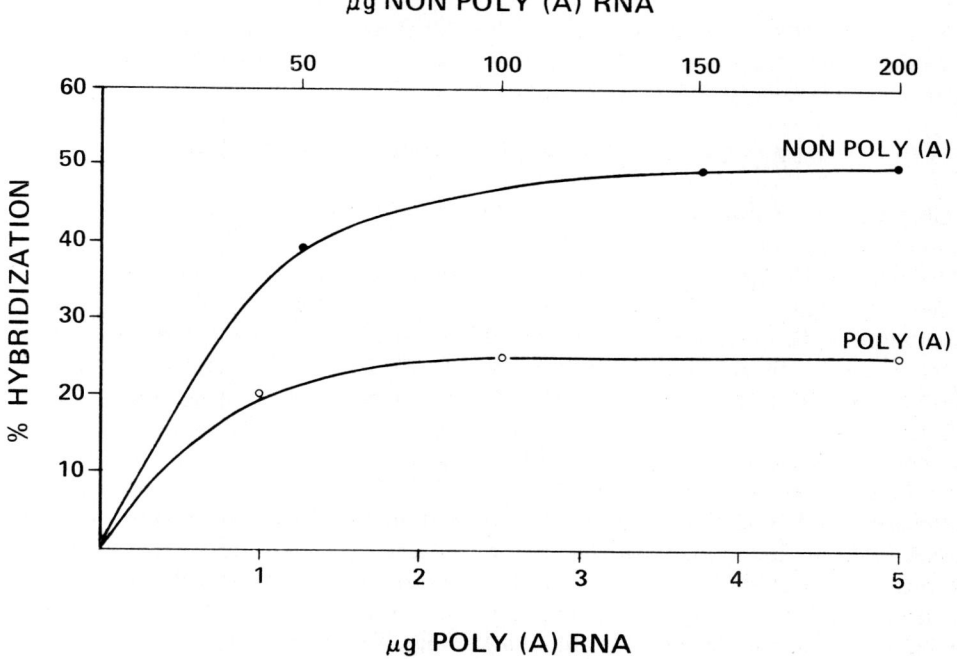

FIGURE 40. Hybridization of poly A- and non poly A-containing RNA to the nick translated ctDNA.

In recent studies, Bedbrook and co-workers[59] identified a plastid gene that is expressed during photoregulated development of plastids. The nonribosomal RNA fraction of the developing plastids and chloroplasts were found to contain RNA which

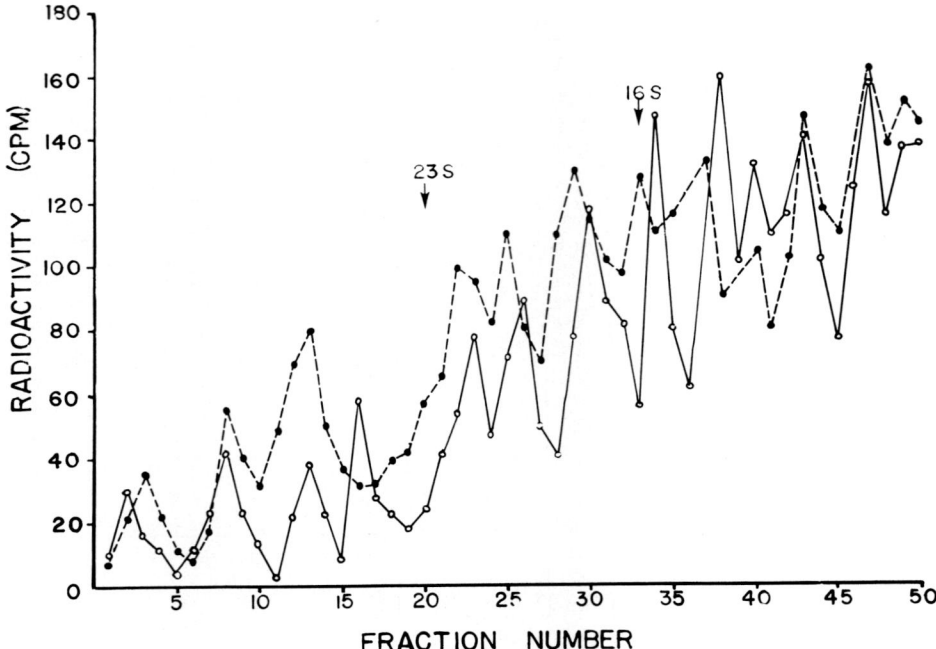

FIGURE 41.   Fraction of ctDNA specific RNA transcripts in 2.5% acrylamide gel electrophoresis.

hybridized to ctDNA Bam fragment 8. This RNA was not found in etioplasts, but appeared during illumination. The RNA which hybridizes to Bam 8 has been found to translate into a 34,500 daltons protein in vitro.

IV. REPLICATION OF CHLOROPLAST DNA

A. DNA Replication In Vivo
1. Displacement Loops in Chloroplast DNA

The ctDNA was isolated from 10-day-old pea or corn leaves, fractionated in CsC1-ethidiam bromide gradients, and collected into three fractions; the lower band of closed circular DNA consisting of 65 to 70% of total ctDNA, and the area between the two bands (middle band) which contained 1% of the total ctDNA. The closed circular DNA from pea was prepared for electron microscopy using the formamide technique. Pea ctDNA molecules containing one or two D-loops were observed (Figure 42). The D-loops could be easily identified because one side of the D-loop appeared to be double stranded while, the other side appeared to be single stranded because of its thinner and kinkier appearance.[60,61] In addition, the D-loops observed in the pea ctDNA molecule exhibited branch migration. The frequency of molecules containing D-loops was not affected by RNase, but brief treatment of ctDNA with alkali released the displacing strand, suggesting that the displacing strand is DNA and hydrogen bonded to the parental molecule. The two D-loops were located at two adjacent sites. The smallest size of D-loops was 820 base pairs and the average distance between the outside edges of the two D-loops was 7.15 kb pairs. The inner distance between the two D-loops was highly variable, indicating that the two D-loops expand towards each other. Small Cairns type of replicative forked structure ranging in size from 7.2 to 10.6 kb have been observed in closed circular pea ctDNA. These Cairn's structures map at the position of the two D-loops and probably result when the two displacing

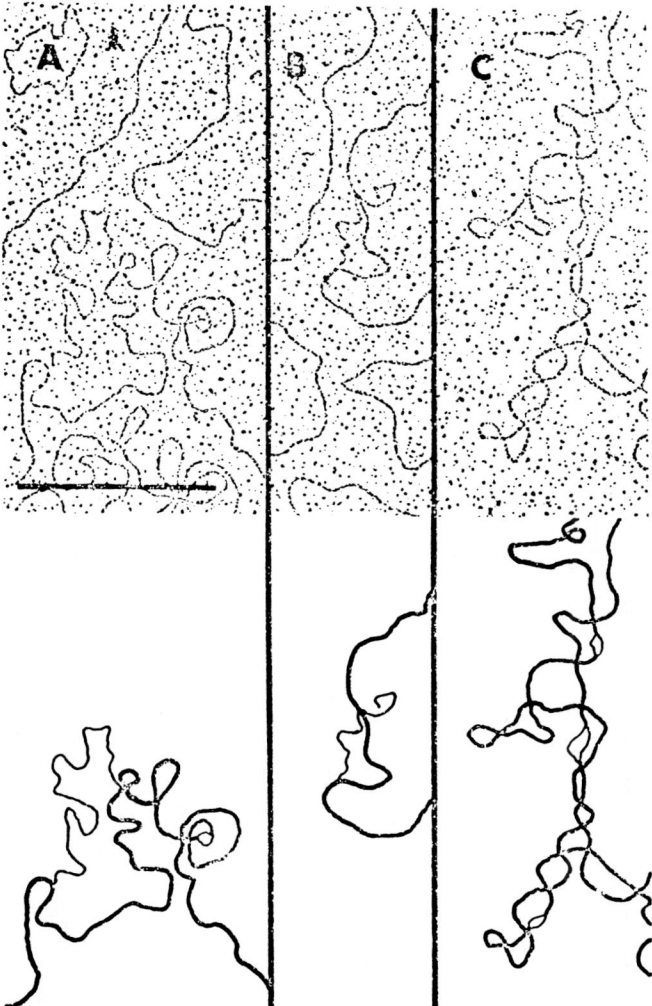

FIGURE 42. High magnification electron micrographs of ctDNA molecules. (A) A molecule containing one D-loop and one Den-loop from the pea ctDNA. (B) A branch migrating D-loop fron corn ctDNA. (C) A pea ctDNA molecule containing three Den-loops. The thin and thick lines in the drawings represent the single-stranded regions, respectively. The bar indicates 1 μm.

strands expand past each other in the opposite strands. This confirms that the two displacing strands are located on the opposite parental strands of the pea ctDNA molecule. The two D-loops in corn ctDNA also are located at two adjacent sites. The size of the corn ctDNA D-loop corresponds to 860 base pairs and the outer distance between the two D-loops was 7.06 kb pairs.

2. Cairns Replicative Intermediates

Replicative forked structures of the Cairns type were found in the DNA from lower, middle, and upper bands (Figure 43). In pea ctDNA, 3% of the circular molecules in the lower band were Cairns' structures and the extent of their replication ranged from 5.2 to 8.2%. The Cairns replicative intermediates made up an average of 5.7% of the

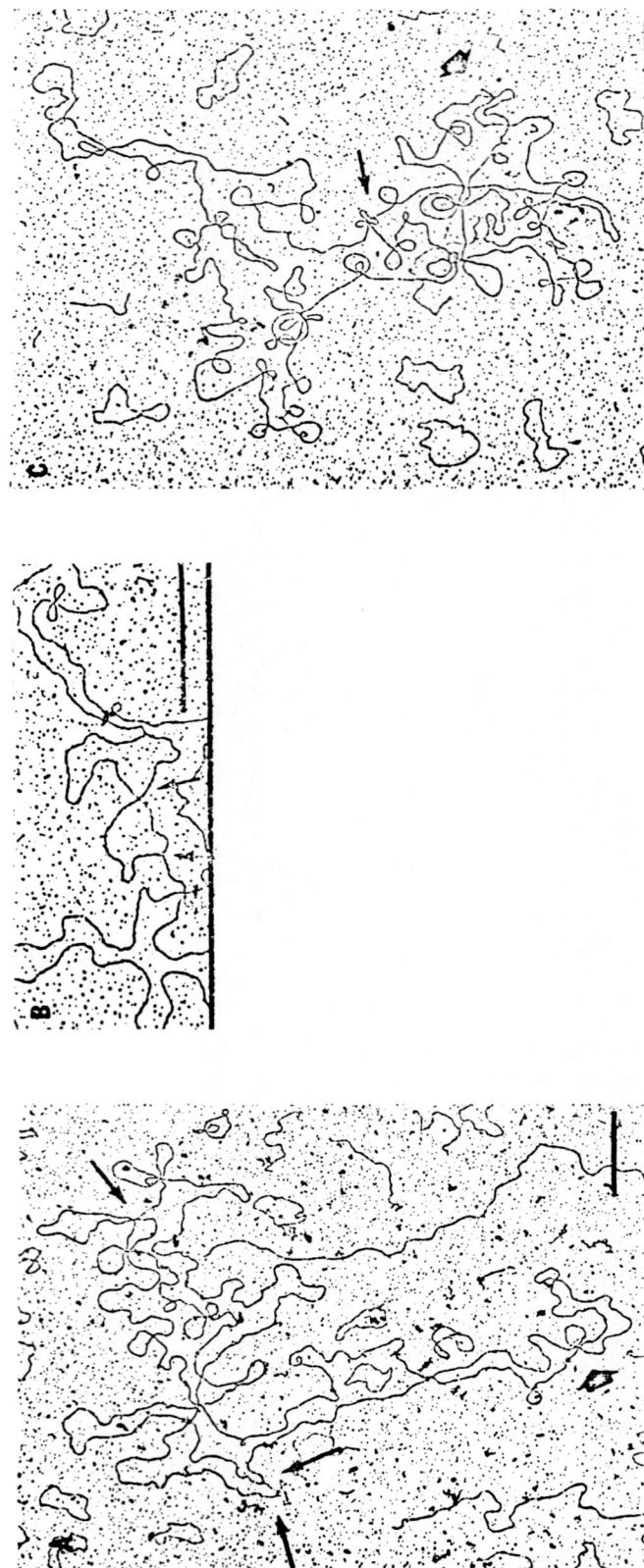

FIGURE 43.   Cairns replicative intermediates.   (A) A molecule that is 37% replicated and contains a single-strand region (two small arrows) at one growing fork and a denatured region (large arrow) in the unreplicated portion of the molecule.   (B) A portion of a molecule that is 5.2% replicated and contains a single strand region (two small arrows) at one growing fork.   (C) A Cairns replicative forked molecule showing branch migration at one of the rereplicative forks (large arrow).   The molecules were mounted for electron microscopy by the formamide technique; the spreading solution contained 50% formamide.   Single-stranded and double-stranded φX DNA molecules were used as internal standards. The bars indicated 1 μm.

circular DNA in the middle band and the extent of their replication ranged from 7 to 50%. The Cairus structures consisted of 2.9% of the circular ctDNA molecules in the upper band and the extent of their replication ranged from 38 to 87%. The finding of Cairns replicative intermediates in the lower and middle bands of the CsC1-ethidium bromide density gradients, and the correlation between higher banding position and larger amount of replication again suggests that replication takes place on a covalently closed circular template and is accompanied by nicking and closing cycles. In corn ctDNA, Cairns replicative intermediates made up 4.5pc of the total circular ctDNA molecules and the extent of their replication ranged from 9.4 to 70%.

3. Rolling Circle Replicative Intermediates

In the pea ctDNA preparations, circular molecules with an attached double stranded tail accounted for 4.9% of the circular molecules in the upper band, and 2.8% of the circular molecules in the lower band (Figure 44). There were no circular molecules with tails in the lower band of pea ctDNA. The lengths of tails in the pea ctDNA ranged from 1.5 to 124% of the length of the attached monomer-length circular molecule. In corn ctDNA, 11% of the total circular molecules had tails which ranged in length from 2 to 140% of the attached monomer-length circular molecule. The finding of tails that were longer than the attached monomer-length circular molecule eliminated the possibility that the tails arose by breakage of a Cairns forked structure at a replicative fork.

4. A Model for the Replication of Chloroplast DNA

Based upon the above results, a model for the replication of pea and corn ctDNA is presented in Figure 45. The ctDNA replication is initiated by the formation of two displacement loops, whose displacing strands are complementary to the opposite parental strands of ctDNA (Figure 45B). The two displacing strands expand toward each other and initiate the formation of Cairns replicative forked structures (Figure 45D). The small Cairns-forked structures (Figure 45D) expand bidirectionally until termination takes place at a site that is 180° around the cicular ctDNA molecule from the initiation site. Separation of the daughter molecules takes place yielding two circular molecules that each have a single-strand break or small gap at the same site located in opposite daughter strands (Figure 45F). In ctDNA, the nicked circles could be sealed to close the circles or the 3′OH of each nicked progeny molecule could be extended by a DNA polymerase molecule. This would displace a single-stranded tail (Figure 45G) from the molecule and this tail could be filled in by discontinuous duplex synthesis to yield a molecule with a double-stranded tail (Figure 45H) The tails might then be converted to circular molecules by an intrastrand recombination event. If boh progeny from a Cairns round of replication initiated rolling circle synthesis, two types of rolling circles would be formed. In each case, the tip of the tail would map at the same site, but the sequence of the two types of tails would extend in the opposite direction from this site. The denaturation maps with pea ctDNA rolling circle confirm that this is the case.

B. DNA Synthesis In Vitro

The above experiments on replication have been carried out with invivo replicative intermediates. In order to further understand the replication process inside the cell, we must resolve and reconstitute each function outside the cell. As a first step, we have studied DNA synthesizing activity in the isolated chloroplasts.[62] Sucrose density gradient purified pea chloroplasts were found to carry out a very active DNA synthesis in the presence of dNTP utilizing endogenous ctDNA. The lysis of chloroplasts with triton, deoxycholate, and EDTA failed to solubilize any significant amount of DNA

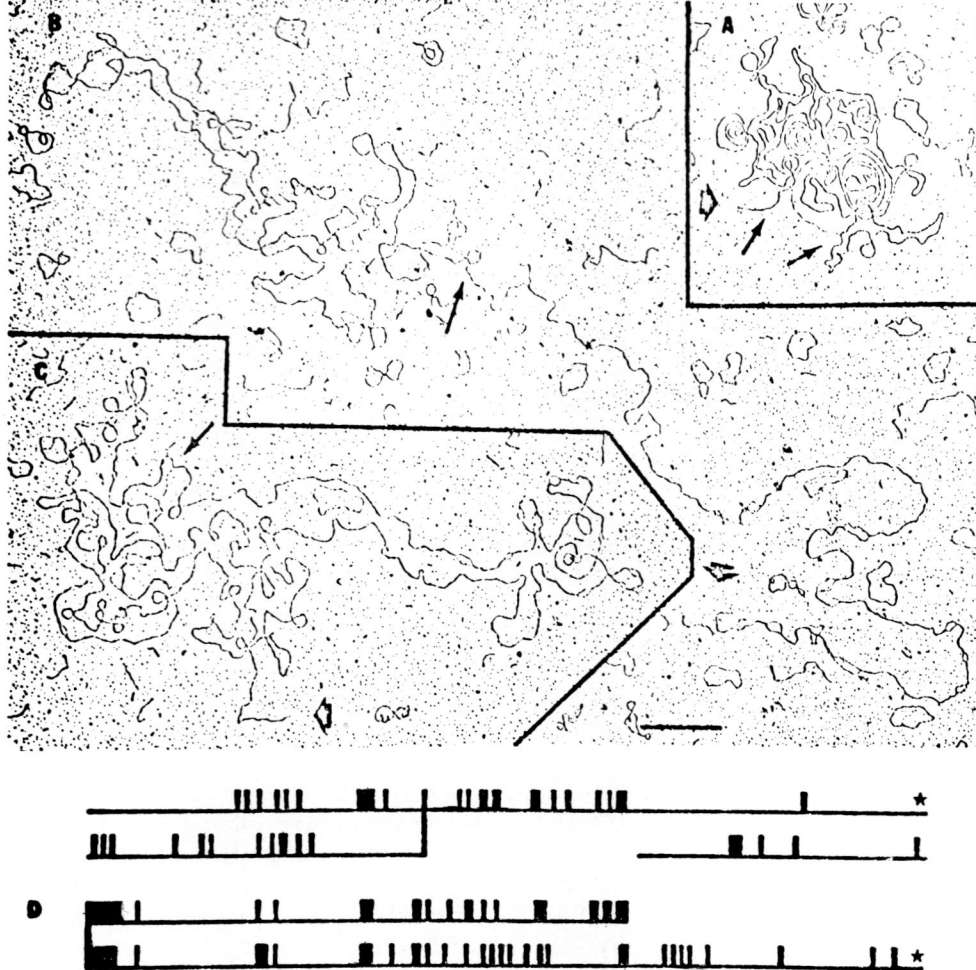

FIGURE 44.   Rolling circle molecules. (A) A molecule with a tail that is 19% of length of the attached circle. There is a single-strand region at the growing fork that is 3500 bases long (indicated by arrows). (B,C) Partially denatured molecules with tails that illustrate the two different denatured patterns at the tip of the tails. (D) Graphical representation of molecules B and C. The boxes indicate the denatured regions, the ★ indicates the circular part of the molecule that has been linearized. The unmarked free end is the tip of each tail. The bars indicate 1 μm.

synthesis activity from chloroplasts. The DNA synthesis activity was completely solubilized when chloroplasts were treated with 0.5% Lubrol.®

The solubilized fraction was adjusted to 0.3 *M* KCL and passed through a DEAE-cellulose column equilibrated with the same buffer. The effluent from the column contained all the activity of the isolated chloroplasts and this activity was totally dependent upon th externally added DNA. This fraction was able to synthesize DNA to the extent of 1 to 2% of the template DNA and was active against both double-stranded and single- stranded DNA. Using single stranded φX174 and G4 as a template the in vitro synthesized product was found to be very close to the molecular size of the template. This fraction was allowed to synthesize DNA in the presence of ctDNA by substituting bromodeoxyuridine triphosphate for thymidine triphosphate in the DNA synthesis reaction system. The isolated product had a buoyant density expected

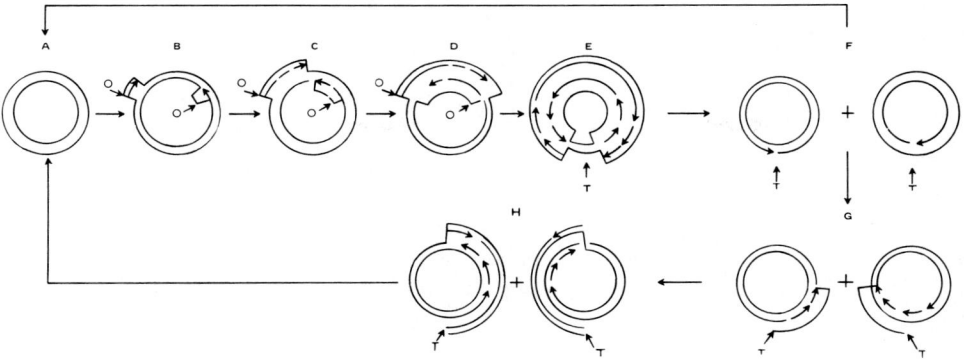

FIGURE 45. A model for the replication of ctDNA. (A) closed circular parental molecule; (B) D-loop containing molecule; (C) expanded D-loop containing molecules; (D,E) Cairn's type of replicative intermediate; (F) nicked progeny molecules; (G,H,) rolling circles. The thin and thick lines mark the opposite strands of a molecule. The lines with the arrows are the daughter strands. The "O" indicates the positions of the two origins of D-loop synthesis which are 5.2% of pea ctDNA apart. The "T" indicates the terminus of the Cairn's round of replication which is 180° opposed to the origins of D-loop synthesis.

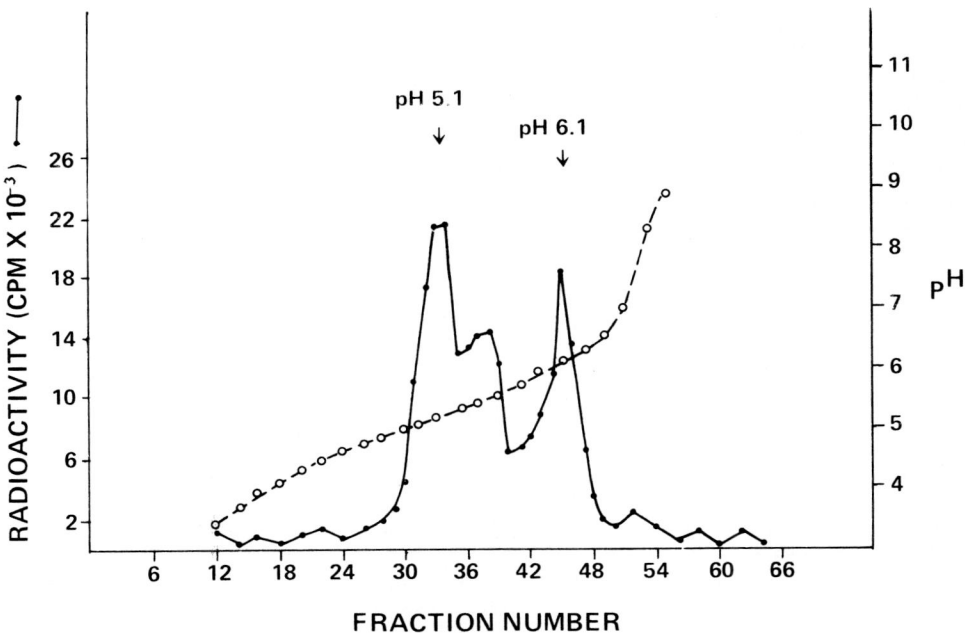

FIGURE 46. Isoelectric focusing of DNA polymerase purified through a phosphocellulose column.

of a DNA whose one entire strand contains bromodeoxyuridine instead of thymidine. This material was further fractionated on a DEAE cellulose column with 0 to 0.3 M linear KCL gradient. DNA synthesis activity was found at three different fractions. The first two peaks eluted at approximately 0.07 and 0.13 M KCL. These two peaks were found to be active in the presence of both double-stranded and single-stranded DNA. The third peak eluted at 0.25 M KCl and was active only on the double-stranded DNA as a template. The above three different fractions were further purified on a phosphocellulose column followed by isoelectric focusing. The enzymatic activity was found to focus in two distinct peaks with isolectric pHs at 5.1 and 6.1, respectively

(Figure 46). The purity of these enzyme fractions and their function in DNA replication is currently under investigation.

V. SUMMARY

The studies described here have contributed significantly to the understanding of the sequence organization of ctDNA, the arrangement of genes in the ctDNA, and the replication of 275ctDNA. With ever increasing sophistication of techniques in the molecular biology of nucleic acids, it is now possible to predict that all of the genes of ctDNA will be identified and mapped in the forseeable future. Studies on the mechanism of transcription of ctDNA and its regulation in vitro and in vivo will add tremendously to our knowledge of the development and differentiation of chloroplasts. Studies on the purification and properties of proteins involved in the replication of ctDNA will add new insight into the mechanism of organelle duplication as well as DNA replication in general.

ACKNOWLEDGMENT

The work presented here reflects the contribution of many of my graduate students. I would like to particularly thank Dr. J. R. Thomas, Dr. R. Kolodner, Dr. M. White, Mr. N. Chu, Mr. R. McKown, Ms. K. Oishi, Mr. T. Summicht, and Mr. W. Dobkin. This chapter could not have been completed without the assistance of Ms. J. Keithe who has been responsible for printing of the figures and Ms. Ann Rule who has been patient and understanding in preparing the manuscript. I am very thankful to the National Science Foundation for supporting this research for the last 10 years. The current grant number is PCM 7817443. I would like to dedicate this chapter to Dr. S. G. Wildman for this pioneering work on chloroplast of higher plants.

REFERENCES

1. **Bauer, E.,** Das Wesen and die Erblichkeitsverhaltnisse der 'Varietates albomarginatae hort' von *Pelargonium zonale, Z. Verebungsl.,* 1, 330, 1909.
2. **Correns, C.,** Verebungsversuche mit blass (gelb) grunen und bluntblattrigen sipper bei *Mirabilis, Uritca,* und Lunaria, *Z. Verebungsl.,* 1, 291, 1909.
3. **Boardman, N. K., Linnane, A. W., and Smillie, R. M., Eds.** *Autonomy and Biogenesis of Mitochondria and Chloroplasts,* North-Holland, Amsterdam, 1971.
4. **San Pietro, A., Greer, F. A., and Army, T. J., Eds.,** *Harvesting the Sun: Photosynthesis in Plant Life,* Academic Press, New York, 1971,
5. **Goodwin, T. W., Ed.,** *Biochemistry of Chloroplasts,* Vol. I and II, Academic Press, New York, 1967,
6. **Gibbs, M., Ed.,** *Structure and Function of Chloroplasts,* Springer-Verlag, Berlin, 1971,
7. **Miller, P. L., Ed.,** Control of organelle development, *Symp. Soc. Expt. Biol.,* Vol. 24, Cambridge University Press, London, 1970,
8. **Kirk, J. T. O. and Tilney- Bassett, R. A. E.,** *The Plastids. Their Chemistry, Structure, Growth and Inheritance,* W. H. Freeman, London, 1967.
9 **Levine, R. P. and Goodenough, V. W.,** The genetics of photosynthesis and of the chloroplast in *Chlamydomonas reinhardi, Annu. Rev. Genet.,* 4, 397, 1970.
10. **Sager, R.,** *Cytoplasmic Genes and Organelles,* Academic Press, New York, 1972.
11. **Kirk, J. T. O.,** Biochemical aspects of chloroplast development, *Annu. Rev. Plant pyhsiol.,* 21, 11, 1970.
12. **Kirk, J. T. O.,** Chloroplast structure and biogenesis, *Annu. Rev. Biochem.,* 40, 161, 1971.

13. Bogorad, L. and Weil, J., Eds., *Proceedings of the N.A.T.O. Advanced Study Institute on Nucleic Acids and Protein Synthesis in Higher Plants,* Plenum Publishing, London, 1977.

14. Bucher, T., Ed., *Genetics and Biogenesis of Chloroplasts and Mitochondrira,* Elsevier, Amsterdam, 1977.

15. Kolodner, R. and Tewari, K. K., The molecular size and conformation of the chloroplast DNA from higher plants, *Biochim. Biophys. Acta,* 402, 375, 1975.

16. Tewari, K. K., *Annu. Rev. Plant Physiol.,* 22, 141, 1971.

17. Kirk, J. T. O., in *Autonomy and Biogenesis of Mitochondria and Chloroplasts,* Boardman, N. K., Linnane, A. W., and Smillie, R. M., Eds., North-Holland, Amsterdam, 1971.

18. Wells, R. and Sager, R., Denaturation and renaturation kinetics of chloroplast DNA from *Chlamydomonas reinhardi, J. Mol. Biol.,* 58, 611, 1971

19. Slavik, N. S. and Hershberger, C. L., The kinetic complexity of *Euglena gracilis* chloroplast DNA, *FEBS Lett.,* 52, 171, 1975.

20. Bayen, M. and Rode, A., Heterogenity and complexity of *Chlorella* chloroplastic DNA, *Eur. J. Biochem.,* 39, 413, 1973.

21. Manning, J. E., Wolstenholme, R., Ryan, R. S., Hunter, J. A., and Richards, O. C., *Proc. Natl. Acad. Sci. U.S.A.,* 68, 1169, 1971.

22. Kleinschmidt, A. K., *Methods in Enzymology,* Vol. 12, Grossman, L., and Moldave, K., Eds., Academic Press, 1968, 361.

23. Davis, R. W., Simon, M., Davidson, N., *Methods in Enzymology,* Vol. 21, Part D., Grossman, L. and Moldave, K., Eds., 1971, 413.

24. Strider, W., Ph. D. thesis, New York University, New York, 1971,

25. Behn, W. and Herrmann, G., Circular molecules in the β-satellite DNA of *Chlamydomonas reinhardii, Mol. Gen. Genet.,* 157, 25, 1977.

26. Rochaix, J. D., *Genetics and Biogenesis of Chloroplasts and Mitochondria,* Bucher, T., Ed., North-Holland, Amsterdam, 1976, 375.

27. Howell, S., Heizmann, P., and Gelvin, S., *Genetics and Biogenesis of Chloroplasts and Mitochondria,* Bucher, T., Ed., North-Holland, Amsterdam, 1976, 625.

28. Kolodner, R., Tewari, K. K., and Warner, R. C., Physical studies on the size and structure of the covalently closed circular chloroplast DNA from higher plants, *Biochim. Biophys. Acta,* 447, 144, 1976.

29. Rush, M. G. and Warner, R. C., Alkali denaturation of covalently closed circular duplex deoxyribonucleic acid, *J. Biol. Chem.,* 245, 2704, 1970.

30. Borst, P. and Kroon, A. M., Mitochondrial DNA: physicochemical properties, replication, and genetic function, *Int. Rev. Cytol.,* 26, 108, 1969.

31. Waring, M., Variation of the supercoils in closed curcular DNA by binding of antibiotics and drugs: evidence for molecular models involving intercalation, *J. Mol. Biol.,* 54, 247, 1970.

32. Wang, J. C., The degree of unwinding of the DNA helix by ethidium I. Titration of twisted PM2 DNA molecules in alkaline cesium chloride density gradients, *J. Mol. Biol.,* 89, 783, 1975.

33. Kolodner, R. and Tewari, K. K., Molecular size and conformation of chloroplast DNA from pea leaves, *J. Biol. Chem.,* 247, 6355, 1972.

34. Bastia, D., Chiang, K. S., Swift, H., and Sievsma, P., Heterogeneity, complexity, and repetition of the chloroplast DNA of *Chlamydomonas reinhardii, Proc. Natl. Acad. Sci. U.S.A.,* 68, 1157, 1971.

35. Britten, R. J. and Kohne, D. E., Repeated sequences in DNA, *Science,* 161, 529, 1968.

36. Inman, R. B., Denaturation maps of the left and right sides of the lambda DNA molecule determined by electron microscopy, *J. Mol. Biol.,* 28, 103, 1967.

37. Davis, R. W. and Hyman, R. W., A study in evolution: the DNA base sequence homology between caliphages T7 and T3, *J. Mol. Biol.,* 62, 287, 1971

38. Davis, R. W., Simon, M., and Davidson, N., *Methods Enzymol.,* 21, 413, 1971.

39. Kolodner, R. D. and Tewari, K. K., Denaturation mapping of pea chloroplast DNA, *J. Biol. Chem.,* 250, 4888, 1975.

40. Kolodner, R., Warner, R. C., and Tewari, K. K., The presence of covalently linked ribonucleotides in the closed circular deoxyribonucleic acids from higher plants, *J. Biol. Chem.,* 250, 7020, 1975.

40a. Kolodner, R. and Tewari, K. K., Inverted repeats in chloroplast DNA from higher plants, *Proc. Natl. Acad. Sci. U.S.A.,* 76, 41, 1979.

41. Bedbrook, J. R. and Bogorad, L., Endonuclease recognition sites mapped on *Zea mays* chloroplast DNA, *Proc. Natl. Acad. Sci. U.S.A.,* 73, 4309, 1976.

42. Herrmann, R. G., Bohnert, H. J., Driesel, A., and Hobom, G., *Genetics and Biogenesis of Chloroplasts and Mitochondria,* Bucher, T., Ed., North-Holland, Amsterdam, 1976, 351.

43. Southern, E. M., Detection of specific sequences among DNA fragments separated by gel electrophoresis, *J. Mol. Biol.,* 98, 503, 1975.

44. Bedbrook, J. R., Kolodner, R., and Bogorad, L., *Zea mays* chloroplast ribosomal RNA genes are part of a 22,000 base pair inverted repeat, *Cell*, 11, 739, 1977.

45. Thomas, J. R. and Tewari, K. K., Ribosomal RNA genes in chloroplast DNAs of higher plants, *Biochim. Biophys. Acta*, 361, 73, 1974.

46. Thomas, J. R. and Tewari, K. K., Conservation of 7OS ribosomal RNA genes in chloroplast DNAs of higher plants, *Proc. Natl. Acad. Sci. U.S.A.*, 71, 3147, 1974.

47. Weil, J. H., Burkard, G., Guillemant, P., Jeanin, G., Martin, R., and Steinmetz, H., *Genetics and Biogenesis of Chloroplasts and Mitochondria*, Bucher, T., Ed., 1976, 667.

48. Hecker, L. I., Egan, J., Reynolds, R. J., Nix, C. E., Schiff, J. A., and Barnett, W. E., The sites of transcription and translation for *Euglena* chloroplastic aminoacyl-tRNA synthetases, *Proc. Natl. Acad. Sci. U.S.A.*, 71, 1910, 1971.

49. Haff, L. A. and Bogorad, L., Hybridization of maize chloroplast DNA with transfer ribonucleic acids, *Biochemistry*, 15, 4105, 1976.

50. Williams, G. R., Williams, A. S., and George, S. A., Hybridization of leucyl-transfer ribonucleic acid isoacceptors from green leaves with nuclear and chloroplast deoxyribonucleic acid, *Proc. Natl. Acad. Sci. U.S.A.*, 70, 3498, 1973.

51. Schwartzbach, S. D., Hecker, L. I., and Barnett, W. E., Transcriptional origin of *Euglena* chloroplastic tRNAs, *Proc. Natl. Acad. Sci. U.S.A.*, 73, 1984, 1976.

52. Chang, S. H., Brun, C. U., Silberklang, M., Rajbhandary, U. L., Hecker, L. I., and Barnett, W. E., The first nucleotide sequence of an organelle transfer RNA: chloroplastic tRNA, *Cell*, 9, 717, 1976.

53. Meeker, R. and Tewari, K. K., tRNA genes in pea chloroplast DNA, in preparation, 1979.

54. Meeker, R., Ph. D thesis, University of California, Irvine, 1978.

55. Haff, L. A. and Bogorad, L., Poly(adenylic acid)-containing RNA from plastids of maize, *Biochemistry*, 15, 4110, 1976.

56. Chelm, B. K., Hallick, R. B., Changes in the expression of the chloroplast genome of *Euglena gracilis* during chloroplast development, *Biochemistry*, 15, 593, 1976.

57. Rawson, J. R. Y. and Boerma, C. L., Hybridization of *Euglena* RNA to the heavy and the light chloroplast DNA components separated inalkaline CsCl gradients, *Biochem. Biophys. Res. Comm.*, 74, 912, 1977.

58. Coen, D. M., Bedbrook, J. R., Bogorad, L., and Rich, A., Maize chloroplast DNA fragment encoding the large subunit of ribulose diphosphate carboxylase, *Proc. Natl. Acad. Sci. U.S.A.*, 74, 5487, 1977.

59. Bedbrook, J. R., Link, G., Coen, D. M., Bogorad, L., and Rich, A., Maize plastid gene expressed during photo regulated development, *Proc. Natl. Acad. Sci. U.S.A.*, 75, 3060, 1978.

60. Kolodner, R. and Tewari, K. K., The presence of displacement loops in the covalently closed circular chloroplast DNA from higher plants, *J. Biol. Chem.*, 250, 8840, 1975.

61. Kolodner, R. and Tewari, K. K., Chloroplast DNA from higher plants replicates by both the Cairns and the rolling circle mechanism, *Nature (London)*, 256, 708, 1975.

62. Tewari, K. K., Kolodner, R. D., and Dobkin, W., *Genetics and Biogenesis of Chloroplasts and Mitochondria*, Bucher, T., Ed., North-Holland, Amsterdam, 1976, 379.

Section II
Plant RNA

RNA POLYMERASES IN PLANTS

W. M. Becker

TABLE OF CONTENTS

I. EUKARYOTIC RNA POLYMERASES: A BIOGRAPHICAL SKETCH

The control of transcription is a topic of key importance in any consideration of the regulatory mechanisms underlying eukaryotic development. Although we are becoming increasingly aware of the multiplicity and significance of regulatory possibilities operating distant from the gene, it nonetheless remains true that the differential transcription of nuclear and organellar genes is a primary mode of control. The detailed mechanisms whereby such control is exerted are still largely unknown, but they obviously must involve an interaction between the RNA polymerases responsible for the actual transcription and the chromosomal template on which these enzymes must function. In theory, of course, transcriptional control could reside primarily or even exclusively in the chromosomal template, with changes in chromatin structure responsible for relative accessibility of specific genes and polymerase enzymes playing the relatively passive role of simply transcribing all available sites. As an alternative, regulation could be accomplished at the level of the polymerase molecule mediated by changes in forms, amounts, or subunit composition of the enzyme and/or by the possible transient involvement of regulatory factors.

While there is already ample and continuously accumulating evidence on hand supporting changes in chromosomal structure as a means of restricting the availability of specific DNA sequences and thereby regulating gene expression, it is also true that eukaryotic cells contain multiple forms of RNA polymerase differing in their localization, subunit composition, and function. Thus, the attention focused on eukaryotic RNA polymerases and on regulatory possibilities raised by their demonstrated multiplicity of form and intracellular locale seems appropriate indeed. Although several enzymes capable of synthesizing RNA hetero- or homopolymers in the absence of a DNA template are known, the present discussion will be restricted to the DNA-dependent RNA polymerases (ribonucleoside triphosphate RNA nucleotidyl transferase, EC 2.7.7.6) which catalyze synthesis of an RNA molecule complementary in base sequence to the transcribed strand of the DNA template.

Several reviews have already appeared on RNA polymerases from prokaryotes[1-3] and eukaryotes,[4-10] but only that of Duda[9] deals specifically with the RNA polymerases of plants. Following a brief historical account of RNA polymerase and its multiple forms in eukaryotic nuclei, this chapter will focus on what is presently known about the plant RNA polymerases and their involvement in developmental phenomena.

A. Beginnings

The first report of an enzyme capable of synthesizing RNA with four ribonucleoside triphosphates as substrates was that of Weiss[11] based on studies with rat liver nuclei. Shortly thereafter, Hurwitz et al.[12] and Stevens[13] demonstrated the same reaction in *Escherichia coli*, which quickly became the organism of choice for further studies due to difficulties initially encountered in the solubilization of the enzyme from mammalian and other eukaryotic sources. Burgess[14] isolated the *E. coli* enzyme and showed it to be a large molecule with five polypeptide subunits: two α subunits (39,000 daltons each), one β subunit (155,000 daltons), one β' subunit (165,000 daltons), and one dissociable molecule of sigma (σ) factor (95,000 daltons). Upon chromatography on phosphocellulose, the holoenzyme ($\alpha_2\beta\beta'\sigma$) could be resolved into a core enzyme ($\alpha_2\beta\beta'$) and the dissociable sigma factor.[15] Sigma factor was shown to be required for specific initiation of RNA synthesis, after which it is released, leaving the core enzyme responsible for further elongation of the RNA chain, and the sigma factor free to assist in initiation in a cyclic and catalytic manner.[16]

Progress in characterization of the transcriptional apparatus of nuclei of eukaryotic cells was hindered initially by the difficulty of solubilizing eukaryotic polymerases from chromatin. A key breakthrough came in 1969 with the report of Roeder and Rutter[17] that the quantitative extraction of RNA polymerase activity from several animal sources could be achieved by sonication of nuclei in the presence of high ionic strength. Upon dilution of the sonicate to lower ionic strength, the histones and DNA reassociate and can be removed by centrifugation, leaving the polymerase activity in solution and amenable to purification by standard enzymological procedures.

B. The Multiplicity of Eukaryotic Nuclear RNA Polymerases

Once eukaryotic RNA polymerase activities were solubilized from chromatin, ion exchange chromatography (on DEAE cellulose or DNA Sephadex®, with linear gradients of KCl or $(NH_4)_2SO_4$) quickly revealed the existence of multiple forms of the enzyme.[17,18] This multiplicity of nuclear RNA polymerases has now been demonstrated for a phylogenetically diverse group of eukaryotic sources and may be presumed a common feature of all eukaryotic cells.

The three distinctively different forms of RNA polymerases present in most eukaryotic nuclei are designated as RNA polymerases I, II, and III, in order of their elution from DEAE Sephadex.[10,17,18] (For the alternative "European" nomenclature, see Chambon.)[8] As indicated in Table 1, the three major forms of RNA polymerases also

TABLE 1

Properties of RNA Polymerases I, II, and III

	Enzyme class		
	I	II	III
Elution from DEAE Sephadex (M $(NH_4)_2SO_4$)	0.1	0.2	0.2—0.3
Elution from DEAE Cellulose (M $(NH_4)_2SO_4$)	0.1	0.2	0.1
Sensitivity to α-amanitin	None	Highly sensitive	Moderately sensitive
Location within nucleus	Nucleolus	Nucleoplasm	Nucleoplasm
Products of transcription	rRNA	hnRNA	tRNAs, 5S rRNA

differ in localization within the nucleus, in the kinds of RNA synthesized, and in sensitivity to the fungal toxin, α-amanitin (a bicyclic octapeptide isolated from the mushroom *Amanita phalloides*). Additionally, RNA polymerases I, II, and III are readily distinguished by ionic strength optima, relative responses to Mn^{++} vs. Mg^{++}, and relative activities with synthetic templates (e.g., poly (dA-dT)) versus native DNA.[4-10]

RNA polymerase I is localized within the nucleolus,[17,19] synthesizes ribosomal RNA,[20,21] and is refractory to α-amanitin.[19,22] RNA polymerase II, by contrast, is found in the nucleoplasm,[17,19] synthesizes heterogeneous nuclear RNA,[23] (generally presumed to serve in part as precursor to cytoplasmic mRNA) and is acutely sensitive to α-amanitin[19,22] with only approximately 0.01 ppm inhibitor required for 50% inhibition of form II activity in most species. RNA polymerase III is also a nucleoplasmic enzyme[17,24] (though it may also be present in the cytoplasm[24,25]), but is responsible for synthesis of transfer (4S) RNAs and 5S ribosomal RNA.[26] Sensitivity of the form III enzyme to α-amanitin seems to be phylogenetically variable: that from higher animals is moderately sensitive[24,25] (20 to 200 ppm α-amanitin required for 50% inhibition), while the plant enzyme is 10- to 100-fold less sensitive[27] (see below). The otherwise comparable enzymes from *Bombyx*,[10] *Mucor*,[28] and yeast[29] appear completely refractory.

In much of the early work on animal RNA polymerases and in most of the plant work to date, only RNA polymerases I and II were detected. For animal systems at least, this was partly because form III is labile and represents only a minor polymerase activity[24] and partly because DEAE cellulose chromatography is unable to resolve polymerases I and III from one another as indicated in Table 1.

Multiple subspecies of RNA polymerases have been demonstrated within each of the three major classes based on chromatographic and electrophoretic heterogeneity.[10]

However, none of the subspecies within a given class can be distinguished on the basis of catalytic properties or α-amanitin sensitivity, leading to the suggestion that the classification of nuclear RNA polymerases as forms I, II, and III is functionally meaningful with only minor structural variations within classes.

C. Subunit Composition of RNA Polymerases

All known nuclear RNA polymerases are complex, multisubunit enzymes with molecular weights in the range of 500,000. Subunit analysis typically involves the electrophoresis of purified enzyme on polyacrylamide gels in the presence of sodium dodecyl sulfate, followed by staining the protein and scanning to quantitate amounts of subunit protein. Subunit analysis has been undertaken for RNA polymerases from a variety of animal and other sources. Some of the most definitive results are those reviewed recently by Roeder et al.[10] from their work on mouse plasmacytoma cells; their findings are summarized here to facilitate comparisons with the subunit compositions of plant RNA polymerases discussed later (see Figure 1).

By a combination of chromatography, sucrose density sedimentation, and electrophoresis under nondenaturing conditions, nine individual enzyme species were resolved from plasmacytoma nuclei, including two form I enzymes, three form II enzymes, and four species of the form III enzyme.[10,24] RNA polymerase Ia contained six polypeptide subunits, with molecular weights ($\times 10^{-3}$) of 195, 117, 61, 52, 29, and 19, respectively. Enzyme Ib had the same subunit composition, except for the absence of the 61,000-dalton component. The three subspecies of RNA polymerase II had seven subunits in common with molecular weights ($\times 10^{-3}$) of 140, 41, 29, 27, 22, 19, and 16, respectively; they differed only in the size of the largest subunit which had a molecular weight of 170,000 for one subspecies, 205,000 for another, and 240,000 for the third. The four form III enzymes (two each of IIIa and IIIb) had ten distinct polypeptides. For the two IIIa enzymes, these had molecular weights ($\times 10^{-3}$) of 155, 138, 89, 70, 53, 49, 41, 32, 29, and 19. The two form IIIb enzymes were almost identical to this in component polypeptides except that the 32,000-dalton subunit of form IIIa was clearly distinguishable from the 33,000-dalton subunit of form IIIb.

From these[10,24] and similar (though usually less definitive) data for RNA polymerases from various sources, three general conclusions can be drawn: (1) the subunit composition of each of the three major classes of RNA polymerases appears to be a distinctive feature of that class of enzyme with most subunits unique to a given class; (2) in no case has it yet been possible to correlate the minor structural heterogeneity within a given class of RNA polymerase with any convincing functional heterogeneity within that class; and (3) the structural features of the corresponding enzymes from different species seem to be strikingly similar, suggesting a remarkable evolutionary conservation.

II. ISOLATION AND PROPERTIES OF PLANT POLYMERASES

The first reports of the isolation of RNA polymerases from plant sources appeared in 1971 and described the partial purification of the chloroplast[30] and nuclear[31] enzymes of maize (*Zea mays*) in Bogorad's laboratory and the characterization of the corresponding enzymes from wheat leaves by Polya and Jagendorf.[32,33] Since then, RNA polymerases have been isolated and characterized from a variety of plants including pea seedlings[34] and buds,[35] soybean hypocotyl,[36-39] sugar beet root,[40] cauliflower inflorescence,[27,41-43] Jerusalem artichoke tubers,[44] lentil roots,[45] coconut endosperm,[46] parsley cell culture,[47] wheat germ,[48-50] and rye embryos.[51]

Generally, plants seem to follow the animal pattern in their possession of three basic forms of RNA polymerase: one eluting from DEAE Sephadex (or DEAE cellulose) at low salt (approximately 0.1 M (NH$_4$)$_2$SO$_4$) and showing resistance to α-amanitin, a

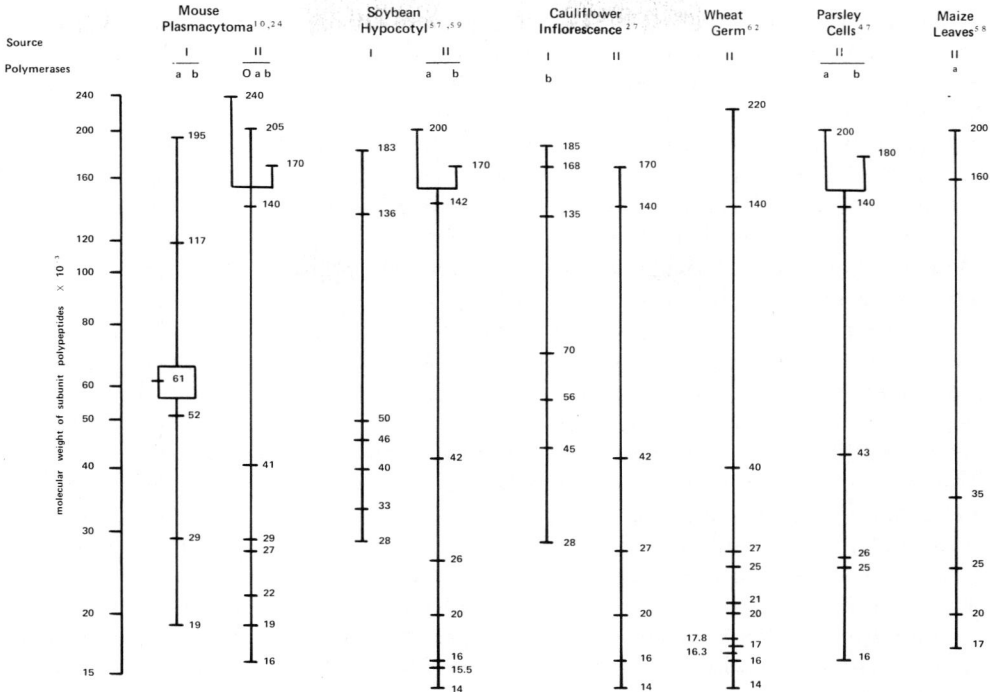

FIGURE 1. Subunit composition of several plant RNA polymerases. RNA polymerases I and/or II have been isolated and purified from soybean hypocotyl,[57,59] cauliflower inflorescence,[27] wheat germ,[62] cultured parsley cells,[47] and maize leaves.[58] Their subunit compositions were determined by sodium dodecyl sulfate-polyacrylamide gel electrophoresis. The compilation is by no means exhaustive, but is intended to include most of the plant RNA polymerases for which the subunit analysis appears reasonably credible at present. Data for the corresponding RNA polymerases of mouse plasmacytoma cells[10,24] are included on the left for comparative purposes. Each vertical line represents an RNA polymerase molecule. Subunits are indicated by the crosslines and their respective molecular weights (in kilodaltons) by the adjacent numbers. Enzyme subspecies differing in subunit composition are indicated by branching of the vertical axis. Subunit molecular weight is on a logarithmic scale to correspond to electrophoretic mobility on SDS-polyacrylamide (direction of migration would be down the page), but the figure is an idealized representation in that it is seldom possible to accomodate all subunits from 14 to 240 kilodaltons on a single gel, and no provision has been made to indicate relative band intensities. Discrepancies in numbers and sizes of subunits, especially in the lower molecular weight range, are probably due more to technical problems than to phylogenetic variability, since it is difficult to resolve all polypeptides and determine their molecular weights accurately based on electrophoresis in a single dimension. Of the results shown, only those of Jendrisak and Burgess[62] for wheat germ RNA polymerase II were obtained using two-dimensional polyacrylamide gel electrophoresis (8 M urea in the first dimension, 0.1% SDS in the second) to resolve all of the polypeptide components. Not surprisingly, theirs is the most complex subunit composition for a plant RNA polymerase to date.

second eluting at higher salt (approximately 0.2 M $(NH_4)_2SO_4$) and displaying great sensitivity to the inhibitor, and a third enzyme which, when detected,[27,51] elutes from DEAE Sephadex at still higher salt and requires an especially high α-amanitin concentration for inhibition.[27] The elution order of the amanitin-insensitive and -sensitive enzymes (forms I and II) is the same for all plant sources thus far examined, except for coconut endosperm. It was suggested[17] that the inverted order of elution reported for coconut endosperm RNA polymerases[46,52] is due to the high salt (2 M NaCl) used to extract the polymerase from chromatin, although the explanation does not seem especially convincing.

RNA polymerase III is not mentioned in most of the published reports on plant RNA polymerases to date. Whether this can be accounted for by the very minor proportion of total activity which this enzyme represents in most plant sources or by the frequency with which DEAE cellulose rather than DEAE Sephadex has been used in the initial separation remains to be ascertained. In any case, RNA polymerase III has not yet been isolated from any plant source in sufficient quantity to allow serious examination of its structure or specific function.

That both RNA polymerases I and II of plants are of nuclear origin is suggested by isolation of both enzymes from chromatin[37,39,45,46,53] and is demonstrated more directly by the presence of both enzymes in nuclei isolated from soybean hypocotyl.[54] More recently, RNA polymerases I, II, and III have all been purified from isolated cauliflower nuclei.[27] By isolating nucleoli from soybean hypocotyl, Lin et al.[54] were able to establish a nucleolar localization for RNA polymerase I. Since RNA polymerase II was present in isolated nuclei, but not in the purified nucleoli, it is apparently a nucleoplasmic enzyme suggesting a similar intranuclear localization of both RNA polymerases I and II in plants and animals. Further evidence of similarities between plant and animal enzymes comes from the demonstration by Gurley et al.[55] that the α-amanitin-resistant RNA polymerase bound to soybean chromatin synthesizes rRNA in vitro, although with the qualification that only about 35% of the total transcript was demonstrably rRNA by competition hybridization to total DNA with and without a 100-fold excess of unlabeled rRNA. Polymerase II (or, more specifically, its apparent equivalent in coconut endosperm, the RI enzyme) appears to synthesize a nonribosomal RNA since rRNA is an ineffective competitor in hybridization to DNA.[46]

Although both RNA polymerases I and II are apparently tightly bound to chromatin in nuclei of animal cells, it seems that, at least for some plant species, chromatin isolated by conventional methodology (i.e., by homogenization at pH 8 by the method of Huang and Bonner)[56] contains mainly RNA polymerase I, with most RNA polymerase II activity remaining behind upon chromatin sedimentation. This is the case for both soybean hypocotyl[37] and wheat germ.[50] Neither requires sonication to render RNA polymerase II soluble. Lin et al.[37] have shown, however, that a several-fold greater (though still relatively small) recovery of chromatin-bound RNA polymerase II can be achieved if the chromatin is isolated at pH 6 instead of pH 8. Recently, Guilfoyle and Key[57] reported the existence of two RNA polymerase II activities in germinating soybean. One was a soluble enzyme present in the ungerminated embryo, and the other was a chromatin-bound form that increased in amount as germination progressed (see further discussion below).

A. Isolation Procedures

Various procedures have been employed in the isolation of RNA polymerase from plant sources. (For a detailed tabulation of extraction procedures used in comparable work with animal sources, see the review by Jacob.[4]) Commonly, the starting material is either chromatin[32,37,39,40,45] or total plant tissue,[31,34-36,41,44,47-49] although isolated nuclei have been used in several recent reports.[27,54] When chromatin is used, it is usually prepared by the method of Huang and Bonner,[56] and the solubilization of DNA is accomplished by subsequent high salt treatment of the chromatin. Chromatin should be used with caution for this purpose, since chromatin prepared from at least some plant sources (e.g., soybean hypocotyl[37,55] and wheat germ[50]) is, as already noted, virtually devoid of RNA polymerase II activity. Sonication in high salt is widely used in the solubilization of animal RNA polymerases and has also been used to prepare polymerases from wheat germ,[48,49] peas,[34,35] soybean,[36,37] and cauliflower,[27] but many workers prefer to depend simply on stirring or blending in the presence of high salt.

Subsequent resolution of the solubilized RNA polymerase activities usually depends on ion exchange chromatography on either DEAE cellulose or DEAE Sephadex. With the former, RNA polymerase III tends to co-elute with polymerase I at low salt concentration. With the latter, the form III enzyme may be obscured by the RNA polymerase II peak.[27] By using both procedures in tandem, RNA polymerases I and III can first be separated from form II on DEAE cellulose and then resolved from each other by rechromatography on DEAE Sephadex.[25,27]

Fractionated polymerase activities as obtained by DEAE chromatography are usually identified as forms I, II, and sometimes III based on their relative sensitivities to α-amanitin. Enzymes at this stage of purity have frequently been used in characterization studies intended to determine their optima with respect to pH, temperature, ionic strength, divalent cation (Mg^{++} or Mn^{++}) concentrations, and their template preference (native DNA vs. denatured DNA vs. synthetic polynucleotides).

In general, further purification of individual RNA polymerase species depends upon chromatography on phosphocellulose or other ion exchange resin, gel filtration, DNA cellulose affinity chromatography, or centrifugation through density gradients of sucrose or glycerol. Three representative procedures successfully used for purification of individual RNA polymerase species from several sources (maize leaves,[58] auxin-treated soybean hypocotyls,[59] and wheat germ[50]) are summarized in Table 2. In each case, the authors provide a tabular record of the step-by-step yields and specific activities in the original reports. Subunit compositions of the three enzymes prepared by these purification procedures are compared in Figure 1.

The procedure of Mullinix et al.[58] begins with a direct homogenization of maize leaves in a Waring Blendor®, followed by filtration of the homogenate and clarification by high-speed centrifugation (100,000 g for 90 min). RNA polymerase activity was precipitated from the supernatant in a 30 to 50% $(NH_4)_2SO_4$ fraction. The protein precipitate was collected by centrifugation, dissolved in a glycerol buffer, and heated to 50°C for 10 min to destroy RNA polymerase I activity.[31] RNA polymerase IIa (one of the two forms into which maize RNA polymerase II can be resolved on DEAE cellulose[31]) was then purified by passage successively through Sepharose 6B, Sepharose 4B, DEAE cellulose, phase partition into polyethylene glycol, and DNA cellulose.

For isolation of RNA polymerase I from soybean hypocotyl, Guilfoyle et al.[59] began by preparing chromatin from auxin-treated tissue because RNA polymerase I is known to be chromatin-bound in this tissue[37,55] and markedly enhanced in activity following auxin treatment of seedlings,[60] as discussed below. Solubilization of the RNA polymerase I activity was achieved by stirring in buffered 0.5 M $(NH_4)_2SO_4$ for 3 hr at 0°C. Following clarification by centrifugation (50,000 rpm for 60 min), RNA polymerase activity was precipitated from the supernatant with $(NH_4)_2SO_4$ (added to 0.35 g/ml). The precipitate was collected by centrifugation (50,000 rpm for 60 min) and resuspended for chromatography. Purification was achieved by successive passages through Agarose A-1.5 m, DEAE cellulose, CM cellulose, phosphocellulose, and finally by sedimentation (41,00 rpm for 60 hr) through a 5 to 20% linear sucrose density gradient. Since the procedure began with a polymerase-enriched source (isolated chromatin), the final purification was only about 300-fold with respect to the chromatin. However, the enzyme was electrophoretically pure, and the overall purification with respect to the starting tissue was calculated as more than 20,000-fold.[59]

The novel approach used by Jendrisak and Burgess[50] for isolation of RNA polymerase II from wheat germ is especially noteworthy for its high yield (63% vs. 5% and 14%, respectively, for the two previous procedures[58,59]) and its use of only two chromatographic steps. The procedure avoids ultracentrifugation and dialysis of large sample volumes, thereby allowing completion of the purification of an RNA polymerase in 2 days. These advantages are achieved by an initial fractionation of the crude homogenate (obtained by homogenization of wheat germ in a Waring Blender®) with Polymin P, a polyethylenimine first used by Zillig et al.[61] in purification of E. coli RNA polymerase. Polymin P is a highly basic polymer that precipitates acidic molecules including nucleic acids, nucleoproteins (such as chromatin and ribosomes), and

TABLE 2

Summary of Several Representative Procedures for Purification of Individual RNA Polymerase Species from Plant Sources

Authors	Mullinix et al.[58]	Guilfoyle et al.[59]	Jendrisak and Burgess[50]
Source	Maize leaves	Soybean hypocotyl	Wheat germ
Enzyme	IIa	I	II
Tissue disruption	Leaves ground in Waring Blendor®; homogenate filtered, then clarified by centrifugation; polymerase precipitated with $(NH_4)_2SO_4$, then heated to inactivate polymerase I	Auxin-treated hypocotyls ground in Polytron®; chromatin recovered by centrifugation; RNA polymerase solubilized in high salt and precipitated with $(NH_4)_2SO_4$	Wheat germ ground in Waring Blendor;® homogenate centrifuged, and filtered. Polymerase precipitated with Polymin P®, extracted with high salt, and precipitated with $(NH_4)_2SO_4$
Purification procedure	Gel filtration on Sepharose 6B and 4B	Gel filtration on Agarose A-1.5m	DEAE cellulose chromatography
	DEAE cellulose chromatography	DEAE cellulose chromatography	Phosphocellulose chromatography
	Phase partition in polyethylene glycol/dextran	CM cellulose chromatography	
	DNA cellulose chromatography	Phosphocellulose chromatography	
		Sucrose density centrifugation	
Analytical assay	Electrophoresis on SDS-polyacrylamide gels	Electrophoresis on polyacrylamide and on SDS-polyacrylamide gels	Gel filtration on Bio-Gel® A-1.5m; electrophoresis on SDS-polyacrylamide gels
Yield	5%	14%	63%
Purification	3,500-fold	300-fold (chromatin) 20,000-fold (tissue)	4,000-fold
Specific activity[a]	90	100—150	250

[a]   Specific activity defined as nanomoles nucleotide incorporated per milligram protein in 15 min. Values adjusted from the 20-min incubation of Mullinix et al.[58] and the 30-min incubation of Guilfoyle[59] to the 15-min incubation used by Jendrisak and Burgess.[50]

acidic proteins of which RNA polymerase is an example. Because RNA polymerase II is only moderately acidic, it can be selectively solubilized from the Polymin P precipitate by extraction with a high salt ($0.2\ M\ (NH_4)_2SO_4$) buffer. RNA polymerase is purified about 30-fold by this initial step, with a quantitative yield. Subsequent purification steps are limited to $(NH_4)_2SO_4$ precipitation and passage through DEAE cellulose and phosphocellulose. By this procedure, it was possible to prepare 30 mg of RNA polymerase II from 1 kg of wheat germ in less than 2 days, with a 4000-fold purification and greater than 60% recovery of the initial activity.[50] The procedure has since been adapted successfully to isolation of RNA polymerase IIa from soybean by Guilfoyle and Key[57] and appears to be applicable to other systems as well.

B. Enzymatic Properties of Plant RNA Polymerases
The properties by which RNA polymerases are characterized have become almost

codified by their repeated, predictable use in various laboratories involved in isolation and characterization of RNA polymerases from many different sources. In general, partially purified enzymes are used (after elution from DEAE cellulose or Sephadex, most commonly), and most (if not all) of the following properties are examined: pH optima, temperature optima, divalent cation requirements, ionic strength optima, response to inhibitors, and template preferences. Some of these data have been tabulated by Jacob[4] for mammalian polymerases and by Duda[9] for plant enzymes. However, caution is necessary when considering these or similar tabulations of RNA polymerase properties since such properties are only rarely determined under the same conditions in different laboratories, and direct comparisons are at best hazardous. Ionic strength optima and the relative stimulatory effects of the divalent cations Mg^{++} and Mn^{++} are especially difficult to compare because of the interdependence of ionic strength and divalent cation effects and because of the dependence of both on the nature of the template (natural or synthetic), and the state of the template (native or denatured) and its concentration in the reaction mixture. These catalytic properties should not be considered as invariant characteristics of the enzymes. Meaningful comparisons of enzyme properties determined in different laboratories presuppose more uniform methods of determining and reporting such properties than presently prevail. In fact, Duda[9] has already made such suggestions with respect to determination and reporting of ionic strength and divalent cation requirements.

Despite these reservations, some generalizations appear valid with respect to the properties of plant RNA polymerases I and II (too little work has yet been done on RNA polymerase III to warrant discussion of its properties). In general, both enzymes appear to function optimally (or at least are routinely assayed) at a pH of approximately 8.0 to 8.1 (see Polya and Jagendorf[33] for actual data on pH dependency) and an incubation temperature in the range of 25 to 30°C (but see Strain et al.[31] for their claim of an optimum at 44°C for both enzymes). Both enzymes have an absolute divalent cation requirement that can be met with either Mg^{++} or Mn^{++} ion. Responses of the two enzymes are quite different, however, and a given enzyme usually displays different responses to the two cations — often a definite preference for one over the other (though with the qualification that such a preference depends in turn on the ionic strength of the medium). In general, RNA polymerase I has a relatively low ionic strength requirement and is quite responsive to both Mg^{++} and Mn^{++}, but often with a slight preference for Mn^{++} ion.[34-36,40,48,53] RNA polymerase II, on the other hand, requires a higher ionic strength for optimal activity and usually displays a definite preference for Mn^{++} ion.[34,35,39,44,45,47,48]

Template preference is also a complicated property to evaluate critically, especially because of the extent to which transcriptional rates are influenced by template size and presence or absence of single-stranded breaks ("nicks") at which initiation can occur. Indeed, reports of template preference ought to be regarded as incomplete without appropriate characterization of the template itself. Reports claiming a preference for DNA from a particular source (especially for homologous DNA rather than more commonly used calf thymus DNA)[31,33,41,53] should be scrutinized with special care to ensure that the alleged preference is not in fact for DNA chemically mistreated during isolation with a resultant increase in frequency of nicks or loose ends at which transcription can be artifactually initiated. Any generalization immediately admits exceptions, but RNA polymerase I tends to prefer native to denatured DNA as template, and RNA polymerase II is usually more active with denatured DNA as template. This has led to the interesting speculation[9] that these template preferences may be linked to the reported tendency of RNA polymerase I, but not II, to be tightly bound to the chromatin template.[37,40,55,60] Of possible further significance is the report of Strain et al.[31] that maize RNA polymerase II could be resolved into two components, one of which (IIa) preferred denatured DNA while the other (IIb) preferred native DNA. Although these workers did not investigate possible preferential association of one or the other of these enzymes with chromatin, Guilfoyle and Key[57] have recently reported that RNA polymerase IIb for soybean is a chromatin-bound enzyme, while IIa is a soluble (indeed, an extranuclear) form of the enzyme (see below). However, unlike RNA polymerases IIa and IIb from maize,[31] the soybean enzymes showed no differences in template preference,[57] and there is, in any case, no assurance that the distinction between IIa and IIb enzyme forms has the same meaning in soybean as in maize.

C. Electrophoretic Characterization of RNA Polymerase Subunits

Further characterization of RNA polymerases beyond the enzymatic properties discussed above generally involves an electrophoretic analysis of enzyme structure under denaturing conditions using sodium dodecyl sulfate-polyacrylamide gels. This obviously requires an enzyme preparation of high purity since contaminating polypeptides are likely to be mistaken for enzyme subunits.

All plant nuclear RNA polymerases described to date have molecular weights in the range 400,000 to 600,000 and are composed of multiple polypeptide subunits. The numbers, sizes, or molar ratios of subunits reported for plant RNA polymerases are not always in good agreement. Some discrepancies are very likely due to technical problems encountered in resolving, detecting, quantitating, and estimating the size of subunits ranging in molecular weight from approximately 15,000 to 240,000, but analyses are almost certainly further complicated by varying degrees of enzyme purity obtained in different laboratories and by difficulties in distinguishing true subunits from contaminating impurities or proteolytic degradation products, especially in the lower molecular weight range. Note that although the term "subunit" is used for convenience, it is not yet possible to distinguish true subunits from polypeptides which are merely tightly bound to the enzyme. For this purpose, reconstitution experiments will eventually be necessary.

Presented in Figure 1 are the subunit compositions of several relatively well-characterized plant RNA polymerases for which the work appears credible. In several cases, the procedure for preparation and purification of these enzymes has already been described and is summarized in Table 2. For comparative purposes, the subunit data for the corresponding RNA polymerases of mouse plasmacytoma cells[10,24] discussed earlier are included in Figure 1. Although frequently included in the original reports, molar ratios of the various polypeptide chains do not appear in Figure 1 primarily because of a lack of conviction that such values are accurate enough to permit useful comparisons and conclusions.

It seems a valid generalization that, like the corresponding enzymes from animal sources,[8] plant RNA polymerases consist of two subunits of high molecular weight (135 to 220 kilodaltons) and several (4 to 6 in most cases[27,47,57-59] but up to 11 for wheat germ[62]) subunits of lower molecular weight. Subunit compositions for RNA polymerase I from two sources are shown in Figure 1. In both cases (cauliflower[27] and soybean[59]), the enzyme has one subunit of approximately 135,000 daltons and another of approximately 185,000 daltons. The cauliflower enzyme (Ib) has an additional component corresponding to 168 daltons, but Guilfoyle[27] suggests that this may be due to contamination of the preparation with RNA polymerase Ia, since the molar ratios of the 168,000- and 185,000-dalton subunits sum to that of the 135,000-dalton polypeptide alone, and the purified RNA polymerase Ib could be resolved into two very closely migrating bands upon electrophoresis under nondenaturing conditions.

The subunit composition of RNA polymerase II has been reported for a greater variety of plant sources than is presently the case for polymerase I. As seen in Figure 1, agreement on the subunit pattern of RNA polymerase II is quite good between species. Like the form I enzyme, polymerase II contains two large subunits and several of lower molecular weight. The presence of a large polypeptide with a molecular weight of approximately 140,000 seems an invariant feature of RNA polymerase II molecules isolated from both plant[27,47,57,62] and animal[8,10,24] sources. A second subunit of still higher molecular weight is also invariably present, but varies in size from approximately 200,000 daltons for form IIa to approximately 170,000 daltons for form IIb. Presence of more than two large subunits in a given preparation probably indicates that the preparation may in fact be a composite of multiple subspecies of the enzyme, especially if the molar ratios of the polypeptides with molecular weights of 170,000 or more sum to that of the 140,000-dalton band. That this accords well with knowledge of the corresponding animal enzymes can be seen from data for the three subspecies of mouse plasmacytoma RNA polymerase II shown in Figure 1 and from Chambon's[8] more general discussion.

Although the data are still quite limited, reports thus far available for plant RNA polymerases confirm and extend the generalization[8,10] that eukaryotic RNA polymerases I and II are distinct in structure and in function and intracellular location with few if any subunits common to both classes (the only possible exceptions being several polypeptides of low molecular weight[10]). Within either class (I or II), however, subunit composition appears to be highly conservative between species. This structural conservatism within RNA polymerase classes is sometimes characterized by authors or reviewers as "remarkable" or "striking", yet it is interesting to reflect that such adjectives are seldom if ever voiced with respect to the equally conservative features of corresponding translational machinery of the cell (ribosomes, tRNAs, etc.). Given the essential and universal nature of both the transcriptional and the translational processes, it might be well to consider whether it would be "remarkable" or "striking" only if it were found that structural features of key components such as the RNA polymerases were not highly conserved.

Only a modest beginning has thus far been made toward understanding specific functions of individual subunits of RNA polymerase molecules. For the *E. coli* enzyme, several subunit properties are known.[1-3,14-16] The σ subunit is involved in strand selection and in specific initiation. The β' subunit may be responsible for DNA binding. The β subunit appears to contain the binding sites for the inhibitors rifampicin and streptolydigin. The α subunit may be responsible for actual formation of the phosphodiester bond based on its ability to synthesize poly(A) from ATP.[63] An additional subunit (ω) has been reported as part of the *E. coli* RNA polymerase molecule,[1,15] but its function is even more obscure. For eukaryotic RNA polymerases, neither essentiality nor specific function has been established definitively for any reported subunits. In the case of maize RNA polymerase IIa, Mullinix et al.[58] concluded that the 160,000-dalton subunit is necessary for activity because enzyme activity loss upon storage could be correlated with disappearance of that subunit and appearance of polypeptides with molecular weights of 70,000 and 90,000, presumably breakdown products. In general, however, further understanding of the importance and function of various eukaryotic RNA polymerase subunits is likely to occur only if and as it becomes possible to dissociate these enzymes into their component parts and then reconstitute active enzymes from selected subunits.

D. Regulatory Factors

The discovery of sigma factor as part of the *E. coli* RNA polymerase[15] and recognition of its cyclic role in specific initiation of transcription[16] triggered not only a prokaryotic "sigma boom"[2] (now largely subsided), but also a frantic scramble to find similar factors in eukaryotic cells. Although hailed by some as a new era in the field of transcription,[7] a more candid appraisal might be that this has been at best a mixed blessing in light of, on the one hand, the ease and frequency with which such stimulatory factors have been reported and, on the other hand, the absence of definitive structural or functional characterization in most such reports.

The term "factor" may apply to any protein molecule specifically stimulating or inhibiting RNA synthesis (usually assayed in vitro with isolated RNA polymerases), whether the mode of factor interaction is with the template (either the DNA directly or the chromatin complex as a whole) or with the polymerase molecule itself. Consideration of factors acting at the template level would involve discussion of chromatin proteins (especially the acidic or nonhistone chromosomal proteins) that would quickly move beyond the intended scope of this chapter. Therefore, attention will be focused primarily on transcriptional factors known or thought to interact directly with the RNA polymerase molecule (although the distinction may be more illusory than real since most such factors, especially those involved in initiation, will, when adequately characterized, almost certainly interact with both the polymerase molecule and its DNA or chromatin template). A further complicating feature of the discussion arises from the difficulty of adequately distinguishing between polymerase subunits (presumably obligatory, indispensable structural elements) and regulatory factors (dissociable polypeptides dispensable to the routine functioning of the enzyme, but capable of modulating its activity when associated with the polymerase). The following discussion assumes that such distinctions can be made and are meaningful, if only because literature on regulatory factors seems to assume so. The cautious reader may, however, choose to view the entire topic of regulatory factors with a good measure of healthy skepticism.

Several groups[64-71] have reported the partial purification from mammalian tissues of factors which stimulate RNA polymerase activities from the same tissue. For example, Lee and Dahmus[66] reported that several protein factors appearing in the eluate upon fractionation of ascites cell polymerases on DEAE cellulose can be resolved by chromatography on CM cellulose and stimulate RNA polymerase II activity only when native DNA is used as template. Similar properties have been reported for factors isolated from HeLa and KB cells.[67] Factors that specifically stimulate activity of polymerase II have also been reported by others.[68,69] Other groups claim the existence of factors specific for polymerase I.[70,71] In plant studies, claims have also been made for factors which stimulate RNA polymerase activities. Several of these reports will be considered later in the context of hormonal effects on polymerase activities. For example, Hardin et al.[72] have described a factor that stimulates chromatin-bound RNA polymerase from control tissue, but not from auxin-treated tissue, although nothing further has been reported to date. In a more recent report from the same laboratory,[73] attention was focused instead on a factor alleged to stimulate RNA polymerase activity upon release by auxin from the plasma membrane. Transcriptional factors have also been reported by Matthysse and Phillips,[74] Matthysse and Abrams,[75] Venis,[76] and Mondal et al.[46,52,53] In the latter work, the factors were purified from the nonhistone component of the chromatin of coconut endosperm nuclei and have been implicated in initiation (factor B) and termination (factor C) of RNA synthesis. More recently, Teissere et al.[77,78] have reported the isolation from lentil root nuclei of four transcription factors that they claim are involved in the initiation rather than the elongation step of transcription.

It is still too early to assess the likely physiological significance of these and other studies of transcriptional factors. Some difficulties lie in the apparent tendency of workers in this area to publish their data in advance of the definitive purification and documentation necessary to support their claims adequately, but a good deal of confusion derives from present ignorance of the actual structure of eukaryotic RNA polymerases, such that it is not yet possible to determine what is factor and what is subunit, as already noted. It is likely that a process as complicated as eukaryotic transcription will be shown to require factors aiding initiation, termination, strand selection, and possibly even recognition of specific cistrons. Mandel and Chambon[79,80] concluded that purified mammalian enzymes resemble *E. coli* core polymerase in that they may lack some specific factor(s) necessary for specific initiation on intact DNA. Several protein factors mentioned above are proposed to interact specifically with the polymerases to increase their rate of transcription of native DNA, but there are presently no data indicating that any of these factors play a specifically sigma-like role in regulating the locus specificity as for the prokaryotic enzyme. Clearly, much more information about the chemistry and physiology of the transcriptional process in eukaryotes is needed before reports on factors possibly involved in its regulation can be intelligently evaluated.

III. RNA POLYMERASES IN PLANT DEVELOPMENT

A. Developmental Changes in RNA Polymerase Activities

This chapter focuses attention not only on the plant RNA polymerases as agents of transcription, but also on their possible involvement in actual regulation of transcription as it underlies eukaryotic development. Accordingly, it becomes relevant to ask whether and to what extent data are available to document such polymerase-mediated regulation.

There is, of course, a great variety of evidence suggesting that both rates and products of transcription as well as actual levels of RNA polymerase activities undergo striking changes during developmental events as diverse as amphibian metamorphosis,[81] liver regeneration,[82,83] lymphocyte transformation,[84] embryonic development,[85-87] seed germination,[88-137] and hormonal responses in animals[138-144] and plants.[145-167] In most cases, however, little is yet known about actual underlying regulatory mechanisms.

Any such developmental modulation in RNA polymerase activities, kinds of RNA synthesized, or rates of synthesis of specific classes of RNA can, in principle, be explained in at least three alternative ways depending on whether transcription is thought to be limited by the amount of a particular RNA polymerase species on hand in the cell, by the relative activities of the polymerase molecules present, or by the availability of transcribable template. Thus, an increase in the rate of synthesis of a particular species of RNA might be due to an increase in (1) the number of RNA polymerase molecules of a specific type present in the cell, (2) the transcriptional activity of existing polymerase molecules, (owing in turn to stimulation of either the initiation frequency or the rate of chain elongation), or (3) template accessibility, whether mediated by changes in chromosomal structure or by availability of specific factors required for selective, efficient initiation.

Distinguishing among these several possibilities is difficult and consequently not undertaken often. Yet it is precisely this kind of information that will be required if one is to understand the means by which transcriptional regulation is effected during eukaryotic development. A significant step in this direction has been taken by Roeder and colleagues, who have investigated RNA polymerase levels in several animal systems in which gross alterations in gene activity have been documented in a developmental or quasidevelopmental context including tumor cell proliferation,[24] lymphocyte response to phytohemagglutinin,[84] *Xenopus* embryogenesis,[86] and adenovirus infection of cultured human cells.[10] In the two former cases[24,84] involving transition to rapid growth and proliferation, increases in RNA synthesis could be correlated in at least a general way with increases in actual enzyme levels within the cell, suggesting that specific RNA polymerases "may in some cases be limiting and that the level of a specific enzyme may in part regulate the rate of transcription of a specific class of genes.''[10] That the transcription of a particular class of genes by a specific RNA polymerase molecule can be affected quantitatively in this way by changes in enzyme level seems especially reasonable in light of structural differences in the three main forms of RNA polymerases. If each enzyme form is assembled from its own unique subunits, it is easy to see how the cellular concentration of a given form of RNA polymerase can be regulated independently of the other forms by different rates of accumulation of the appropriate subunits.

However, one can point to contrasting cases, such as early embryonic development in *Xenopus*[86] and lytic viral infection of human KB cells,[10] in which both qualitative and quantitative changes in gene expression and hence in RNA polymerase activities occur without significant changes in actual enzyme levels. Such findings, plus the limited number of different RNA polymerases available to effect transcription, almost inevitably imply involvement of factors capable of modifying enzyme activity or selectivity and/or template accessibility.

B. Changes in RNA Polymerase Activities During Plant Development

We now consider several developmental phenomena in plants, looking at what is known presently about the nature and control of the transcriptional process and about the changes in levels, activities, and specificities of plant RNA polymerases in three quite different developmental settings: seed germination, hormonal responses (especially to auxin), and photomorphogenesis. Although little is known about these phenomena in molecular detail, it will be instructive to keep the preceding several alternatives in mind when considering the following sections since each transcriptional modulation described in these systems must ultimately be explained in terms of the underlying molecular control mechanisms.

IV. RNA POLYMERASES IN SEED GERMINATION

Seed germination involves resumption of active growth and development by a previously quiescent or dormant angiosperm embryo.[88] Upon imbibition of water, the embryo undergoes a rapid, often dramatic transition from an extended period of low metabolic activity in a highly dessicated state to a phase of active metabolism with resumption of growth, differentiation, and morphogenesis. Among the more striking physiological features of seed germination are rapid onset of macromolecular biosynthesis and sudden, marked increases in activities of a variety of enzymes. The latter is hardly surprising in a system undergoing resumption of metabolism and dependent for its initial nourishment on food reserves, the mobilization of which requires enzymes (such as amylases,[89,90] proteases,[91,92] or glyoxylate-cycle enzymes[92]) absent from the ungerminated embryo.

A. Protein Synthesis During Germination

Most such enzymes appear to be synthesized *de novo* during the germination process,[90-92,94] which correlates well with initiation of protein synthesis known to occur very early in germination. In the wheat embryo, for example, initial water uptake by isolated, germinating embryos is essentially complete within 30 min, and the protein synthesizing capacity of the embryo also rises rapidly during the same time period, at least as measured by the amino acid incorporating ability of ribosomes in vitro.[95-97] The actual in vivo demonstration of protein synthesis during very early stages of germination is complicated by difficulties encountered in getting radioactively labeled amino acids into the system quickly enough, but an early onset of protein synthesis has been convincingly documented in embryos of wheat,[98] barley,[99] rye,[100] peas,[101,102] and bean.[103]

This early initiation of protein synthesis has been correlated in several systems with rapid conversion of preexisting monoribosomes into polyribosomes,[95,96,100,104,105] repeatedly claimed[95,96,105-108] to occur in the absence of transcription. If so, then it must be presumed[95,96,106,107,109] that the messenger RNA needed for polyribosome formation and initiation of translation preexists in the ungerminated embryo, as is known to be the case for the other major components of the protein synthesizing system.[95-97] The existence of stable, preformed messenger RNA in ungerminated embryos and its subsequent utilization during germination has in fact been suggested for a variety of plant species.[99,101,105,109-111] However, early evidence supporting the concept of preformed messenger RNA and its involvement in germination was often taken quite uncritically and was usually indirect. For example, the apparent insensitivity of early protein synthesis to actinomycin D and to other inhibitors of RNA synthesis was often unsupported by data indicating that the inhibitor was taken up and actually penetrated to the site of RNA synthesis. Thus, the conclusion of Barker et al.,[112] stating that much of the early evidence was not reliable and the presence of such preformed messenger RNA in ungerminated seed should be regarded as tentative, appeared refreshingly critical.

Since then, further reports have been published and the concept of preformed messenger RNA in ungerminated embryos has become better substantiated, though still not rigorously proven. By direct labeling of the RNA of developing *Phaseolus* embryos, for example, Walbot[113] established that RNA synthesized during embryogenesis is functional during germination before detectable RNA synthesis is resumed and that preexisting ribosomes are utilized for formation of polyribosomes which accompanies onset of protein synthesis. Also relevant is the recent demonstration by Spiegel and Marcus[114] that despite the messenger RNA synthesis detectable at very early stages in germinating wheat embryos,[115] initial polyribosome formation is independent of both transcription (i.e., insensitive to α-amanitin) and polyadenylation (insensitive to cordycepin, an adenosine analogue). Additional (and, on the surface, quite convincing) evidence in support of preformed messenger RNA in ungerminated seeds comes from the demonstration by Dure and colleagues[91,94,105] with cotton and more recently by Tester[116] with soybeans that the appearance during germination of specific enzymes (carboxypeptidase and isocitrate lyase) of obvious physiological relevance to the germinating embryo depends on *de novo* synthesis of the enzymes, but apparently not of corresponding messenger RNAs. Insensitivity of enzyme appearance to actinomycin D provides the basis for their claim that the messenger RNAs for these and perhaps other "germination enzymes" are synthesized at a characteristic point during embryogenesis and maintained thereafter in an untranslated state during the remainder of embryogenesis by abscisic acid supplied by surrounding maternal tissue.[91,94] Though internally consistent and convincing, these conclusions depend almost entirely on inhibitor data and have been questioned (at least for isocitrate lyase in cotton) by findings (also based on inhibitor effects) of Smith et al.[117] and of Radin and Trelease.[118]

B. RNA Synthesis During Germination

Much of the above discussion can be summarized by noting that protein synthesis is an early and obligatory feature of seed germination; that many enzymes specifically required for postgerminative metabolism are synthesized *de novo* by the germinating embryo; and that, in some species at least, the complete requisite machinery for protein synthesis, possibly including messenger RNA, appears to be present in the dry seed. In a discussion intended to focus on RNA polymerases, however, these are only preliminary observations. Similar inquiries ought to be made about RNA synthesis and the transcriptional apparatus. Relatively little is yet known about the developmental profiles of actual RNA polymerase concentrations or activities in the germinating embryo, but a considerable (though not very conclusive) literature exists concerning the importance, onset, and temporal pattern for reactivation of RNA synthesis during germination.

As already suggested, evidence concerning the indispensability of RNA synthesis during early germination is presently contradictory and inconclusive, based as it is on inhibitor studies. Reports[91,94-96,105-107,109,114,116,119] claiming lack of dependence of early germinative events (protein synthesis or enzyme appearance, most commonly) on RNA synthesis must be balanced by others[117,118,120-122] showing a sensitivity of germination to inhibitors such as actinomycin D or 6-methyl purine and therefore presumably a requirement for RNA synthesis during early germination. (See the report of Black and Richardson[119] for the claim that actinomycin D, while markedly reducing RNA synthesis, stimulated rather than inhibited germination!) Furthermore, it should be noted that even though metabolic inhibitors can be used to demonstrate that protein synthesis, enzyme appearance, or other physiological or even morphological events during early germination can occur in the absence of transcription, such experiments probably do not bear directly on the question of the actual occurrence or role of RNA synthesis during germination, since the ability of a system to withstand temporary suspension or delay of RNA synthesis is not likely to be a helpful criterion for determining the actual importance or pattern of transcription during normal germination.

More directly relevant to an understanding of the role and significance of RNA polymerase activities during seed germination would be an inquiry into the temporal sequence with which the synthesis of various RNA species (and hence presumably the activity of the specific RNA polymerases) is resumed during germination. This, too, is a topic over which there is controversy and no good general agreement. Some workers claim that RNA synthesis is not detectable during early germination.[106,107,123] but is triggered only after 12 hr[113,123] or even longer periods[106] of imbibition. Taken at face value, this strengthens the data indicating an insensitivity to inhibitors of RNA synthesis and supports the importance of preformed translational machinery, including messenger RNAs. Others, however, have reported that the synthesis of RNA is resumed immediately[98,124] or shortly after[108,120,124-129] exposure of the seed to favorable germination conditions. Strong supporting evidence for this viewpoint is provided by recent work of Spiegel et al.[115] demonstrating incorporation of label into both ribosomal and messenger RNA within the first hour following transfer from 0 to 25°C. Further confirmatory evidence for early activation of RNA synthesis is provided for rye embryos by Sen et al.[130] who found all major species labeled within 2 hr. These data clearly establish an earlier onset of transcription than that generally recognized previously and raise the possibility (though obviously do not prove) that newly synthesized RNA may be available as template for even the proteins synthesized in the early hours of germination.

Conflicting claims have often been made about the order in which synthesis of the major kinds of RNA begins even within a given species. In wheat embryo studies, Chen et al.[106,125] claimed that no detectable RNA synthesis occurs until after 2 or 3 hr and that only ribosomal RNA is synthesized during the first 12 hr. No new messenger RNA appears until after 24 hr. Buchowicz and colleagues,[98,124,131] on the other hand, report that RNA synthesis begins immediately upon germination of preimbibed wheat embryos, with a small "messenger-like" fraction highly labeled after 3 hr and tRNA and rRNA not labeled until much later. They conclude, therefore, that synthesis of messenger-like RNA actually precedes protein synthesis in early germination and that the order of initiation of macromolecular synthesis initiation in the germinating wheat embryo is mRNA, protein, DNA, rRNA, and tRNA.[124] Presuming that what Dobrzanska et al.[124] regard as "mRNA" can be equated with heterodisperse nuclear RNA, their findings are roughly in accord with those of Van de Walle et al.[132] for maize, since "stage one" of reactivation of RNA synthesis, as identified by Dobrzanska et al.,[124] is characterized by synthesis of heterodisperse nuclear RNA, while nucleolar rRNA synthesis does not commence until "stage two." These findings[124,132] appear to contradict the data of Spiegel et al.[115] showing synthesis of both messenger and ribosomal RNA at the earliest times assayed. A possible explanation is offered by observations of these latter workers revealing that mRNA synthesis remains essentially constant throughout the first 18 hr of germination while the rate of ribosomal RNA synthesis increases after the onset of cell expansion, eventually accounting for more than 90% of the high molecular weight RNA synthesized after 16 hr. As the authors suggest,[115] this could prompt the erroneous conclusion that rRNA synthesis is negligible at early germination times — an impression that might be further enhanced by the requirement for processing of ribosomal RNA before it appears in ribosomes. As already noted, however, Sen et al.[130] found all major classes of RNA labeled at early stages — within 2 hr in rye and within 1 hr. in wheat. In fact, they detected label in 4S and 5S RNA and in what might be heterodisperse nuclear RNA within 20 min and in rRNA within 40 min.

C. Changes in RNA Polymerase Activities During Germination

The rapid onset of synthesis of all major species of RNA in the germinating seed strongly implies that all components necessary for RNA synthesis (most notably the several classes of RNA polymerases) are present in the ungerminated embryo. This is further supported by incorporation of low levels of ATP into acid-precipitable form by chromatin prepared from dry pea seeds,[104] by autoradiographic demonstration of RNA synthesis in sections of both dry and imbibed onion seeds,[133] and by direct measurement of RNA polymerase activity in a crude extract from wheat seeds.[134] Since then, both RNA polymerases I and II have been isolated from wheat germ[48,49] (isolated wheat embryos) shown to be an especially rich source of eukaryotic RNA polymerase II (25 to 30 mg of enzyme from 1 kg of starting material,[50] see Table 2). Ungerminated embryonic soybean axes have also been reported[57] rich sources of RNA polymerase II, which apparently exists in the ungerminated embryo as a soluble enzyme unassociated with nuclei or chromatin. Thus, in addition to its well-documented potential for protein synthesis, the ungerminated angiosperm embryo appears to contain the RNA polymerases (especially the Class II enzyme) required for RNA synthesis. Whether the complete transcriptional machinery is on hand in functional form is not yet known, however, and would be somewhat difficult to assess at present because of our inadequate understanding of eukaryotic transcription and the factors involved in its initiation and regulation. As yet, little has been published concerning modulations of polymerase levels or activities during the germinative process. Mazus and Buchowicz[134] originally reported a twofold increase in polymerase activity of whole seeds during the first 48 hr of germination. Since then, however, Mazus[135] has claimed that isolated wheat embryos undergo a gradual decrease in RNA polymerase activity beginning about 6 hr after initiation of germination. She further reported that ungerminated embryos contained both RNA polymerases I and II, but that only RNA polymerase II could be detected after 48 hr of germination — a rather puzzling observation.

Of considerable interest is the recent report of Guilfoyle and Key[57] that germination in soybean is accomplished by the progressive conversion of RNA polymerase II from its soluble form (IIa) to a chromatin-bound form (IIb) that actively synthesizes RNA in vitro in isolated nuclei or chromatin. Form IIa is the dominant form (greater than 95%) in the ungerminated embryo. Form IIb is essentially absent until the onset of germination. Since the two enzymes appear to differ (see Table 1) only in size of the largest subunit (200,000 daltons for form IIa and 170,000 daltons for form IIb), the conversion from IIa to IIb may involve a specific preteolytic cleavage. The authors speculate that soluble, extranuclear form IIa enzyme may represent a storage or precursor form of RNA polymerase II that is activated upon conversion to the chromatin-bound form IIb enzyme. They suggest that this dichotomy of both function and subcellular localization may be a general characteristic of the two forms of RNA polymerase II, at least in higher plants.[57] It will be of considerable interest to ascertain whether this conversion of soluble RNA polymerase II into a chromatin-bound, transcriptionally active form in germinating soybean[57] can be correlated with the rapid onset of heterodisperse nuclear RNA and/or messenger RNA synthesis reported for other species.[98,115,124,130-132]

D. Changes in RNA Polymerase Activities During Postgerminative Seedling Growth

Only two papers have been published to date dealing specifically with RNA polymerase activities in seedlings at later stages of postgerminative development. Rizzo and Cherry[136] determined the relative levels of RNA polymerases I and II (distinguished both by α-amanitin sensitivity in crude extracts and by resolution on DEAE Sephadex) in several regions of the soybean hypocotyl corresponding to different stages of development. More total activity and a higher form I/form II ratio were found in the meristematic region (upper) than in the fully elongated (lower) region. The hypocotyl hook was especially high in polymerase I and contained a polymerase activity which corresponded to RNA polymerase III in elution position but not in amanitin sensitivity. In a more detailed study with the same tissue, Lin et al.[137] also found a higher form I/form II activity ratio (as measured in isolated nuclei) in meristematic tissue than that in mature tissue. A marked decrease in RNA polymerase I activity of isolated nuclei was found to accompany the normal growth transition from the meristematic to the fully elongated state, while the nuclear activity of RNA polymerase II decreased only negligibly. Auxin treatment resulted in a striking increase in nuclear activity of RNA polymerase I (but not of polymerase II) in fully elongated hypocotyl tissue commensurate with the large increase in rRNA synthesis and accumulation known to occur in that region in response to auxin. Therefore, the authors[137] concluded that RNA polymerase I activity changed much more than did polymerase II activity during both normal and auxin-induced growth transitions, with a generally strong correlation between the pattern of RNA (especially rRNA) accumulation in vivo and levels of RNA polymerase I activity expressed in isolated nuclei. Lesser variations were seen when polymerase activities from different stages were first solubilized, fractionated on DEAE cellulose, and assayed on a free DNA template, suggesting that the RNA polymerase I of hypocotyl nuclei is subject to greater modulations in activity than in actual enzyme concentration during these growth transitions.

E. Summary

Data bearing on the significance and temporal ordering of RNA synthesis and regulation of polymerase activities during germination are at present still fragmentary and inconclusive, but some patterns are beginning to emerge. With some speculative license, the following working model is suggested, but is is probably more valuable in formulating testable hypotheses than in providing a clear concensus of literature opinion:

1. The typical quiescent angiosperm embryo possesses the complete metabolic machinery for both protein synthesis and RNA synthesis (and is an especially rich source of both ribosomes and RNA polymerases, especially polymerase II), the

components of which are held in abeyance either by the dehydrated state of the seed alone or by other, more specific control factors.

2. Protein synthesis is initiated almost immediately upon imbibition and probably depends on preexisting messenger RNA (as well as other components) for initial polyribosome formation, although the extent to which mRNAs unique to the germination process (and therefore of the greatest developmental interest) preexist in the ungerminated seed remains an open question.

3. RNA synthesis is also initiated early in germination — certainly within the first few hours following imbibition. However, the amount of RNA synthesized in advance of the onset of protein synthesis is probably quite limited.

4. Informational RNA (heterodisperse nuclear RNA and messenger RNA) is synthesized at the earliest assayable times almost certainly as the result of the activation of RNA polymerase II, quite possibly upon conversion of the inactive, soluble form of the enzyme (IIa) to the transcriptionally active, chromatin-bound form (IIb). However, the extent to which this newly synthesized RNA is required for early protein synthesis is unclear.

5. Synthesis of rRNA (polymerase I activity) and tRNA (polymerase III activity) also begins at an early stage, although these species are presumably present in adequate supply in the dry embryo to meet needs of early germination, since (for ribosomal RNA anyway) a marked increase in its synthesis does not occur until the onset of cell expansion.

6. Developmental changes in RNA polymerase activities (especially of the nucleolar enzyme) accompanying postgerminative growth and development may depend more upon modulation in activity than on actual enzyme levels, suggesting involvement of subunits or regulatory factors capable of modifying either enzyme activity or template availability.

V. HORMONAL EFFECTS ON PLANT RNA POLYMERASES

A. Plant Hormones and Nucleic Acid Synthesis

Beginning with pioneering studies in Skoog's laboratory on the auxin-stimulated increased in RNA content of cultured tobacco pith tissue,[145,146] a large literature has accumulated relating effects of the plant hormones to nucleic acid metabolism in general and to RNA synthesis in particular. Much of the early (pre-1969) work was summarized in the very competent review by Key.[147] Hormonal responses such as cell elongation in response to auxin, gibberellic acid-induced synthesis of hydrolases, and ethylene-stimulated abscission generally involve enhanced synthesis and accumulation of RNA, although in no case has it yet been possible to establish definitively a primary effect of any plant hormone on nucleic acid metabolism in general or on the transcriptional process in specific.

Most early work on hormonal stimulation of RNA synthesis in plant systems[72,74,148-157] involved chromatin-bound RNA polymerases simply because procedures for isolation and purification of the enzymes were unavailable or, in some cases, because it was not yet realized that the enzymes of interest already existed in soluble form in the cell.[50,57,158] This made it difficult, if not impossible, to distinguish between changes in template activity (availability of greater numbers or kinds of genetic loci) and direct modulation of RNA polymerase levels or activities. Attempts were made to resolve this problem indirectly by using saturating levels of exogenous bacterial RNA polymerase as means of assaying for template availability.[151-156] In some cases, it was asserted that hormonal stimulation of RNA synthesis involved enhanced activities or levels of chromatin-bound RNA polymerase[152-155] rather than (or at least before[154]) template changes, while other reports alleged an increased template availability as the main,[156] or at least the initial,[151] effect of hormonal treatment. At best, conclusions based on such an approach ought to be regarded only as suggestive in light of present knowledge of differences in structure and specificity of the prokaryotic and the eukaryotic RNA polymerases and consequent uncertainty concerning what is being measured qualitatively or quantitatively when a bacterial polymerase is used to transcribe chromatin.

The following discussion focuses primarily on enhancement of RNA synthesis by auxin, since this is the only plant hormone for which effects on RNA synthesis and accumulation[145,147] have been extended competently to the isolated enzyme level.[60] Effects of other hormones on RNA synthesis and RNA polymerase activities will be summarized briefly.

B. Effects of Auxin on RNA Polymerase

Auxin treatment has long been known to stimulate RNA synthesis at the level of the whole plant, excised tissue, isolated nuclei, and chromatin.[145-147,152,154,157] Using chromatin isolated from soybean hypocotyls treated with the synthetic auxin, 2,4-dichlorophenoxyacetic acid (2,4-D) O'Brien et al.[152] found the activities of endogenous RNA polymerase to be enhanced by the in vivo treatment with auxin, but not by 2,4-D added to the assay incubation. Since saturation with added *E. coli* RNA polymerase yielded similar response curves with auxin-treated and control chromatin, they concluded that the enhanced RNA synthesis due to auxin resulted from an increase in the level of chromatin-bound RNA polymerase rather than from a greater template availability (see the reservation of Key[147] concerning this interpretation). Johnson and Purves[154] concluded that the effect of auxin (indoleacetic acid) treatment was to stimulate endogenous RNA polymerase (within 4 hr), but that this was followed (after 12 hr) by an increase in template availability. The report of Arens and Stout[158] on the auxin enhancement of RNA polymerase activity in maize seedlings generally agrees with both of these findings. Since the maize polymerase with which they worked is a soluble enzyme, their demonstration of enhanced polymerase activity rather than template availability was more direct and convincing.

Accompanying and immediately following publication of these and similar reports on auxin stimulation of polymerase activity, there appeared in the literature a spate of reports describing regulatory factors alleged to mediate auxin effects on RNA synthesis. Matthysee and Phillips,[74] for example, isolated a factor from tobacco or soybean nuclei which they claimed could interact with auxin to enhance template activities of chromatin as assessed by incubation with *E. coli* RNA polymerase. Mondal et al.[159,160] found a similar factor in coconut endosperm nuclei and suggested that modification of template capacity was the result of binding to the DNA of an auxin-factor complex which could then associate with an initiation factor and an RNA polymerase, giving rise to synthesis of additional species of RNA. A recent preliminary report from the same laboratory claims involvement of an auxin receptor protein in this enhancement of transcription.[161] O'Brien et al.[152] reported isolation of a factor from soybean cotyledons which was asserted to stimulate chromatin-bound RNA polymerase from control soybean hypocotyl tissue, but not from auxin-treated tissue. The same group later reported[73] that auxin (2,4-D) treatment of plasma membrane could cause release of a transcriptional factor capable of stimulating activity of RNA polymerase solubilized from soybean chromatin. Since the enzyme activity was amanitin sensitive, it was assumed to be RNA polymerase II, although this is somewhat difficult to reconcile with findings of others[37,60] stating that little RNA polymerase II is associated with chromatin isolated from either auxin-treated or control soybean hypocotyls.

Venis[76] reported isolation, by affinity chromatography of crude extracts from pea or corn shoots on agarose-linked 2,4-D, of a protein factor capable of stimulating RNA synthesis by *E. coli* RNA polymerase on purified DNA. The same technique (but using 2,4-D Sepharose) has recently been used by Rizzo et al.[162] to isolate a similar but more active soluble transcriptional factor from soybean hypocotyl. In addition to stimulating *E. coli* RNA polymerase two- to sevenfold, their factor showed a preferential stimulation (25 to 80%) of solubilized soybean RNA polymerase I, but only if 2,4-D was included in the assay mixture. A further contribution to the auxin factor literature of the early 1970s was the report of Teissere et al.[163] claiming enhanced chromatin activity as an early (1.5 hr) response of lentil root to auxin, followed (by 14 hr) by increases in nucleolar (Ia and Ib) but not in nucleoplasmic (IIa and III) RNA polymerase activities. The enhanced activity of RNA polymerase I in this system has since been attributed[77,78] to a transcription factor (γ) thought to be specific for the nucleolar polymerase, active at the level of chain initiation, and auxin dependent in its cellular level.

In general, these claims for protein factors which apparently mediate transcriptional effects of auxin do not integrate well with each other and present too little data to substantiate the broad claims often made. They usually have not been followed up by adequate characterization of either the factor itself or its mode of action. It seems incontrovertible that such stimulatory factors do exist and that more adequate characterization may, in at least some cases, eventually be brought to bear. However, presently, it is still difficult to avoid the conclusion that most reports to date have done little to enhance either our understanding of the molecular basis of auxin-stimulated RNA synthesis or the general credibility of work in this area.

Of more immediate substance is the competent work of Guilfoyle and colleagues[59,60,164,165] concerning specific effects of auxin on activity of soybean RNA polymerase I. Having confirmed the earlier findings of O'Brien et al.[152] stating that auxin treatment of soybean seedlings results in enhanced activity of chromatin-bound RNA polymerase activity, Guilfoyle and Hanson[164] established that the enzyme activity was due almost exclusively to RNA polymerase I. This correlates well with earlier observations that the abnormal proliferation of mature hypocotyl tissue in response to auxin[137] is preceded by a large increase in ribosomal RNA and a copious synthesis of ribosomes.[147] By fractionating RNA polymerases I and II from both chromatin and nuclei of auxin-treated and control hypocotyl tissue, Guilfoyle et al.[60] were able to demonstrate directly that RNA polymerase I activity increases several-fold in response to auxin treatment; RNA polymerase II activity is not significantly affected. This enhancement of RNA polymerase I activity was maintained after solubilization of the enzyme (from either chromatin or nuclei) and assay on exogenous DNA, strongly suggesting that it is the enzyme rather than the template that responds to auxin. Based on the prior observation[165] stating that actual numbers of amanitin-insensitive RNA polymerase molecules involved in active transcription of the chromatin template did not appear to change appreciably in response to auxin treatment, it was concluded that the augmented rate of RNA synthesis by chromatin from auxin-treated tissue is likely due to a greater rate of transcription by each molecule rather than actual synthesis or mobilization of more polymerase molecules.[60] We have already noted earlier the suggestion of Lin et al.[137] that the RNA polymerase I of hypocotyl nuclei is subject to greater modulations in activity than in actual enzyme concentration during growth transitions such as that induced by auxin treatment. Teissere et al.[77,78,163] have also shown in their work with lentil root that auxin treatment results in selective enhancement of nucleolar RNA polymerase, but, in contrast with the conclusions of Guilfoyle and co-workers,[60,165] Teissere et al. suggest that the main effect of auxin in RNA polymerase I is at the chain initiation level (mediated by the transcription factor γ) rather than chain elongation. Thus, there appears to be growing support for the view that auxin selectively enhances activity of RNA polymerase I at the level of the enzyme itself[60,162,163] but disagreement on the mechanism which is, in any case, not yet understood.

C. Effects of Other Plant Hormones on RNA Polymerases

Effects on RNA synthesis have also been described for each of the other major classes of plant hormones, although in lesser detail to date than for auxin. Gibberellic acid has been reported to enhance RNA synthesis in several systems known to respond physiologically to exogenously supplied gibberellic acid. In dwarf peas, the hormone stimulates internode elongation and has been reported to enhance RNA synthesis by nuclei isolated in its presence.[148] The RNA product was shown by nearest-neighbor analysis to differ qualitatively from that synthesized by untreated nuclei. Using exogenous *E. coli* RNA polymerase, McComb et al.[153] attributed this stimulation to an increased level of chromatin-associated RNA polymerase. A similar finding of enhanced RNA polymerase activity has been reported for chromatin isolated from gibberellic acid-treated soybean hypocotyl tissue.[155] This is in contrast to what has been observed in hazel seed, a system in which gibberellic acid can replace the usual chilling requirement for breaking dormancy. Jarvis et al.[149,151] reported that the marked rise in RNA synthesis by the embryonic axis in response to gibberellic acid involves an increase in template availability, followed eventually by increased RNA polymerase activity. Both effects were also observed with cucumber chromatin,[154] but in this case, stimulation of endogenous RNA polymerase was reported to precede increases in template availability. Since gibberellic acid has no effect on chromatin-bound RNA polymerase in vitro,[148,149,155] several authors[148,149] postulated the existence of a factor which mediates the effect of gibberellic acid on RNA synthesis. However, no direct evidence for such a factor has yet been published. Especially surprising with respect to gibberellic acid is the paucity in the literature of information concerning specific effects of the hormone on the RNA polymerase activities of barley aleurone cells, since this is in many other respects the best-characterized gibberellic acid response system.

Abscisic acid effects on RNA polymerase activities have only rarely been reported. Pearson and Wareing[150] found an inhibition of chromatin activity by abscisic acid in radish, provided that the hormone was included in the grinding medium during chromatin preparation. Bex[166] reported that total RNA synthesis in maize coleoptile was inhibited within 3 hr of treatment with abscisic acid, but that a decrease in the specific activity of soluble RNA polymerase activity was not seen until after 6 hr, suggesting that the hormone's effect was not on the enzyme directly. A similar conclusion was reached by Mondal and Biswas,[167] who postulated further that inhibition of RNA synthesis results from a direct interaction of abscisic acid with the DNA.

In general, the specific effects of gibberellic acid, abscisic acid, and cytokinins (for which reports of effects on RNA polymerase activities are even more scarce)[75,154] must be regarded as open questions certainly deserving more competent and detailed experimental consideration they have thus far been accorded.

VI. EFFECTS OF LIGHT ON CHLOROPLAST RNA POLYMERASE

A. Chloroplast RNA Polymerase

Thus far, we have only been concerned with the multiple forms of RNA polymerase found in the nucleus or (in some cases) in the cytosol of eukaryotic cells. RNA synthesis is, however, also known to occur in the mitochondrion and the chloroplast. Both contain DNA and are genetically semiautonomous organelles. A discussion of eukaryotic RNA polymerases is therefore incomplete without reference to organellar enzymes.[5] Since very little is yet known about the mitochondrial polymerase of plants, attention here will focus on the chloroplast enzyme especially relevant in a developmental context because of the RNA synthesis known to accompany light-induced plastid development.

The existence of a DNA-dependent, RNA-synthesizing system in the chloroplasts of higher plants was first suggested in 1964 by the incorporation of labeled precursors (^{14}C-nucleotides) into RNA by isolated chloroplasts.[168,169] The RNA polymerase activity was shown to be tightly bound with DNA to the thylakoid membranes.[170-172] Early attempts to solubilize the enzyme from the membrane were unsuccessful; RNA polymerase activity (and all of the chloroplast DNA) remained associated with a chloroplast membrane fraction despite hypotonic treatment, homogenization, freezing and thawing, salt extraction, or detergent treatment.[30,171]

Solubilization of chloroplast RNA polymerase was first achieved in 1971 by Bottomley et al.[30] for maize and by Polya and Jagendorf[32] for wheat. Critical to the extraction procedure of the former group were a low magnesium concentration (decreased by addition of EDTA) and a high temperature. The latter workers depended upon high salt (1 M KCl) extraction of subchloroplast membrane fragments prepared from chloroplasts disrupted with 1% Triton® X-100. The RNA polymerase of maize chloroplasts was found to differ from the nuclear[31] (forms I and II) or soluble[174] (form II?) enzymes not only in subcellular localization but also in its unusually high affinity for phosphate groups, as evidenced by the high salt concentration necessary for elution from phosphocellulose. Bottomley et al.[30] have speculated that this may account for the strong binding of the enzyme to DNA and for some of the difficulties in solubilizing the enzyme.

Possible similarities between chloroplast and bacterial RNA polymerases have been examined using rifampin, a potent inhibitor of bacterial RNA polymerase,[3] but not of the eukaryotic nuclear polymerases.[4,9] Surzycki[172] reported that the RNA polymerase activity of isolated *Chlamydomonas* chloroplasts was completely inhibited by rifampin. However, Bottomley et al.[173] found both chloroplast and nuclear RNA synthesis insensitive to rifamycin in peas, maize, and radish. This has since been confirmed for the solubilized chloroplast enzyme as well,[30,32] suggesting that the chloroplast RNA polymerase of higher plants differ from the prokaryotic enzyme (and perhaps also from the algal enzyme) in at least one feature despite general similarities otherwise noted between the genetic systems of bacteria and the DNA-containing organelles of eukaryotes.

Availability of chloroplast RNA polymerase in soluble form made possible the elucidation of its subunit structure by SDS-polyacrylamide gel electrophoresis. Smith and Bogorad[175] reported that the RNA polymerase of maize chloroplasts contains at least two polypeptides with molecular weights of approximately 180,000 and 140,000. Polypeptides of about 100,000, 95,000, 85,000, and 40,000 daltons were also associated with enzymatic activity in relatively constant ratios, but each of these could be removed by at least one specific purification step. No bands of less than 40,000 daltons were detected. No evidence for multiple forms of chloroplast RNA polymerase was seen in their work or in the initial solubilization and characterization studies.[30,32,33] However, Joussaume[176] reported the presence, in pea chloroplasts, of two forms of RNA polymerase, differing in localization (one membrane bound, the other in the stroma), pH optima, template response, and G + C content of the reaction product. She suggested that failure of others to detect the stromal enzyme may be related to the procedures used to isolate and disrupt the chloroplasts, a possibility which Bottomley et al.[30] had already acknowledged could not be excluded.

B. Transcriptional Effects of Light

Illumination of dark-grown seedlings initiates a complex sequence of changes whereby the etioplasts develop into mature, functional chloroplasts.[177-179] A key feature of this response to light is the rapid and specific increase in RNA synthesis and in activity of RNA polymerase.[178] Harel and Bogorad[180] reported that most of the radioactive label administered to etiolated maize seedlings in the dark was incorporated into etioplast ribosomal RNA. This incorporation was strongly and preferentially stimulated by light during the first 2 hr of illumination, and the effect persisted after return of the leaves to darkness. Preliminary evidence suggested a light-dependent synthesis of messenger-like RNA (or at least not rRNA nor tRNA), but only after 2.5 hr or more of illumination.

RNA polymerase activity increases substantially upon illumination of etiolated plants; within 16 hr, the enzyme activity in isolated maize plastids had increased three- to fourfold according to Apel and Bogorad.[181] However, this stimulation was not paralleled by a comparable increase in the amount of enzyme protein, nor were qualitative differences observed upon electrophoresis of the isolated plastid enzymes from 16-hr illuminated vs. control plants. This suggests[181] that the plastid RNA polymerase may be subject to greater modulations in activity than in actual enzyme concentration during greening, a situation reminiscent of similar conclusions with respect to nuclear RNA polymerase I during growth transitions in soybean hypocotyl.[137] Whether the light-induced increase in plastid RNA polymerase activity can be correlated with the enhanced synthesis of specific kinds of RNA remains to be determined.

VII. CONCLUSION

Our understanding of the multiple RNA polymerases of plant cells and especially of their involvement in regulation of transcription during developmental phenomena is still very much in its infancy. Two (sometimes three) forms of nuclear RNA polymerases have been isolated from a variety of plant sources and characterized to varying degrees, including electrophoretic analysis of sununit composition. Polymerases I, II, and III appear to be responsible for synthesis of ribosomal, informational, and transfer RNAs, respectively, in both plant and animal nuclei. Subspecies of these forms are known, differing in subunit composition, but not in enzymatic properties. Less work has thus far been done on the organellar RNA polymerases of plants, especially for the mitochondrial enzyme.

Developmental phenomena such as seed germination, hormonal responses, and light-induced chloroplast development are characterized by substantial, often dramatic changes in the patterns of RNA synthesis and accumulation. In each case, it eventually should be possible to relate these changes to specific modifications in (1) the number of RNA polymerase molecules of a specific type present at a particular location in the cell, (2) the transcriptional activity of existing polymerase molecules, and/or (3) the extent of template availability. Regulatory factors are almost certain to be involved in the interaction of a polymerase molecule with a specific template locus, but such have not yet been well characterized.

A great deal remains to be explored and explained concerning the RNA polymerases and their role in regulation of transcription, but the heightened interest and increased research activity in this area augur well for the future.

REFERENCES

1. Burgess, R. R., RNA polymerase, *Annu. Rev. Biochem.,* 40, 711, 1971.
2. Bautz, E. K. F., Regulation of RNA synthesis, *Prog. Nucleic Acid Res. Mol. Biol.,* 12, 129, 1972.
3. Chamberlin, M. J., Bacterial DNA-dependent RNA polymerase, in *The Enzymes,* 3rd ed., Boyer, P. D., Ed., Academic Press, New York, 1974, 333.
4. Jacob, S. T., Mammalian RNA polymerases, *Prog. Nucleic Acid Res.Mol. Biol.,* 13, 93, 1973.
5. Chambon, P., Eucaryotic RNA polymerases, in *The Enzymes,* 3rd ed., Boyer, P. D., Ed., Academic Press, New York, 1974, 261.
6. Chambon, P., Gissinger, F., Kedinger, C., Mandel, J. L., and Meilhac, M., in *The Cell Nucleus,* Busch, H., Ed., Academic Press, New York, 1974, 270.
7. Biswas, B. B., Ganguly, A., and Das, A., Eukaryotic RNA polymerases and factors that control them, *Prog. Nucleic Acid Res. Mol. Biol.,* 15, 145, 1975.
8. Chambon, P., Eukaryotic nuclear RNA polymerases, *Annu. Rev. Biochem.,* 44, 613, 1975.
9. Duda, C. T., Plant RNA polymerases, *Annu. Rev. Plant Physiol.,* 27, 119, 1976.
10. Roeder, R. G., Schwartz, L. B., and Sklar, V. E. F., Function, structure, and regulation of eukaryotic nuclear RNA polymerases, in *The Molecular Biology of Hormone Action,* Papaconstantinou, J., Ed., Academic Press, New York, 1976, 29.
11. Weiss, S. B., Enzymatic incorporation of ribonucleoside triphosphates into the interpolynucleotide linkages of ribonucleic acid, *Proc. Natl. Acad. Sci. U.S.A.,* 46, 1020, 1960.
12. Hurwitz, J., Bresler, A., and Diringer, R., The enzymatic incorporation of ribonucleotides into polyribonucleotides and the effect of DNA, *Biochem. Biophys. Res. Commun.,* 3, 15, 1960.
13. Stevens, A., Incorporation of the adenine ribonucleotide into RNA by cell fractions from *E. coli* B, *Biochem. Biophys. Res. Commun.,* 3, 92, 1960.
14. Burgess, R. R., Separation and characterization of the subunits of ribonucleic acid polymerase, *J. Biol. Chem.,* 244, 6168, 1969.
15. Burgess, R. R., Travers, A. A., Dunn, J. J., and Bautz, E. K. F., Factor stimulating transcription by RNA polymerase, *Nature, (London),* 221, 43, 1969.
16. Travers, A. A. and Burgess, R. R., Cyclic re-use of the RNA polymerase sigma factor, *Nature, (London),* 222, 537, 1969.
17. Roeder, R. G. and Rutter, W. J., Multiple forms of DNA-dependent RNA polymerase in eukaryotic organisms, *Nature, (London),* 224, 234, 1969.
18. Blatti, S. P., Ingles, C. J., Lindell, T. J., Morris, P. W., Weaver, R. F., Weinberg, F., and Rutter, W. J., Structure and regulatory properties of eukaryotic RNA polymerases, *Cold Spring Harbor Symp. Quant. Biol.,* 35, 649, 1970.
19. Jacob, S. T., Sajdel, E. M., and Munro, H. N., Different responses of soluble whole nuclear RNA polymerase and soluble nucleolar RNA polymerase to divalent cations and to inhibition by α-amanitin, *Biochem. Biophys. Res. Commun.,* 38, 765, 1970.
20. Roeder, R. G., Reeder, R. H., and Brown, D. D., Multiple forms of RNA polymerase in *Xenopus laevis*: their relationship to RNA synthesis in vivo and their fidelity of transcription in vitro, *Cold Spring Harbor Symp. Quant. Biol.,* 35, 727, 1970.
21. Reeder, R. H. and Roeder, R. G., Ribosomal RNA synthesis in isolated nuclei, *J. Mol. Biol.,* 67, 433, 1972.
22. Lindell, T. J., Weinberg, F., Morris, P. W., Roeder, R. G., and Rutter, W. J., Specific inhibition of nuclear RNA polymerase II by α-amanitin, *Science,* 170, 447, 1970.
23. Zylber, E. A. and Penman, S., Products of RNA polymerases in HeLa cell nuclei, *Proc. Natl. Acad. Sci. U.S.A.,* 68, 2861, 1971.
24. Schwartz, L. B., Sklar, V. E. F., Jaehning, J. A., Weinmann, R., and Roeder, R. G., Isolation and partial characterization of the multiple forms of deoxyribonucleic acid-dependent ribonucleic acid polymerase in the mouse myeloma, MOPC 315, *J. Biol. Chem.,* 249, 5889, 1974.
25. Weil, P. A. and Blatti, S. P., Partial purification and properties of calf thymus deoxyribonucleic acid-dependent RNA polymerase III, *Biochemistry,* 14, 1636, 1975.
26. Weinmann, R. and Roeder, R. G., Role of DNA-dependent RNA polymerase III in the transcription of the tRNA and 5S RNA genes, *Proc. Natl. Acad. Sci. U.S.A.,* 71, 1790, 1974.
27. Guilfoyle, T. J., Purification and characterization of DNA-dependent RNA polymerase from cauliflower nuclei, *Plant Physiol.,* 58, 453, 1976.
28. Young, H. A. and Whiteley, H. R., Deoxyribonucleic acid-dependent ribonucleic acid polymerases in the dimorphic fungus *Mucor rouxii, J. Biol. Chem.,* 250, 479, 1975.
29. Adman, R., Schultz, L. D., and Hall, B. D., Transcription in yeast: separation and properties of multiple RNA polymerases, *Proc. Natl. Acad. Sci. U.S.A.,* 69, 1702, 1972.
30. Bottomley, W., Smith, H. J., and Bogorad, L., RNA polymerase of maize: partial purification and properties of the chloroplast enzyme, *Proc. Natl. Acad. Sci. U.S.A.,* 68, 2412, 1971.

31. Strain, G. C., Mullinix, K. P., and Bogorad, L., RNA polymerase of maize: nuclear RNA polymerases, *Proc. Natl. Acad. Sci. U.S.A.*, 68, 2647, 1971.
32. Polya, G. M. and Jagendorf, A. T., Wheat leaf tRNA polymerases. I. Partial purification and characterization of nuclear, chloroplast, and soluble DNA-dependent enzymes, *Arch. Biochem. Biophys.*, 146, 635, 1971.
33. Polya, G. M. and Jagendorf, A. T., Wheat leaf RNA polymerases. II. Kinetic characterization and template specificities of nuclear chloroplast, and soluble enzymes, *Arch. Biochem. Biophys.*, 146, 649, 1971.
34. Glicklich, D., Jendrisak, J. J., and Becker, W. M., Separation and partial characterization of two ribonucleic acid polymerases from pea seedlings, *Plant Physiol.*, 54, 356, 1974.
35. Sasaki, Y., Sasaki, R., Hashizume, T., and Yamada, Y., The solubilization and partial characterization of pea RNA polymerases, *Biochem. Biophys. Res. Commun.*, 50, 785, 1973.
36. Horgen, P. A. and Key, J. L., The DNA-directed RNA polymerases of soybean, *Biochim. Biophys. Acta*, 294, 227, 1973.
37. Lin, C. Y., Guilfoyle, T. J., Chen, Y. M., Nagao, R. T., and Key, J. L., The separation of RNA polymerases I and II achieved by fractionation of plant chromatin, *Biochem. Biophys. Res. Commun.*, 60, 498, 1974.
38. Hardin, J. W. and Cherry, J. H., Solubilization and partial characterization of soybean chromatin-bound RNA polymerase, *Biochem. Biophys. Res. Commun.*, 48, 299, 1972.
39. Rizzo, P. J., Cherry, J. H., Pedersen, K., and Dunham, V. L., Separation and partial characterization of multiple ribonucleic acid polymerases from soybean hypocotyl, *Plant Physiol.*, 54, 349, 1974.
40. Dunham, V. L. and Cherry, J. H., Solubilization and characterization of RNA polymerase from a higher plant, *Phytochemistry*, 12, 1897, 1973.
41. Fukasawa, H. and Mori, S., Cauliflower RNA polymerase: partial purification and preliminary characterization of DNA-dependent enzymes, *Plant Sci. Lett.*, 2, 391, 1974.
42. Sasaki, Y., Goto, H., Wake, T., and Sasaki, R., Purine ribonucleotide homopolymer formation activity of RNA polymerase from cauliflower, *Biochim. Biophys. Acta*, 366, 443, 1974.
43. Mizuochi, T. and Fukasawa, H., Characterization of an α-amanitin insensitive RNA polymerase from cauliflower, *Phytochemistry*, 14, 1707, 1975.
44. Gore, J. R. and Ingle, J., Ribonucleic acid polymerase activities in Jerusalem-artichoke tissue, *Biochem. J.*, 143, 107, 1974.
45. Teissere, M., Penon, P., and Ricard, J., Hormonal control of chromatin availability and the activity of purified RNA polymerases in higher plants, *FEBS Lett.*, 30, 65, 1973.
46. Mondal, H., Ganguly, A., Das, A., Mandal, R. K., and Biswas, B. B., Ribonucleic acid polymerase from eukaryotic cells. Effects of factors on the activity of RNA polymerase from chromatin of coconut nuclei, *Eur. J. Biochem.*, 28, 143, 1972.
47. Link, G. and Richter, G., Properties and subunit composition of RNA polymerase II from plant cell cultures, *Biochim. Biophys. Acta*, 395, 337, 1975.
48. Jendrisak, J. and Becker, W. M., Isolation, purification, and characterization of RNA polymerases from wheat germ, *Biochim. Biophys. Acta*, 319, 48, 1973.
49. Jendrisak, J. J. and Becker, W. M., Purification and subunit analysis of wheat germ ribonucleic acid polymerase II, *Biochem. J.*, 139, 771, 1974.
50. Jendrisak, J. J. and Burgess, R. R., A new method for the large-scale purification of wheat germ DNA-dependent RNA polymerase II, *Biochemistry*, 14, 4639, 1975.
51. Fabisz-Kijowska, A., Dullin, P., and Walerych, W., Isolation and purification of RNA polymerases from rye embryos, *Biochim. Biophys. Acta*, 390, 105, 1975.
52. Mondal, H., Mandal, R. K., and Biswas, B. B., Factors and rifampicin influencing RNA polymerase isolated from chromatin of eukaryotic cell, *Biochem. Biophys. Res. Commun.*, 40, 1194, 1970.
53. Mondal, H., Mandal, R. K., and Biswas, B. B., RNA polymerase from eukaryotic cells. Isolation and purification of enzymes and factors from chromatin of coconut nuclei, *Eur. J. Biochem.*, 25, 463, 1972.
54. Lin, C. Y., Guilfoyle, T. J., Chen, M. M., and Key, J. L., Isolation of nucleoli and localization of ribonucleic acid polymerase I from soybean hypocotyl, *Plant Physiol.*, 56, 850, 1975.
55. Gurley, W. B., Lin, C. Y., Guilfoyle, T. J., Nagao, R. T., and Key, J. L., Analysis of plant RNA polymerase I transcript in chromatin and nuclei, *Biochim. Biophys. Acta*, 425, 168, 1976.
56. Huang, R. C. and Bonner, J., Histone, a suppressor of chromosomal RNA synthesis, *Proc. Natl. Acad. Sci. U.S.A.*, 48, 1216, 1962.
57. Guilfoyle, T. J. and Key, J. L., The subunit structures of soluble and chromatin-bound RNA polymerase II from soybean, *Biochem. Biophys. Res. Commun.*, 74, 308, 1977.
58. Mullinix, K. P., Strain, G. C., and Bogorad, L., RNA polymerases of maize. Purification and molecular structure of DNA-dependent RNA polymerase II, *Proc. Natl. Acad. Sci. U.S.A.*, 70, 2386, 1973.

59. Guilfoyle, T. J., Lin, C. Y., Chen, Y. M., and Key, J. L., Purification and characterization of RNA polymerase I from a higher plant, *Biochim. Biophys. Acta,* 418, 344, 1976.
60. Guilfoyle, T. J., Lin, C. Y., Chen, Y. M., Nagao, R. T., and Key, J. L., Enhancement of soybean RNA polymerase I by auxin, *Proc. Natl. Acad. Sci. U.S.A.,* 72, 69, 1975.
61. Zillig, W., Zechel, K., and Halbwachs, H. J., A new method of large scale preparation of highly purified RNA-polymerase from *Escherichia coli, Z. Physiol. Chem.,* 351, 221, 1970.
62. Jendrisak, J. J. and Burgess, R. R., Studies on the subunit structure of wheat germ ribonucleic acid polymerase II, *Biochemistry,* 16, 1959, 1977.
63. Ohasa, S. and Tsugita, A., Poly A synthesizing activity in a constitutive subunit of RNA polymerase, *Nature, (London), New Biol.,* 240, 35, 1972.
64. Seifart, K. H., A factor stimulating the transcription on double-stranded DNA by purified RNA polymerase from rat liver nuclei, *Cold Spring Harbor Symp. Quant. Biol.,* 35, 719, 1970.
65. Stein, H. and Hausen, P., A factor from calf thymus stimulating DNA-dependent RNA polymerase isolated from this tissue, *Eur. J. Biochem.,* 14, 270, 1970.
66. Lee, S. C. and Dahmus, M. E., Stimulation of eukaryotic DNA-dependent RNA polymerase by protein factors, *Proc. Natl. Acad. Sci. U.S.A.,* 70, 1383, 1973.
67. Sugden, B. and Keller, W., Mammalian deoxyribonucleic acid-dependent ribonucleic acid polymerases. I. Purification and properties of an α-amanitin-sensitive ribonucleic acid polymerase and stimulatory factors from HeLa and KB cells, *J. Biol. Chem.,* 248, 3777, 1973.
68. Lentfer, D. and Lezius, A. G., Mouse-myeloma RNA polymerase B template specificities and the role of a transcription-stimulating factor, *Eur. J. Biochem.,* 30, 278, 1972.
69. Seifart, K. H., Juhasz, P. P., and Benecke, B. J., A protein factor from rat-liver tissue enhancing the transcription of native templates by homologous RNA polymerase B, *Eur. J. Biochem.,* 33, 181, 1973.
70. Higashinakagawa, T., Onishi, T., and Muramatsu, M., A factor stimulating the transcription by nucleolar RNA polymerase in the nucleolus of rat liver, *Biochem. Biophys. Res. Commun.,* 48, 937, 1972.
71. Froehner, S. C. and Bonner, J., Ascites tumor ribonucleic acid polymerases. Isolation, purification, and factor stimulation, *Biochemistry,* 12, 3064, 1973.
72. Hardin, J. W., O'Brien, T. J., and Cherry, J. H., Stimulation of chromatin-bound RNA polymerase activity by a soluble factor, *Biochim. Biophys. Acta,* 224, 667, 1970.
73. Hardin, J. W., Cherry, J. H., Morre, D. J., and Lembi, C. A., Enhancement of RNA polymerase activity by a factor released by auxin from plasma membrane, *Proc. Natl. Acad. Sci. U.S.A.,* 69, 3146, 1972.
74. Matthysse, A. G. and Phillips, C., A protein intermediary in the interaction of a hormone with the genome, *Proc. Natl. Acad. Sci. U.S.A.,* 63, 897, 1969.
75. Matthysse, A. G. and Abrams, M., A factor mediating interaction of kinins with the genetic material, *Biochim. Biophys. Acta,* 199, 511, 1969.
76. Venis, M. A., Stimulation of RNA transcription from pea and corn DNA by protein retained on Sepharose coupled to 2,4-dichlorophenoxyacetic acid, *Proc. Natl. Acad. Sci. U.S.A.,* 68, 1824, 1971.
77. Teissere, M., Penon, P., van Huystee, R. B., Azou, Y., and Ricard, J., Hormonal control of transcription in higher plants, *Biochim. Biophys. Acta,* 402, 391, 1975.
78. Teissere, M., Penon, P., Azou, Y., and Ricard, J., Mode of action of transcription factors in higher plants, *Plant Sci. Lett.,* 6, 49, 1976.
79. Mandel, J. L. and Chambon, P., Animal DNA-dependent RNA polymerases. Studies on the reaction parameters of transcription in vitro of simian virus 40 DNA by mammalian RNA polymerases AI and B, *Eur. J. Biochem.,* 41, 367, 1974.
80. Mandel, J. L. and Chambon, P., Animal DNA-dependent RNA polymerases. Analysis of the RNA synthesized on simian virus 40 superhelical DNA by mammalian RNA polymerases AI and B, *Eur. J. Biochem.,* 41, 379, 1974.
81. Griswold, M. D. and Cohen, P. P., Alteration of DNA-dependent RNA polymerase activities in amphibian liver nuclei during thyroxine-induced metamorphosis, *J. Biol. Chem.,* 247, 353, 1972.
82. Organtini, J. E., Joseph, C. R., and Farber, J. L., Increases in the activity of the solubilized rat liver nuclear RNA polymerases following partial hepatectomy, *Arch. Biochem. Biophys.,* 170, 485, 1975.
83. Yu, F. L., Increased levels of rat hepatic nuclear free and engaged RNA polymerase activities during liver regeneration, *Biochem. Biophys. Res. Commun.,* 64, 1107, 1975.
84. Jaehning, J. A., Stewart, C. C., and Roeder, R. G., DNA-dependent RNA polymerase levels during the response of human peripheral lymphocytes to phytohemagglutinin, *Cell,* 4, 51, 1975.
85. Renart, J. and Sebastian, J., Characterization and levels of the RNA polymerases during the embryogenesis of *Artemia salina, Cell Differ.,* 5, 97, 1976.

86. Roeder, R. G., Multiple forms of deoxyribonucleic acid-dependent ribonucleic acid polymerases in *Xenopus laevis*: levels of activity during oocyte and embryonic development, *J. Biol. Chem.*, 249, 249, 1974.

87. Versteegh, L. R., Hearn, T. F., and Warner, C. M., Variations in the amounts of RNA polymerase forms I, II, and III during preimplantation development in the mouse, *Dev. Biol.*, 46, 430, 1975.

88. Mayer, A. M. and Poljakoff-Mayber, A., *The Germination of Seeds*, MacMillan, New York, 1963.

89. Chrispeels, M. J. and Varner, J. E., Gibberellic acid-enhanced synthesis and release of α-amylase and ribonuclease by isolated barley aleurone layers, *Plant Physiol.*, 42, 398, 1967.

90. Filner, P. and Varner, J. E., A test for de novo synthesis of enzymes: density labeling with H_2O^{18} of barley α-amylase induced by gibberellic acid, *Proc. Natl. Acad. Sci. U.S.A.*, 58, 1520, 1967.

91. Ihle, J. N. and Dure, L., III, Synthesis of a protease in germinating cotton cotyledons catalyzed by mRNA synthesized during embryogenesis, *Biochem. Biophys. Res. Commun.*, 36, 705, 1969.

92. Chrispeels, M. J., Baumgartner, B., and Harris, N., Regulation of reserve protein metabolism in the cotyledons of mung bean seedlings, *Proc. Natl. Acad. Sci. U.S.A.*, 73, 3168, 1976.

93. Beevers, H., Glyoxysomes of castor bean endosperm and their relation to gluconeogenesis, *Ann. N. Y. Acad. Sci.*, 168, 313, 1969.

94. Ihle, J. N. and Dure, L. S., The developmental biochemistry of cottonseed embryogenesis and germination. III. Regulation of the biosynthesis of enzymes utilized in germination, *J. Biol. Chem.*, 247, 5048, 1972.

95. Marcus, A. and Feeley, J., Activation of protein synthesis in the imbibition phase of seed germination, *Proc. Natl. Acad. Sci. U.S.A.*, 51, 1075, 1964.

96. Marcus, A., Feeley, J., and Volcani, T., Protein synthesis in imbibed seeds. III. Kinetics of amino acid incorporation, ribosome activation, and polysome formation, *Plant Physiol.*, 41, 1167, 1966.

97. Marcus, A., Seed germination and the capacity for protein synthesis, *Symp. Soc. Exp. Biol.*, 23, 143, 1969.

98. Rejman, E. and Buchowicz, J., Sequence of initiation of RNA, DNA, and protein synthesis in wheat grains during germination, *Phytochemistry*, 10, 2951, 1971.

99. Stoddart, J., Thomas, H., and Robertson, A., Protein synthesis patterns in barley embryos during germination, *Planta*, 112, 309, 1973.

100. Hallam, N. D., Roberts, B. E., and Osborne, D. J., Embryogenesis and germination in rye (*Secale cereale* L.). II. Biochemical and fine-structural changes during germination, *Planta*, 105, 293, 1972.

101. Sieliwanowicz, B. and Chmielewska, I., Studies on the initiation of protein synthesis in the course of germination of pea seeds, *Bull. Acad. Pol. Sci. Ser. Sci. Biol.*, 21, 399, 1973.

102. Sieliwanowicz, B., Kolanowska, E., and Chmielewska, I., Studies on the initiation of protein synthesis in the course of germination of pea seeds. Quantitative changes of RNA fractions from embryo axes during initiation of germination, *Acta Biochim. Pol.*, 21, 107, 1974.

103. Gillard, D. F. and Walton, D. C., Germination of *Phaseolus vulgaris*. IV. Patterns of protein synthesis in excised axes, *Plant Physiol.*, 51, 1147, 1973.

104. Barker, G. R. and Rieber, M., The development of polysomes in the seeds of *Pisum arvense*, *Biochem. J.*, 105, 1195, 1967.

105. Waters, L. C. and Dure, L. S., III, Ribonucleic acid synthesis in germinating cotton seeds, *J. Mol. Biol.*, 19, 1, 1966.

106. Chen, D., Sarid, S., and Katchalski, E., Studies on the nature of messenger RNA in germinating wheat embryos, *Proc. Natl. Acad. Sci. U.S.A.*, 60, 902, 1968.

107. Barker, G. R. and Hollinshead, J. A., Nucleotide metabolism in germinating seeds. The ribonucleic acid of *Pisum arvense*, *Biochem. J.*, 93, 78, 1964.

108. Walton, D. C. and Soofi, G. S., Germination of *Phaseolus vulgaris*. III. The role of nucleic acid and protein synthesis in the initiation of axis elongation, *Plant Cell Physiol.*, 10, 307, 1969.

109. Dure, L. and Waters, L., Long-lived messenger RNA: evidence from cotton seed germination, *Science*, 147, 410, 1965.

110. Cherry, J. H., Nucleic acid metabolism in ageing cotyledons, *Symp. Soc. Exp. Biol.*, 21, 247, 1967.

111. Jachymczyk, W. J. and Cherry, J. H., Studies on messenger RNA from peanut plants: in vitro polyribosome formation and protein synthesis, *Biochim. Biophys. Acta*, 157, 368, 1968.

112. Barker, G. R., Bray, C. M., and Detlefsen, M. A., An examination of the evidence for stable messenger ribonucleic acid in seed, *Biochim. J.*, 124, 5P, 1971.

113. Walbot, V., RNA metabolism during embryo development and germination of *Phaseolus vulgaris*, *Dev. Biol.*, 26, 369, 1971.

114. Spiegel, S. and Marcus, A., Polyribosome formation in early wheat embryo germination independent of either transcription or polyadenylation, *Nature, (London)*, 256, 228, 1975.

115. Spiegel, S., Obendorf, R. L., and Marcus, A., Transcription of ribosomal and messenger RNAs in early wheat embryo germination, *Plant Physiol.*, 56, 502, 1975.

116. Tester, C. F., Control of the formation of isocitrate lyase in soybean cotyledons, *Plant Sci. Lett.*, 6, 325, 1976.
117. Smith, R. H., Schubert, A. M., and Benedict, C. R., The development of isocitric lyase activity in germinating cotton seed, *Plant Physiol.*, 54, 197, 1974.
118. Radin, J. W. and Trelease, R. N., Control of enzyme activities in cotton cotyledons during maturation and germination. I. Nitrate reductase and isocitrate lyase, *Plant Physiol.*, 57, 902, 1976.
119. Black, M. and Richardson, M., Germination of lettuce induced by inhibitors of protein synthesis, *Planta*, 73, 344, 1967.
120. Frankland, B., Jarvis, B. C., and Cherry, J. H., RNA synthesis and the germination of light-sensitive lettuce seeds, *Planta*, 97, 39, 1971.
121. Fujisawa, H., Role of nucleic acid and protein metabolism in the initiation of growth at germination, *Plant Cell Physiol.*, 7, 185, 1966.
122. Chakravorty, A. K., Ribosomal RNA synthesis in the germinating black eye pea (*Vigna unguiculata*). I. The effect of cycloheximide on RNA synthesis in the early stages of germination, *Biochim. Biophys. Acta*, 179, 67, 1969.
123. Chen, D. and Osborne, D. J., Hormones in translational control of early germination in wheat embryos, *Nature, (London)*, 226, 1157, 1970.
124. Dobrzanska, M., Tomaszewski, M., Grzelczak, Z., Rejman, E., and Buchowicz, J., Cascade activation of genome transcription in wheat, *Nature, (London).* 244, 507, 1973.
125. Chen, D., Schultz, G., and Katchalski, E., Early ribosomal RNA transcription and appearance of cytoplasmic ribosomes during germination of wheat embryos, *Nature, (London), New Biol.*, 231, 69, 1971.
126. Deltour, R., Synthése et translocation de RNA dans les cellules radiculaires de *Zea mays* au début de la germination, *Planta*, 92, 235, 1970.
127. Van de Walle, C. and Bernier, G., The onset of cellular synthetic activity in roots of germinating corn, *Exp. Cell Res.*, 55, 378, 1969.
128. Watanabe, A., Nitta, T., and Shiroya, T., RNA synthesis in germinating red bean seeds, *Plant Cell Physiol.*, 14, 29, 1973.
129. Tanifuji, S., Asamizu, T., and Sakaguchi, K., DNA-like RNA synthesized in pea embryos at very early stage of germination, *Bot. Mag.*, 82, 56, 1969.
130. Sen, S., Payne, P. I., and Osborne, D. J., Early ribonucleic acid synthesis during the germination of rye (*Secale cereale*) embryos and the relationship to early protein synthesis, *Biochem. J.*, 148, 381, 1975.
131. Rejman, E. and Buchowicz, J., RNA synthesis during the germination of wheat seed, *Phytochemistry*, 12, 271, 1973.
132. Van de Walle, C., Bernier, G., Deltour, R., and Bronshart, R., Sequence of reactivation of ribonucleic acid synthesis during early germination of the maize embryo, *Plant Physiol.*, 57, 632, 1976.
133. Payne, J. F. and Bal, A. K., RNA polymerase activity in germinating onion seeds, *Phytochemistry*, 11, 3105, 1972.
134. Mazus, B. and Buchowicz, J., RNA polymerase activity in resting and germinating wheat seeds, *Phytochemistry*, 11, 2443, 1972.
135. Mazus, B., RNA polymerase activity in isolated *Triticum aestivum* embryos during germination, *Phytochemistry*, 12, 2809, 1973.
136. Rizzo, P. J. and Cherry, J. H., Developmental changes in multiple forms of deoxyribonucleic acid-dependent ribonucleic acid polymerase in soybean hypocotyl, *Plant Physiol.*, 55, 574, 1975.
137. Lin, C. Y., Chen, Y. M., Guilfoyle, T. M., and Key, J. L., Selective modulation of RNA polymerase I activity during growth transitions in the soybean seedlings, *Plant Physiol.*, 58, 614, 1976.
138. Smuckler, E. A. and Tata, J. R., RNA polymerase — changes in hepatic nuclear DNA-dependent RNA polymerase caused by growth hormone and triiodothyroxine, *Nature, (London)*, 234, 37, 1971.
139. Yu, F. L. and Feigelson, P., Cortisone stimulation of nucleolar RNA polymerase activity, *Proc. Natl. Acad. Sci. U.S.A.*, 68, 2177, 1971.
140. Glasser, S. R., Chytil, F., and Spelsberg, T. C., Early effects of oestradiol-17β on the chromatin and activity of the deoxyribonucleic acid-dependent ribonucleic acid polymerases (I and II) of the rat uterus, *Biochem. J.*, 130, 947, 1972.
141. Cox, R. F., Haines, M. E., and Carey, N. H., Modification of the template capacity of chick-oviduct chromatin for form B RNA polymerase by estradiol, *Eur. J. Biochem.*, 32, 513, 1973.
142. Webster, R. A. and Hamilton, T. H., Comparative effects of estradiol-17β and estiol on uterine RNA polymerases I, II, and III in vivo, *Biochem. Biophys. Res. Commun.*, 69, 737, 1976.
143. Tsai, M., Towle, H. C., Harris, S. E., and O'Malley, B. W., Effect of estrogen on gene expression in the chick oviduct. Comparative aspects of RNA chain initiation in chromatin using homologous versus *Escherichia coli* RNA polymerase, *J. Biol. Chem.*, 251, 1960, 1976.

144. **Natori, S.,** Selective activation of RNA polymerase I in fat body nuclei of *Sarcophaga peregrina* larvae by β-ecdysone, *Dev. Biol.,* 50, 395, 1976.

145. **Silberger, J., Jr. and Skoog, F.,** Changes induced by indoleacetic acid in nucleic acid contents and growth of tobacco pith tissue, *Science,* 118, 443, 1953.

146. **Skoog, F.,** Substances involved in normal growth and differentiation of plants, *Brookhaven Symp. Biol.,* 6, 1, 1953.

147. **Key, J. L.,** Hormones and nucleic acid metabolism, *Annu. Rev. Plant Physiol.,* 20, 449, 1969.

148. **Johri, M. M. and Varner, J. E.,** Enhancement of RNA synthesis in isolated pea nuclei by gibberellic acid, *Proc. Natl. Acad. Sci. U.S.A.,* 59, 269, 1968.

149. **Jarvis, B. C., Frankland, B., and Cherry, J. H.,** Increased nucleic acid synthesis in relation to the breaking of dormancy of hazel seed by gibberellic acid, *Planta,* 83, 257, 1968.

150. **Pearson, J. A. and Wareing, P. F.,** Effect of abscisic acid on activity of chromatin, *Nature, (London),* 221, 672, 1969.

151. **Jarvis, B. C., Frankland, B., and Cherry, J. H.,** Increased DNA template and RNA polymerase associated with the breaking of seed dormancy, *Plant Physiol.,* 43, 1734, 1968.

152. **O'Brien, T. J., Jarvis, B. C., Cherry, J. H., and Hanson, J. B.,** Enhancement by 2,4-D of chromatin RNA polymerase in soybean hypocotyl tissue, *Biochim. Biophys. Acta,* 169, 35, 1968.

153. **McComb, A. J., McComb, J. A., and Duda, C. T.,** Increased ribonucleic acid polymerase activity associated with chromatin from internodes of dwarf pea plants treated with gibberellic acid, *Plant Physiol.,* 46, 221, 1970.

154. **Johnson, K. D. and Purves, W. K.,** Ribonucleic acid synthesis by cucumber chromatin. Developmental and hormone-induced changes, *Plant Physiol.,* 46, 581, 1970.

155. **Hou, G. C. and Pillay, D. T. N.,** Chromatin RNA polymerase activity from soybean hypocotyls treated with gibberellic acid and AMO-1618, *Phytochemistry,* 14, 403, 1975.

156. **Duda, C. T. and Cherry, J. H.,** Chromatin- and nuclei-directed ribonucleic acid synthesis in sugar beet root, *Plant Physiol.,* 47, 262, 1971.

157. **Holm, R. E., O'Brien, T. J., Key, J. L., and Cherry, J. H.,** The influence of auxin and ethylene on chromatin-directed RNA synthesis in soybean hypocotyl, *Plant Physiol.,* 45, 41, 1970.

158. **Arens, M. Q. and Stout, E. R.,** Enhanced activity of the soluble ribonucleic acid polymerase from 2,4-dichlorophenoxyacetic acid-treated maize seedlings, *Plant Physiol.,* 50, 640, 1972.

159. **Mondal, H., Mandal, R. K., and Biswas, B. B.,** RNA stimulated by indole acetic acid, *Nature, (London), New Biol.,* 240, 111, 1972.

160. **Mondal, H., Mandal, R. K., and Biswas, B. B.,** The effect of indole acetic acid on RNA polymerase in vitro, *Biochem. Biophys. Res. Commun.,* 49, 306, 1972.

161. **Roy, P. and Biswas, B. B.,** A receptor protein for indoleacetic acid from plant chromatin and its role in transcription, *Biochem. Biophys. Res. Commun.,* 74, 1597, 1977.

162. **Rizzo, P. J., Pedersen, K., and Cherry, J. H.,** Stimulation of transcription by a soluble factor isolated from soybean hypocotyl by 2,4-D affinity chromatography, *Plant Sci. Lett.,* 8, 205, 1977.

163. **Teissere, M., Penon, P., and Ricard, J.,** Hormonal control of chromatin availability and of the activity of purified RNA polymerases in higher plants, *FEBS Lett.,* 30, 65, 1973.

164. **Guilfoyle, T. J. and Hanson, J. B.,** Increased activity of chromatin-bound ribonucleic acid polymerase from soybean hypocotyl with spermidine and high ionic strength, *Plant Physiol.,* 51, 1022, 1973.

165. **Guilfoyle, T. J. and Hanson, J. B.,** Greater length of ribonucleic acid synthesized by chromatin-bound polymerase from auxin-treated soybean hypocotyls, *Plant Physiol.,* 53, 110, 1974.

166. **Bex, J. H. M.,** Effects of abscisic acid on the soluble RNA polymerase activity in maize coleoptiles, *Planta,* 103, 11, 1972.

167. **Mondal, H. and Biswas, B. B.,** Abscisic acid as an inhibitor of RNA synthesis by RNA polymerase in vitro, *Plant Cell Physiol.,* 13, 965, 1972.

168. **Kirk, J. T. O.,** DNA-dependent RNA synthesis in chloroplast preparations, *Biochem. Biophys. Res. Commun.,* 14, 393.

169. **Semal, J., Spencer, D., Kim, Y. T., and Wildman, S. G.,** Properties of a ribonucleic acid synthesizing system in cell-free extracts of tobacco leaves, *Biochim. Biophys. Acta,* 91, 205, 1964.

170. **Spencer, D. and Whitfeld, P. R.,** Ribonucleic acid synthesizing activity of spinach chloroplasts and nuclei, *Arch. Biochem. Biophys.,* 121, 336, 1967.

171. **Tewari, K. K. and Wildman, S. G.,** Function of chloroplast DNA. II. Studies on DNA-dependent RNA polymerase activity of tobacco chloroplasts, *Biochim. Biophys. Acta,* 186, 358, 1969.

172. **Surzycki, S. J.,** Genetic functions of the chloroplast of *Chlamydomonas reinhardi:* effect of rifampin on chloroplast DNA-dependent RNA polymerase, *Proc. Natl. Acad. Sci. U.S.A.,* 63, 1327, 1969.

173. **Bottomley, W., Spencer, D., Wheeler, A. M., and Whitfeld, P. R.,** The effect of a range of RNA polymerase inhibitors on RNA synthesis in higher plant chloroplasts and nuclei, *Arch. Biochem. Biophys.,* 143, 269, 1971.

174. **Stout, E. R. and Mans, R. J.,** Partial purification and properties of RNA polymerase from maize, *Biochim. Biophys. Acta,* 134, 327, 1967.
175. **Smith, H. J. and Bogorad, L.,** The polypeptide subunit structure of the DNA-dependent RNA polymerase of *Zea mays* chloroplasts, *Proc. Natl. Acad. Sci. U.S.A.,* 71, 4839, 1974.
176. **Joussaume, M.,** Mise en evidence de deux formes de RNA polymerase dependante du DNA dans les chlorplastes isoles de feuilles de Pois, *Physiol. Veg.,* 11, 69, 1973.
177. **Bogorad, L.,** Biosynthesis and morphogenesis in plastids, in *Biochemistry of Chloroplasts,* Vol. 2, Goodwin, T. W., Ed., Academic Press, New York, 1967, 612.
178. **Bogorad, L.,** Control mechanisms in plastid development, in *Control Mechanisms in Developmental Processes,* Locke, M., Ed., Academic Press, New York, 1967, 1.
179. **Kirk, J. T. O. and Tilney-Bassett, R. A. E.,** *The Plastids,* W. H. Freeman, London, 1967.
180. **Harel, E. and Bogorad, L.,** Effect of light on ribonucleic acid metabolism in greening maize leaves, *Plant Physiol.,* 51, 10, 1973.
181. **Apel, K. and Bogorad, L.,** Light-induced increase in the activity of maize plastid DNA-dependent RNA polymerase, *Eur. J. Biochem.,* 67, 615, 1976.

CYTOPLASMIC AND ORGANELLAR tRNAS IN PLANTS

J. H. Weil

TABLE OF CONTENTS

I. INTRODUCTION*

In 1955, Francis Crick proposed the need of an "adaptor" to fit the DNA code on one hand and the amino acid on the other. By 1957, evidence that amino acids could be covalently linked to a small RNA (25,000 to 30,000 mol wt) was obtained.[1,2] This RNA was more soluble than most cellular RNAs, contained rare nucleotides,[3] and was a metabolic intermediate between the activated amino acid and the ribosome (where protein synthesis occurs). First called sRNA (soluble RNA), this type of RNA was later called tRNA (transfer RNA) and found to consist of a mixture of different tRNA molecules, each of which is able to bind (and to transfer) only one amino acid.

According to the generally accepted mechanism, the first two steps of protein synthesis are: (1) amino acid activation and (2) amino acid attachment to the cognate tRNA (called tRNA aminoacylation or tRNA charging). This two-step mechanism of tRNA esterification (aminoacylation) is summarized in Figure 1. The attachment of the correct amino acid to each tRNA is essential for the fidelity of translation because in the later steps of protein biosynthesis, the aminoacyl-tRNA is recognized through its polynucleotide moiety, especially through its anticodon which interacts with the corresponding codon in the messenger RNA. Thus, any mistake in the nature of the amino acid attached to a tRNA will result in an error in the protein synthesized, as shown by the experiments of Chapeville et al.[5]

These two critical steps in the correct amino-acylation of a tRNA are catalyzed by an enzyme appropriately named "activating enzyme" (referring to the first step), "aminoacyl-tRNA synthetase" or "aminoacyl-tRNA ligase" (referring to the second step). In a bacterial cell, there are 20 different aminoacyl-tRNA synthetases (one specific for each amino acid), but there are about 40 to 50 different tRNAS (obviously more than one for each amino acid). The various tRNAs (usually two or three) which can accept the same amino acid are called "isoacceptors." Although they are recognized by the same aminoacyl-tRNA synthetase, their nucleotide sequences differ. Sometimes only one or two nucleotides are different (out of a total of approximately 70 to 90). Sometimes the structural differences are more important. Two isoacceptors may have the same anticodon. In other cases, their anticodons are different so that they can recognize different triplets coding for the same amino acid.

The situation becomes somewhat more complicated in eukaryotic cells. Here, as shown by the pioneering studies of Barnett et al.,[6,7] mitochondria contain tRNAs and aminoacyl-tRNA synthetases different from the cytoplasmic counterparts. The situation becomes even more complex in plant cells because in addition they contain chloroplast tRNAs and aminoacyl-tRNA synthetases that (as seen below) can differ from the cytoplasmic and mitochondrial counterparts.

Plants contain mitochondria and chloroplasts, and each type of organelle has its own DNA and its own protein synthesizing system. In many respects, protein synthesis in the organelles resembles protein synthesis in prokaryotic organisms (especially bacteria) rather than that taking place in the surrounding cytoplasm. This is illustrated by the small size of the organellar ribosomes, the formylation of the initiation methionyl-tRNA, and by the sensitivity of organellar protein synthesis to antibiotics such as chlor-

---

* Symbols used in text and figures: A = adenosine, *A = modified adenosine, m^1A = 1 methyladenosine, i^6A = N^6-Δ^2-isopentenyladenosine, ms^2i^6A = 2 methylthio N^6-Δ^2-isopentenyl-adenosine, C = cytidine, m^5C = 5 methylcytidine, Cm = 2'O-methylcytidine, D = dihydrouridine, G = guanosine, *G = modified guanosine, m^2G = N^2 methylguanosine, m^2_2G = N^2 dimethylguanosine, m^7G = 7 methylguanosine, Gm = 2'O-methylguanosine, Pu = purine, *Pu = hypermodified purine, Py = pyrimidine, *Py = hypermodified pyrimidine, U = uridine, *U = modified uridine, s^2U = 2 thiouridine, s^4U = 4 thiouridine, acp^3U = 3-(3 amino-3-carboxypropyl)uridine, T = thymidine, ψ = pseudouridine.

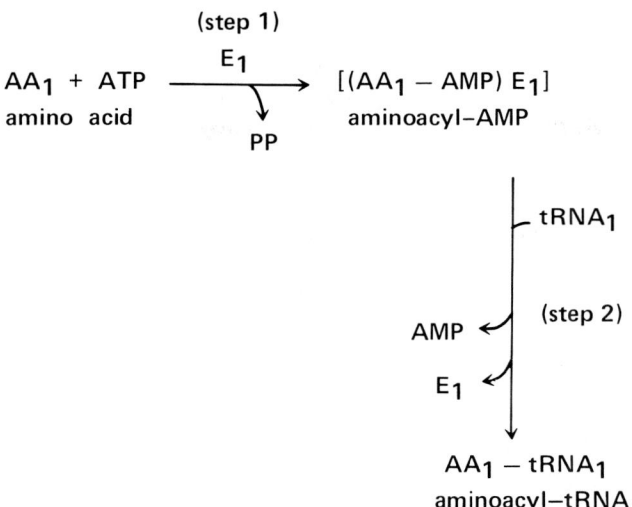

FIGURE 1.   The first steps of protein synthesis: amino acid activation and amino acid attachment to the cognate tRNA.

amphenicol.[8,373.] These observations have been considered by several authors as supporting the theory that the organelles have an endosymbiotic origin and have evolved from prokaryotic ancestors.[9] Studies on plant organellar tRNAs and aminoacyl-tRNA synthetases revealed additional similarities between organelles and prokaryotes (see below).

Many studies have been devoted to tRNAs and aminoacyl-tRNA synthetases in bacteria, in fungi (especially yeast), and in animal cells. Several review articles have been published during the past few years.[10-19,382,395] Plant tRNAs and aminoacyl-tRNA synthetases have been studied less, partly because of difficulties encountered in their extraction and purification. However, the number of laboratories engaged in research on their structure, synthesis, and functions has recently increased. Much progress has been accomplished, judged by comparing the recent review on plant tRNAs and aminoacyl-tRNA synthetases written by Lea and Norris[20] to their previous review published 5 years ago.[21] More information is needed to give better knowledge of the mechanisms of protein synthesis and to understand the interrelations between the various compartments of the plant cells. Some organellar enzymes consist of polypeptides coded in the nucleus and made on cytoplasmic ribosomes combined with polypeptides coded by the organellar DNA and made on organellar ribosomes.[8,22] Furthermore, in addition to their pivotal role in protein biosynthesis, tRNAs have been found in various organisms to have other biological functions, particularly in regulatory processes (see below). This may also be the case in plants, but can only be established by further investigation.

II. EXTRACTION, FRACTIONATION AND PURIFICATION OF tRNAs

A. Extraction of Total tRNA

To extract plant tRNAs, methods originally developed to obtain tRNAs from other sources, especially microorganisms (e.g., *E. coli*[23] and yeast[24]), had to be modified to give satisfactory results. As an example, a general scheme for the extraction of total tRNA from bean hypocotyls is outlined in Figure 2.

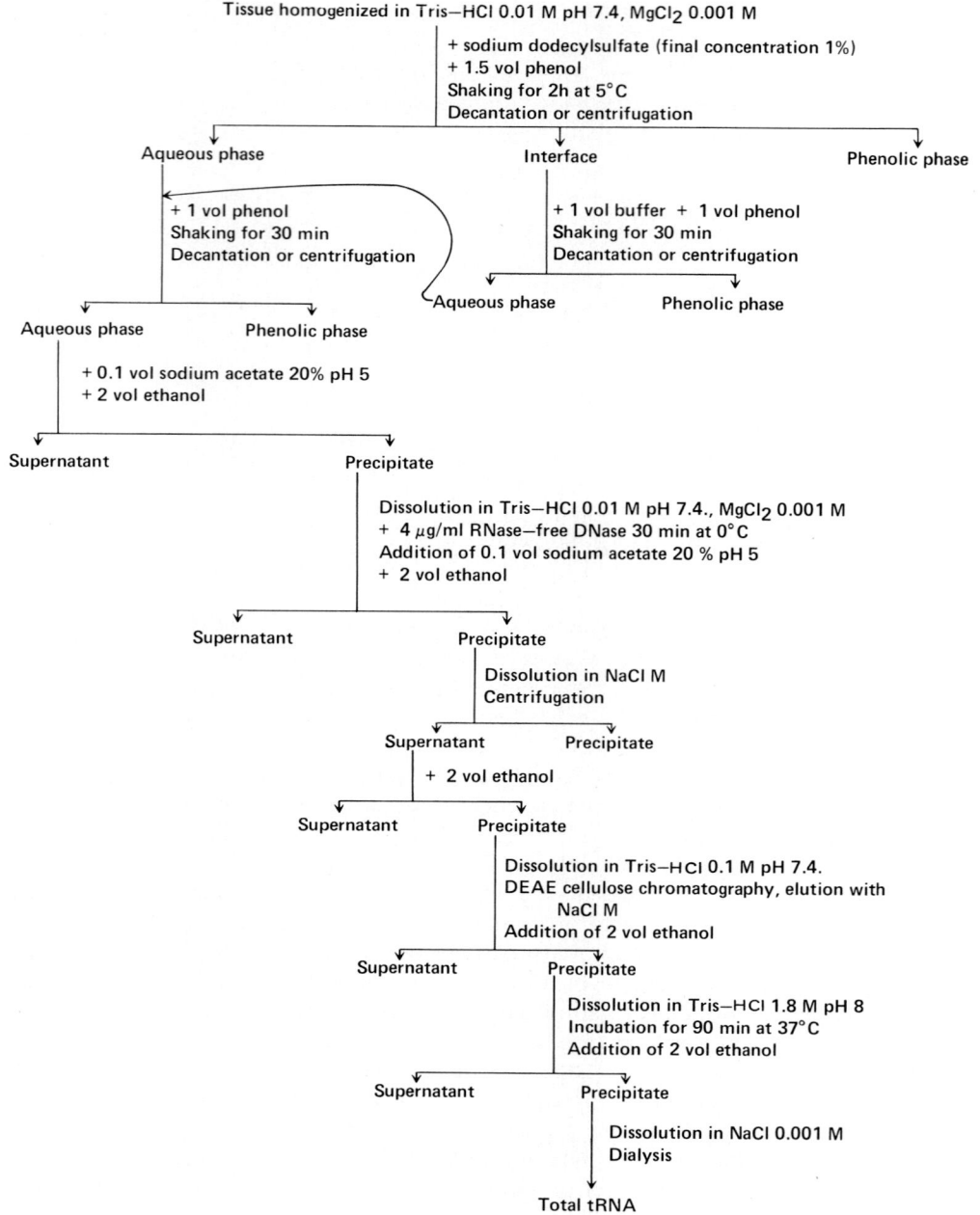

FIGURE 2.   General scheme for the extraction of total tRNA from bean hypocotyls.

Grinding techniques depend on texture of the plant material used (leaves, hypocotyls, roots, seeds, etc.). The cell homogenate is usually deproteinized by phenol treatment,[25] but *m*-cresol is sometimes added to the phenol as it improves deproteinization and prevents crystallization of the phenol when the mixture is cooled to 5°C. The cell homogenate is generally mixed with 1 volume of phenol and shaken, but the amount of phenol added and the time of contact (shaking) with phenol may vary depending on the plant material used. The aqueous phase is separated by decantation or centrifugation, collected, and after addition of sodium acetate (final concentration 0.2 M, pH 4.5), the nucleic acids are precipitated by addition of 2 or 2.5 volumes of cold ethanol. In many cases, a poorly defined interface is present between the aqueous and phenolic phases after centrifugation. Because this can lead to a loss of tRNAs, it is recommended that fresh buffer and phenol be added to this interface, that the mixture be shaken, and after separation of the two phases, the second aqueous phase be combined with the first before ethanol precipitation. The combined aqueous phases are sometimes submitted to one or two more phenol extractions to remove contaminating material (proteins, pigments, etc.).

To prevent action of ribonucleases, bentonite can be used to adsorb these enzymes.[26] Increased yields of tRNAs are obtained when sodium dodecyl sulfate (sodium lauryl sulfate) or other detergents are added during deproteinization.[27] This is probably due not only to prevention of nuclease action, but also to a better disruption of the cellular structures and to a better dissociation of the nucleic acids from complexes with proteins (in polysomes, for instance). Addition of 8-hydroxyquinoline to the phenol has also been suggested to decrease protein contamination and lessen ribonuclease action.[25]

Diethylpyrocarbonate, a very active inhibitor of nucleases, has been proposed to obtain pure and undegraded RNAs from plant tissues.[28] It has been reported that diethylpyrocarbonate can cause opening of the adenine ring[29] and loss of amino acid accepting activity in tRNAs.[30] However, other authors have obtained good yields of tRNA having high activity,[31] and it has been claimed that diethylpyrocarbonate does not degrade RNAs.[32]

Carbohydrates, which often contaminate tRNA preparations, can be removed by treatment with 2-methoxyethanol.[33] Polyphosphate contaminants accumulated by some organisms under certain conditions can be eliminated upon selective precipitation of tRNAs by trimethylhexadecylammonium bromide, dissolution in an organic solvent (such as ethanol), and regeneration of the sodium ribonucleates by NaCl.[34,35] This treatment does not affect the accepting activity of tRNAs.[36,37]

Further purification of tRNAs can then be obtained by fractionation with isopropanol[23,38] to remove DNA (which can also be hydrolyzed by incubation with RNase-free DNase) and high molecular weight RNA. Deproteinization can be achieved by use of a chloroform-isoamyl alcohol mixture.[39] To separate tRNAs from high molecular weight RNAs (essentially ribosomal RNAs, which can represent over 80% of the total cellular RNAs), high salt concentrations in which tRNAs are soluble and high molecular weight RNAs insoluble are generally used. Examples are 1 M to 3 M NaCl,[23,38,40,41] and 3 M potassium or sodium acetate.[42,43] Recently a 2 M ammonium sulfate treatment has been recommended to precipitate ribosomal RNAs and proteins. This is followed by chromatography of the supernatant on Sepharose® CL 4B. In the presence of high concentrations of ammonium sulfate, DNA, carbohydrates, phenol, and other UV absorbing material are not retained, while tRNAs are retained on the Sepharose column and are eluted by a buffer without ammonium sulfate.[44]

Separation of tRNAs from high molecular weight RNAs can also be achieved by chromatography on Sephadex,[45,46] on methylated albumin-kieselguhr (MAK),[47,48] or on diethylaminoethyl cellulose (DEAE cellulose).[23,24,41,43,49] For a review, see Reference 50.

It is possible to separate high molecular weight RNAs and tRNAs by density gradient centrifugation[51] or by electrophoresis,[52] a method allowing separation of 5S RNA and tRNAs (which sediment at about 4S) if the acrylamide concentration is raised (see below). These techniques are generally used for analytical purposes rather than for preparative purposes. In most cases tRNAs are required free from esterified amino acids. This is achieved by alkaline treatment, usually with 1.8 M Tris, pH 8, at 37°C for 90 min.[53] However, chromatography of tRNA on a Sephadex G50 column at pH

7.6 was shown to remove over 90% of the endogenously bound amino acids.[54] A method developed for yeast cells, but applicable to plant cells, permits determination of the percentage of tRNAs esterified in vivo with the corresponding amino acids has been described.[55]

Concentrations of tRNAs are usually determined by simply measuring absorbance of the solution at 260 nm, assuming that one OD_{260} (1 cm cell) corresponds to 40 μg tRNA/ml. Colored substances (often oxidized polyphenols) frequently contaminate plant tRNA preparations (especially from leaves) and interfere with the spectrophotometric determination of tRNA concentration. These can be removed upon Sepharose 4B chromatography, as they are eluted early in the decreasing salt concentration gradient used to fractionate tRNAs by this method (see Section II. B.).

Yields obtained for total tRNA preparations vary appreciably depending on the starting plant material, the extraction procedure used, and the degree of purification achieved. For instance, 35 mg of crude tRNA were obtained from 100 g lupin seeds;[39] 5 to 7 mg tRNA from 100 g (2 days old, dark-grown) bean seedlings,[44] 2 to 3 mg tRNA from 100 g *Parthenocissus* cells (grown in vitro for 10 days in the dark);[44] and 6 mg, 13 mg, and 33 mg tRNA from 100 g of, respectively, the maturing, elongating, and meristematic regions of the pea root.[78]

The extraction procedures mentioned above yield total tRNA preparations containing all isoacceptor tRNAs present in the cell. Such complex mixtures can be further fractionated to separate the individual tRNA species, and especially to distinguish between various isoacceptors present in the different cellular compartments (see Section II, B).

The purity of a total tRNA preparation can be judged by electrophoresis on polyacrylamide gel.[52] On a 2.2 to 3% gel, for instance, one can check for presence of high molecular weight RNAs; on a 7.5 to 8% gel it is possible to separate tRNAs (which sediment at about 4S) from 5S RNA. The molecular weight of 5S (ribosomal) RNA is close to that of the tRNAs, and it is a frequent contaminant of tRNA preparations. The purity can also be estimated by measuring the accepting activity of the tRNA preparation, namely, the attachment of the amino acids under conditions as close as possible to the optimal, especially in regard to the Mg^{2+}P ratio. The best ratio for the aminoacylation reaction varies depending on the amino acid considered.[41,44] Assuming that tRNAs have an average molecular weight of 25,000, 1 mg of tRNA (40 nmol) should accept a total of 40 nmol of amino acids. Thus, if 1 mg of a tRNA preparation accepts a total of only 32 nmol of amino acids, that preparation can be considered as being approximately 80% pure.

Aminoacylation of a total tRNA preparation by one single amino acid allows determination of the relative concentration of the isoacceptor tRNAs specific to that amino acid and evaluation of the purity of an individual tRNA following fractionation (see the fractionation methods below). Conversely, the aminoacylation of a tRNA with one amino acid can also be used to titrate the corresponding aminoacyl-tRNA synthetase in an enzymatic preparation by introducing limiting amounts of protein in the reaction mixture. The aminoacylation assays are based on the reaction scheme shown in Figure 1. A radioactive (^{14}C, ^{3}H, or ^{35}S) amino acid is used, and the radioactive aminoacyl-tRNA formed in the reaction is precipitated with acid (usually trichloroacetic acid), washed (to remove the free amino acid), and its amount determined by measurement of the acid insoluble radioactivity. The aminoacylation reaction is sensitive to salts.[41,56] For instance, it is inhibited by phosphate ions possibly present in the tRNA or enzyme fractions tested after chromatographic separations using a phosphate gradient for elution. It is sensitive to organic solvents,[57,58] and to enzymes such as proteases and nucleases[43,59] which may be present especially when crude preparations are used as a source of aminoacyl-tRNA synthetases. Optimal reaction conditions, such as pH, temperature, and especially Mg^{2+}P ratio vary depending on the tRNA and the amino acid studied.[41,43,44,60,61] As an example, however, the composition of the reaction mixture used to study[41] the esterification of leucine to *Phaseolus vulgaris* (bean) cytoplasmic tRNA by the homologous leucyl-tRNA synthetase can be given: sodium cacodylate (pH 7.4), 5.5 μmol; ATP, 1 μmol; Mg Cl$_2$, 1.5 μmol; KCl 3, μmol; glutathione, 0.12 μmol; ^{14}C-leucine, 0.01 μmol; (0.2 μCi), tRNA 15 to 20 μg; bovine serum albumin, 10 μg, enzymatic proteins, 30 μg (in a final volume of 0.1 ml).

In addition to its analytical uses, the aminoacylation reaction is employed for preparation of (radioactive or unlabeled) aminoacyl-tRNAs. These are needed for studies of the later steps of protein biosynthesis. They are also required for the fractionation and purification of individual tRNA species where they can be used for identification and quantification of radioactive aminoacyl-tRNA in fractions eluting from a chromatography column. Sometimes the fractionation procedure is based on differences in properties between one aminoacyl-tRNA and all other uncharged tRNAs (see below).

B. Fractionation and Purification of Individual tRNA Species

Partitioning of tRNAs between two liquid phases by counter-current distribution has been successfully used in tRNA fractionation and purification,[62-66] even in the case of plant tRNAs,[67] however, this method requires specialized equipment available in only a few laboratories, and tRNA fractionation is usually achieved using various types of column chromatography.

Adsorption chromatography has been employed to fractionate plant tRNAs[67,68] using as a support hydroxylapatite,[67-73] methylated albumin-kieselguhr (MAK)[74-76,133] or methylated albumin silicic acid.[77]

Ion exchange chromatography, essentially on DEAE cellulose, has given good results in tRNA purification and fractionation including plant tRNA.[78,370] For a review, see Reference 50. To take advantage of the differences in the melting temperatures (Tm) of the various tRNAs, elution (by a salt gradient) from the column can be performed at elevated temperature or using a temperature gradient.[50] DEAE Sephadex has also been used by several authors and has furnished better resolution in some cases.[50]

Substitution of hydroxyl groups of DEAE cellulose by aromatic groups increases the nonionic interactions with tRNAs, as shown by Gillam et al.[79] who recommended benzoylated DEAE cellulose (BD cellulose) columns to fractionate isoaccepting tRNAs. Some tRNAs are more strongly held on the column because they contain an aromatic (hydrophobic) group called "Y" base.* Their elution is retarded, and they are, therefore, easily isolated and purified. This is the case, for instance, for cytoplasmic tRNAs ^{Phe}.[79,80] BD cellulose chromatography can in fact be used to fractionate and purify any tRNA species, either after esterification of the tRNA with the corresponding amino acid if it has an aromatic ring (Phe, Tyr, Trp),[80,81] or else after aminoacylation of the tRNA and subsequent derivatization of the amino acid by attachment of an aromatic substituent (usually a phenoxyacetyl or a naphthoxyacetyl group).[82] Esterification with an aromatic amino acid or derivatization with an aromatic substituent causes delay in the elution of the corresponding tRNA (while there is no effect on the mobility of the other tRNAs in the mixture), thus resulting in its purification. BD cellulose chromatography has been successfully used for fractionation of tRNAs from many sources. For a review, see Reference 83. Examples of its use with plant tRNAs include fractionation of methionyl-tRNAs from wheat germ[84,85] and from the broad (horse) bean, *Vicia faba*,[86] the purification of wheat germ phenylalanyl-tRNA,[87] and the separation of *Euglena gracilis* cytoplasmic phenylalanyl-tRNA (which

* This base is now called "wye", and the corresponding nucleoside "wyosine".

contains a Y base and is eluted late in the ethanol fraction) from chloroplastic phenylalanyl-tRNA having no Y base and therefore eluted early in the salt gradient.[88]

Partition chromatography has also been used to fractionate tRNAs taking advantage of the solubility of quaternary ammonium salts of tRNAs in organic solvents.[36,37] The first partition techniques described used a stationary aqueous phase bound to the support and a mobile organic phase. A variety of solvent systems and supports have been employed, such as Sephadex G25,[89-92] silica gel,[93] or cellulose.[94] Unfortunately, it is difficult to prevent alterations in the volume of the column due to dehydration of the gel by the organic solvents even if they are saturated with the aqueous phase.

Better results have been obtained by reversed phase chromatography (RPC), in which the organic phase is bound to the support, and the aqueous phase is the eluting solution. An inert support is coated with a thin film of a high molecular weight quaternary ammonium salt having a limited solubility in water or dilute salt solutions and very good ion exchange properties under neutral or weakly acidic conditions. The first system described by Kelmers et al.,[95] called RPC-1, was followed by several other RPC systems that differ in the support and in the quaternary ammonium salt used. For a review, see Reference 96. These systems have been very efficient in the fractionation of tRNAs from a great variety of sources, including plants. For instance, RPC-2 columns[97] have been used at pH 4.5 or 7.5 to purify wheat germ phenylalanyl-tRNA;[98] to characterize light-induced (chloroplastic) phenylalanyl-, glutamyl-, and isoleucyl-tRNAs in *Euglena*;[99] to separate cytoplasmic, chloroplastic, and mitochondrial leucyl- and valyl-tRNAs of *Phaseolus vulgaris*[41] or leucyl-tRNAs of *Nicotiana tabaccum*;[100] and to study soybean leucyl-tRNAs in cotyledons and seedlings during development.[40,102-105] RPC-5 columns[106,107] have been used to fractionate *Phaseolus vulgaris* cytoplasmic, chloroplastic, and mitochondrial leucyl-tRNAs;[108] lysyl-tRNAs,[109] phenylalanyl-RNAs,[388] and prolyl-tRNAs.[109] Methionyl-tRNAs have been separated using either RPC-5 (in some cases in the presence of 6 *M* urea to achieve a better resolution) or RPC-6 columns.[110,111] As an example, the fractionation of bean leaf tRNAs[Leu] by RPC-5 chromatography is shown in Figure 3.

Very good resolution of tRNAs can also be obtained by two recently described methods. These are chromatography on Sepharose 4B in which the tRNAs are adsorbed to the support at a high concentration of ammonium sulfate and eluted with a linear decreasing gradient of this salt[112] and chromatography on Aminex® A-28 which is a polystyrene anion exchanger.[113]

Advantage has been taken of the specific antigen-antibody reaction to purify tRNA[Phe] species containing a Y nucleoside on a column where antibodies against this nucleoside had been immobilized.[114]

After immobilizing yeast tRNA[Phe] (anticodon GmAA) by covalent linkage (through its 3′ end) to polyacrylamide (Biogel® P20), Grosjean et al. were able to retain specifically *E. coli* tRNA[Glu] (which has the complementary anticodon s[2]UUC) and to obtain on such a column a 19-fold enrichment of this tRNA[Glu] after one passage of total *E. coli* tRNA.[115] The same principle has been applied to purify ^{32}P-labeled tRNA precursors.[116]

Several fractionation methods are based on the aminoacylation of the desired tRNA by the corresponding amino acid and on the separation of this aminoacyl-tRNA from the other (uncharged) tRNAs. Such a strategy usually leads to isolation not of a single tRNA species, but of the various isoacceptors specific of that amino acid. These isoacceptors must then be further fractionated if pure individual tRNA species are desired. Among these methods, the isolation on BD cellulose of an aminoacyl-tRNA derivatized using an aromatic substituent (if the amino acid itself has no aromatic ring) has already been discussed. See Section II.B.

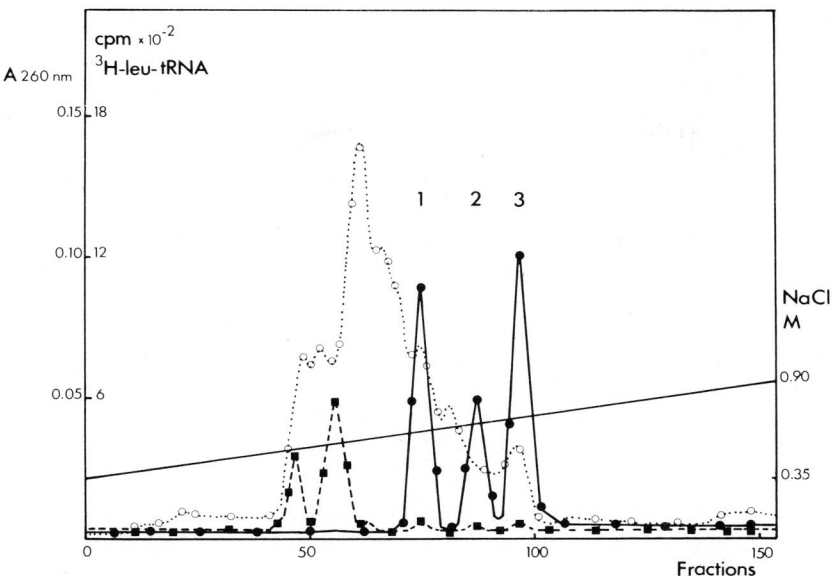

FIGURE 3. Fractionation of bean leaf tRNAs^{Leu} by RPC-5 chromatography. Column size : 60 × 0.6 cm. NaCl gradient from 0.35 to 0.9 M (2 × 85 ml) in Tris-HCl 0.05 M pH 7.4, MgCl$_2$ 0.01 M. Fractions of 1 ml were collected. O . . . . . . O A$_{260}$nm; ■- - - -■ ^{3}H-leu-cyl-tRNAs charged by the cytoplasmic enzyme; ■——■ ^{14}C-leucyl-tRNAs charged by the *E. coli* enzyme (these are the three chloroplast-specific leucyl-tRNAs).

Other chemical modifications of the aminoacyl moiety of the aminoacyl-tRNA have been used.[117,118] More recently, the reaction of a bifunctional reagent (*p* chloromercuribenzenesulfonyl chloride or PAMBSYL) with the aminoacyl group has been described, which allows the subsequent attachment of the derivatized (pambsylated) aminoacyl-tRNA to a sulfhydryl containing resin and thus its separation from the bulk of all other (uncharged) tRNAs. It can then be released from the resin by deacylation at alkaline pH.[119]

Conversely, instead of modifying the aminoacyl-tRNA, the specific chemical modification of the free 3′-end of all uncharged tRNAs has been recommended to allow their removal from the mixture.[38,120,121] These approaches suffer from two drawbacks. First, there is the danger that the chemical reagent used might cause structural alterations of the tRNAs in addition to the desired modification of the 3′-end or the esterified amino acid. Secondly, tRNA purification depends on the efficiency of the chemical modification and on the efficiency of the esterification reaction when the chemical modification is performed on the aminoacyl-tRNA.

A fractionation method exploiting the difference between charged and uncharged tRNAs, but requiring no chemical modification has recently been described. According to this method, uncharged tRNAs bind (through the free *cis*-diol group located at the 3′-end) to the dihydroxyboryl groups attached to aminoethyl cellulose and are retained on the column (called DBAE cellulose) while the aminoacyl-tRNA molecules are not retained.[122,123]

In the presence of GTP and protein synthesis elongation factor EF-Tu from *E. coli*, a ternary complex with aminoacyl-tRNA is formed that can be separated from all uncharged tRNAs and from free aminoacyl-tRNA by gel filtration on Sephadex G-100. This method has been used to purify chicken liver tRNAs, but appears of general use and yields highly purified tRNAs.[124]

An aminoacyl-tRNA can also be purified by specific enzyme-substrate complex formation on a column where the cognate aminoacyl-tRNA synthetase has been bound to a modified cellulose matrix. The tRNA is liberated by hydrolysis of the ester bond between the amino acid and the 3′-terminal adenosine of the tRNA.[125]

In addition to these various chromatographic techniques, ^{32}P-labeled tRNAs can also be purified by two-dimensional electrophoresis on polyacrylamide gels.[126-128] This method has the advantage of allowing the simultaneous purification of many tRNAs in one step, and has also been used to fractionate tRNA precursors.[129] Although this is a very powerful technique, the actual amounts of tRNAs obtained are very small. However enough material can be obtained to allow, after elution of the purified individual tRNAs from the gel, identification of each tRNA by aminoacylation and mapping of its gene upon hybridization of the tRNA (labled in vitro with ^{125}I) with DNA fragments obtained by action of a specific restriction endonuclease.[419,428,429]

Many techniques concerning tRNAs (including tRNA extraction and purification) and enzymes acting on them (especially aminoacyl-tRNA synthetases) can be found in two volumes of "Methods in Enzymology" devoted to nucleic acids and protein synthesis.[129a,130a]

III. MULTIPLICITY AND INTRACELLULAR LOCALIZATION OF PLANT tRNAs

The fractionation methods described above have been applied to the tRNAs from a large variety of plants including green algae such as *Euglena*. Isoacceptors have been characterized for the tRNAs specific of almost all 20 amino acids (see Table 1). In some instances, it has been shown that certain isoaccepting tRNA species are specifically located in one compartment of the plant cell, namely the cytoplasm, the chloroplasts, or the mitochondria (see Table 1). This raises the question of their origin or localization of the genes coding for these isoacceptors, a problem discussed in Section V. Quite often, distinct isoacceptors can be found within the same subcellular com-

TABLE 1

Multiplicity and Intracellular Localization of Plant tRNAs

tRNAs accepting:	Plant and tissue	Number of isoacceptors and comments	Organellar isoacceptors	Ref.
Ala	Cotton seedlings	4		130
	Tobacco tissue cultures	1—2		132
Arg	Wheat embryo and seedlings	3		133
	Cotton seedlings	7		130
	Tobacco tissue cultures	3		132
Asn	Cotton seedlings	1		130
Asp	Cotton seedlings	4		130
Glu	Wheat embryos and seedlings	3		133
	Euglena gracilis	3	1 chloro	99
	Tobacco tissue cultures	2		132
Gly	Wheat embryos and seedlings	1		133
	Wheat germ	2 tRNA,[Gly] sequenced		158
	Cotton seedlings	3		130

TABLE 1 (continued)

Multiplicity and Intracellular Localization of Plant tRNAs

tRNAs accepting:	Plant and tissue	Number of isoacceptors and comments	Organellar isoacceptors	Ref.
	Soybean seedlings	2 shifts on chilling		134
	Tobacco tissue cultures	2		132
His	Wheat embryos and seedlings	2		133
	Cotton seedlings	2	1 chloro	130
Ile	Lupin seeds	5		67, 368
	E. gracilis	3—5	1 chloro	99, 135, 136, 149, 420
		5	1 mito	
	Cotton seedlings	4	2 chloro	130
	Tobacco tissue cultures	2		132
Leu	Lupin seeds	3		67
	Wheat embryos and seedlings	2		133
	Soybean cotyledons and seedlings	6 shifts during development		40, 102, 103, 104, 134, 137, 138, 430
	Pea cotyledons and leaves	4 shifts during development		102, 139, 140
	Pea pods	2		140
	Pea roots	5—6		141, 423
	Bean leaves	9	3 chloro, 4 mito	41, 108, 218
	Cotton seedlings	6	3 chloro	130
	Tobacco leaves	6	2 chloro 2 mito	100
	Tobacco tissue cultures	5		132
	E. gracilis	6	2 chloro	142, 396
	Mercurialis annua flowers	6		143
	Tomato fruits	3 (+ 2) shifts on ripening		144
	Apple and pear fruits	3		367
	Barley embryos	5		422
	Barley seedlings	4	shifts upon greening	422
Lys	Wheat embryos and seedlings	5 shifts during development		133
	Pea roots	4		141
	Vigna sinensis	3		145
	Cotton seedlings	4	1 chloro	130
	Wheat grains	3—4		146
	Tomato fruits	5 shifts upon ripening		144
	Tissue cultures	3		132
	Bean leaves	4	1 chloro, 1 mito	109
	Apple and pear fruits	4		367
	Lupin seeds and cotyledons	5 shifts during development	1 chloro, 1 mito	369, 403

TABLE 1 (continued)

Multiplicity and Intracellular Localization of Plant tRNAs

tRNAs accepting:	Plant and tissue	Number of isoacceptors and comments	Organellar isoacceptors	Ref.
Met	Bean leaves	8	3 chloro (1 formylated) 3 mito (2 formylated)	110, 111, 147, 148
	Pea roots	3		141
	Wheat leaves	5	2 chloro (1 formylated)	85
	Cotton seedlings	3	2 chloro (1 formylated)	130
	E. gracilis	3	1 chloro	149
	Tomato fruits	2 shifts upon ripening		144
	Tobacco tissue cultures	2		132
	Scenedesmus obliquus	5 (2 formylated) one sequenced		101, 404
Phe	Wheat embryos and seedlings	1		133
	Wheat germ	2 one species sequenced		87, 157
	E. gracilis	2	1 chloro (sequenced)	99, 135, 150
	Pea roots	4 one species sequenced		141, 151
	Cotton seedlings	4	2 chloro	130
	Barley seedlings	9	2 chloro	152
	Barley embryos	2 one species sequenced		153, 424
	Lupin seeds	2 one species sequenced		39, 154
	Bean leaves	4	2 chloro (1 sequenced) 1 mito	156, 388, 155
	Tobacco tissue cultures	1 one extra species in crown-gall tumors		132
Pro	Wheat embryos and seedlings	2 shifts during development		133
	Wheat grains	3		146
	Pea roots	4		141
	Bean leaves	5	1 chloro, 2 mito	109
Ser	Wheat embryos and seedlings	2 shifts during development		133
	Pea roots	3		141
	Soybean cotyledons	3		138
	M. annua flowers	6		143
Thr	Wheat embryos and seedlings	2		133
	Pea roots	3		141
	E. gracilis	2	1 chloro	396
Trp	Cotton seedlings	3	2 chloro	130
	Chlamydomonas reinhardi	2	1 chloro	159
Tyr	Pea roots	4 differences between dividing and nondividing cells	1 mito	141
	Soybean seedlings	3 shifts during development		103
	Soybean cotyledons	4	2 chloro	138, 387

TABLE 1 (continued)

Multiplicity and Intracellular Localization of Plant tRNAs

tRNAs accepting:	Plant and tissue	Number of isoacceptors and comments	Organellar isoacceptors	Ref.
	Tobacco tissue cultures	2		132
	M. annua flowers	5		143
	Tomato fruits	2		144
Val	Wheat embryos and seedlings	3		133
	Lupin seeds	2		67
	Cotton seedlings	5	1 chloro	130
	Bean leaves	4	1 chloro	41
	Tobacco tissue cultures	3		132
	M. annua flowers	3		143
	Barley embryos and seedlings	6		

partment. However, it should be pointed out that in a chromatographic profile, each peak does not necessary correspond to a distinct tRNA species because there are several possibilities for artefacts which can lead to the appearance of an extra peak. Depending on the chromatographic procedure being used, these include changes in the tridimensional configuration of the tRNA molecule, aggregation (formation of tRNA dimers for instance), breaks in the polynucleotide chain or partial degradation of the 3′-terminal CCA sequence due to the action of nucleases, and incomplete posttranscriptional modification or loss of a rare base. The Y base, for instance, can be missing in a tRNA*Phe*. A trivial cause of error is presence of other labeled amino acids contaminating the amino acid studied and esterifying their cognate tRNAs during the aminoacylation reaction, thus causing the appearance of extra peaks when the aminoacyl-tRNAs are fractionated. In order to avoid fractionation of false species generated by nuclease nicking of tRNAs during aminoacylation, isoaccepting species can be visualized by DEAE cellulose or CM cellulose chromatography of the ^{14}C-aminoacyl-oligonucleotides resulting from a T_1 RNAse digestion of the ^{14}C-aminoacyl-tRNAs.[130,131] It should be pointed out, however, that two isoaccepting tRNA species can be visualized by this technique only if they differ in their nucleotide sequence between the 3′-end and the last guanylic residue (as T_1 RNase cleaves phosphodiester bonds when a guanylic residue is present on the 3′-side of the bond).

IV. STRUCTURE OF tRNAs

A. Primary Structure of tRNAs

Nucleotide sequence determination involves partial hydrolysis of the purified tRNA by pancreatic or T_1 RNase and separation of the resulting mono- and oligo-nucleotides either by column chromatography or by paper electrophoresis.[160] The nucleotides are identified by their UV absorption spectra, a procedure requiring large amounts of purified tRNA, but not requiring labeled tRNA sometimes difficult to obtain, especially from higher plants. If ^{32}P-labeled tRNA is available, separation and identification of the oligonucleotides can be achieved starting from much less tRNA.[161] Here, the main drawbacks are (1) the necessity of handling large amounts of radioactive material in the growth medium and during tRNA extraction and purification and (2) the impossibility of characterizing minor nucleotides by their UV spectra because of the small amount of material present.

Recently in vitro end-group labeling techniques have been developed. An unlabeled tRNA is used as starting material, hydrolysed, and the resulting mono- and oligonucleotides labeled to obtain radioactive fingerprints after two-dimensional electrophoresis. Much smaller amounts of starting material (nonradioactive tRNA) are needed to determine the sequence. These methods include:

1. ^{32}P-labeling of the 5'-end of the oligonucleotides (obtained after pancreatic or T$_1$ RNase digestion) using ^{32}P-labeled ATP and polynucleotide kinase.[162,163] The nucleotide sequence can be deduced from shifts in mobility observed upon successive removal of nucleotides from the 3'-end using snake venom phosphodiesterase.[162,164]
2. ^{32}P-labeling of the 3'-end of the oligonucleotides using polynucleotide phosphorylase.[165] The oligonucleotide sequence is deduced from shifts in mobility observed upon successive removal of nucleotides from the 5'-end using spleen phosphodiesterase.[166]
3. ^{3}H-labeling of the 3'-end of oligonucleotides after oxidation of the 2'-3' diol endgroup by periodate and reduction of the resulting 2'-3' dialdehyde end-group by [^{3}H] sodium borohydride into a ^{3}H-labeled dialcohol.[167,168]

The first primary structure of a tRNA, that of yeast tRNA,[41a] was published in 1965 by Holley et al.[169] Today, approximately 100 tRNAs have been sequenced. Most of these nucleotide sequences can be found in handbooks,[170,171] and more structures are being reported every year. The list includes tRNAs from a great variety of sources: mycoplasma, phages, bacteria, fungi, plants, animals (including mammals), and even man.)[172,382,409]

B. Secondary Structure of tRNAs

1. General Features

All tRNA sequences can be arranged to fit the cloverleaf model proposed in 1965 by Holley et al.[169] shown in Figure 4. The polynucleotide chain is folded upon itself forming double helical regions called stems held together by base pairs usually of the Watson-Crick type except for occasional G-U pairs and single stranded loops. A stem and its adjacent loop form what is called an arm. Five regions are usually distinguished in a tRNA structure:

1. The amino acid (or acceptor) stem containing the 5' phosphate and the 3' C-C-A$_{OH}$ end (which accepts the cognate amino acid) and consisting of seven base pairs and four unpaired nucleotides at the 3'-end of the molecule.
2. The D arm, or dihydrouridine (hU) arm consisting of the D stem (3 or 4 base pairs long) and the D loop or loop I (7 to 11 nucleotides long).
3. The anticodon arm consisting of the anticodon stem (five base pairs) and the anticodon loop, or loop II (seven nucleotides long, three of which will bind to the complementary codon).
4. The variable arm or extra arm (loop III) which can be short (four to five nucleotides) or long (13 to 21 nucleotides), thus accounting (together with the D arm) for the rather large variation in the length of tRNAs (73 to 93 nucleotides).
5. The TψC arm, consisting of the TψC stem (five base pairs) and of the TψC loop, or loop IV (seven nucleotides), as in the anticodon arm.

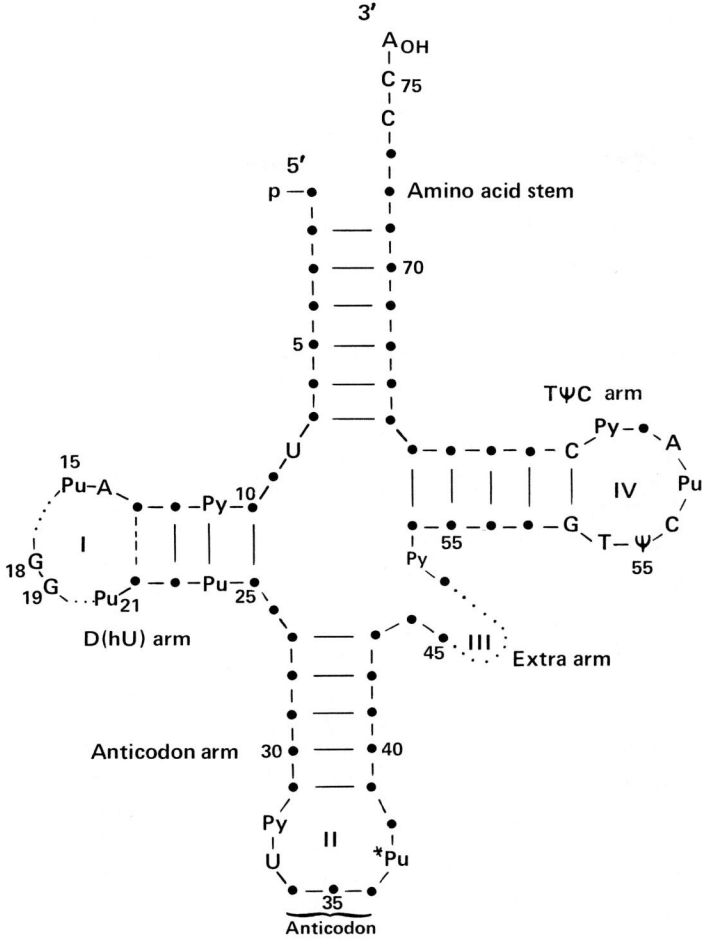

FIGURE 4. General diagram representing the cloverleaf form characteristic of most tRNAs (for exception, see Section IV.B.2 and showing the position of invariant and semi-invariant nucleotides (for symbols, see footnote on page 1). Dotted lines in the D-arm and the extra-arm correspond to regions of variable length. The numbering system is that of yeast tRNA^{Phe}.

There are always two bases between the amino acid stem and the D stem, and only one base between the D stem and the anticodon stem.

2. Invariant and Semi-invariant Nucleotides

In addition to the above-mentioned general features of the cloverleaf structure, most tRNAs have invariant (conserved) or semi-invariant (semi-conserved) nucleotides located at the same position in the molecule. However, there are exceptions.

The invariant nucleotides represented on Figure 4 by the corresponding symbol (see Footnote on the first page of this chapter) are characteristic of almost all tRNAs involved in protein biosynthesis. Starting from the 5'-end of the tRNA molecule (using subscript numbers to denote position), one finds 14 invariants: U_8, A_{14}, G_{18}, G_{19}, U_{33}, G_{53}, T_{54}, ψ_{55}, C_{56}, A_{58}, C_{61} and C_{74}, C_{75}, A_{76} at the amino acid acceptor (3') end. There are exceptions, for instance in the case of the staphylococcal tRNAs^{Gly} involved in cell wall (not in protein) synthesis, where G_{18} and G_{19} are replaced by two U, and ψ_{55} by

G. Exceptions exist even in the case of tRNAs active in protein biosynthesis. In some cases, U_8 is replaced by s^4U, G_{18} by Gm, T_{54} by U, ψ, or another pyrimidine derivative; and A_{58} can be replaced by m^1A. For other exceptions, see Reference 172.

Nine nucleotides are usually considered semi-invariant and are present in most tRNAs active in protein biosynthesis: Py_{11}, Pu_{15}, Pu_{21}, Pu_{24}, Py_{32}, $*Pu_{37}$, Py_{48}, Pu_{57}, and Py_{60}. Three more could be added to that list: Pu_9, Pu_{10} (usually a G or $a*G$), and Pu_{26}. The term "correlated invariants," used for Py_{11} and Pu_{24} (base paired in the D stem), refers to the fact that in tRNAs in which Py_{11} is U, Pu_{24} is A, and in tRNAs in which Py_{11} is C, Pu_{24} is G.

3. Special Features of Initiator tRNAs

Although prokaryotic and eukaryotic initiator tRNAs have a cloverleaf structure and most of the mentioned invariant and semi-invariant nucleotides, they possess some unusual structural features.

Prokaryotic initiator tRNAs have no Watson-Crick base pair between the first nucleotide at the 5′-end and the fifth nucleotide from the 3′-end (in the acceptor stem.)[173,176] Furthermore, instead of the Py_{11}-Pu_{24} correlated invariant base pair (see above), they have a A_{11}-U_{24} base pair in their D stem.

Eukaryotic cytoplasmic initiator tRNAs have no T_{54} ψ_{55} sequence (in the TψC loop), but have an AU (or *U) sequence instead. In all those examined so far, the sequence $AUCGm^1AAA$ (where U can be *U) has been found.[177-180]

4. Plant tRNAs

Very few plant tRNA structures have been determined so far. Four plant cytoplasmic tRNAs^{Phe} from wheat germ,[87] pea,[151] lupin,[154] and barley[424] have been studied. Their sequences were found to be identical (as shown in Figure 5) except for a G-C base pair present in wheat germ, barley, and pea tRNAs^{Phe} which is replaced by an A-U base pair in 80% of the lupin tRNA^{Phe} molecules.

Another plant cytoplasmic tRNA which has been sequenced is wheat-germ tRNA$_1^{Gly}$, which has some unusual features:[158] T is absent (and replaced by U), there is a 2′-O-methylcytidine in the acceptor stem (which is usually devoid of methylated nucleotides) and the D stem has only one Watson-Crick base-pair (A-U) in addition to three weaker basepairs (G-U, A-C, and G-ψ).

The initiator methionine tRNA from wheat embryo cytoplasm has been sequenced.[180] As in other eukatyotic cytoplasmic initiator tRNAs, the trinucleotide sequence TΨC found in most tRNAs (in the TΨ loop, also called loop IV) is absent and replaced by AΨC ; but unlike other eukaryotic (animal) cytoplasmic initiator tRNAs^{Met}, wheat embryo initiator tRNA cannot be formylated by the *E. Coli* transformylase. The initiator methionine tRNA from *Scenedesmus obliquus* (a green alga) has also been sequenced;[404] it has 85% homology with wheat cytoplasmic initiator tRNA, but can be formylated by *E. Coli* transformylase.

Up to 1976, no information was available on the structural features of organelle tRNAs, because of the great difficulties encountered in obtaining sufficient amounts of purified tRNAs from organelles. Recently, nucleotide composition analyses were published for *Euglena* Chloroplast tRNA^{Phe},[181] and for *Phaseolus* chloroplast tRNA^{Phe}.[156] Bean chloroplastic tRNA^{Phe} was found to resemble *E. coli* tRNA^{Phe} in that they have the same two hypermodified nucleotides,[156] namely ms^2i^6A and acp^3U, and that they both lack m^2G, m^1A, and Y which are usually present in cytoplasmic (including plant) tRNAs^{Phe} (the structures of the hypermodified nucleotides ms^2i^6A, acp^3U, and Y are shown on page 108 in Reference 182). This resemblance between *E. Coli* and chloroplastic tRNAs^{Phe} was confirmed when the first chloroplastic tRNA structure,

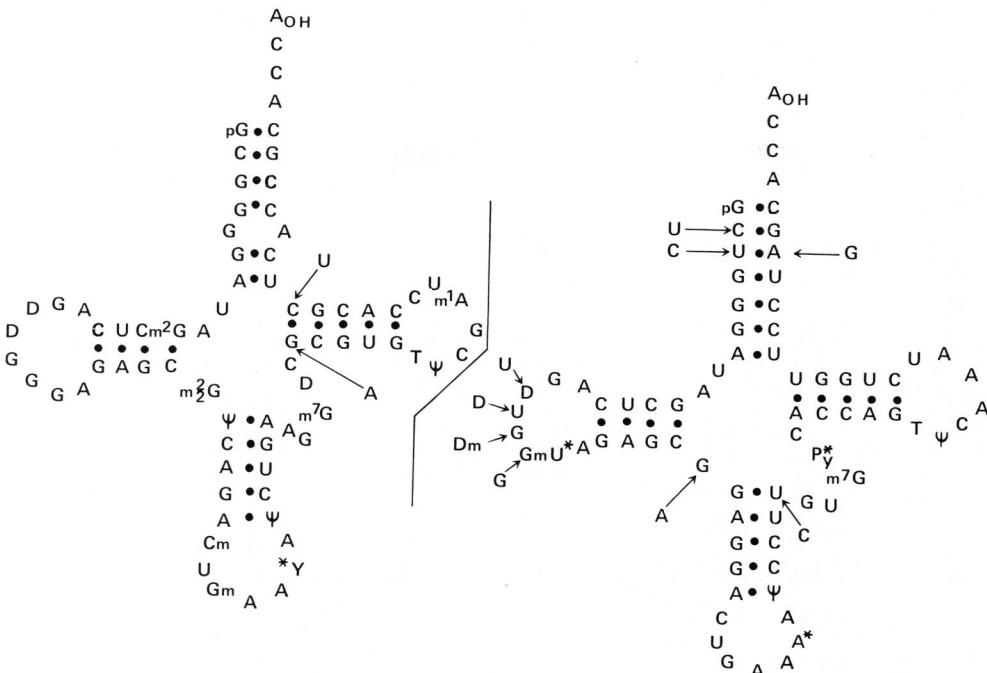

FIGURE 5. Structure of plant tRNAs^{Phe}. Left: Structure of plant cytoplasmic tRNAs^{Phe} from wheat germ,[87] pea,[151] lupin,[154] and barley[424] which are identical except for an A-U base-pair replacing a G-C base-pair in 80% of the lupin tRNA^{Phe} molecules (as shown by the arrows). Right structure of plant chloroplastic tRNAs^{Phe} from *Euglena*[150] and bean,[155] which are similar (the cloverleaf structure shown is that of *Euglena* chloroplast tRNA^{Phe}, while the nucleotides which are different in bean chloroplast tRNA^{Phe} are shown with an arrow).

that of *Euglena* chloroplast tRNA^{Phe} was determined by Chang et al.[150] These authors compared the eight positions in the tRNA^{Phe} molecule where invariant nucleotides in prokaryotes differ from the invariants in eukaryotes, and found that at five of these eight positions (*U$_{20}$, G$_{44}$, U$_{45}$, U$_{60}$, and C$_{68}$) *Euglena* chloroplastic tRNA^{Phe} has the same nucleotide as that present in prokaryotic tRNAs^{Phe}, while only at three positions (G$_4$, U$_{17}$, and *G$_{26}$) are the nucleotides similar to those found in eukaryotic cytoplasmic tRNAs^{Phe}. Bean chloroplastic tRNA^{Phe} differs slightly from that of *Euglena*:[155] only five nucleotides are different in addition to differences in post-transcriptional modifications, as shown in Figure 5. A blue-green alga tRNA^{Phe} has recently been sequenced and shown to have 87% homology with bean chloroplast tRNA^{Phe}.[406]

On the other hand, *Euglena* cytoplasmic tRNA^{Phe} has been sequenced and shown to resemble mammalian cytoplasmic tRNAs^{Phe} rather than plant cytoplasmic tRNA^{Phe}.[406]

C. Three-dimensional Conformation of tRNAs

X-ray diffraction is a very powerful method to study the three-dimensional structure of macromolecules, and it has been successfully used to study crystalline proteins. Its application to tRNAs could only be considered after 1968, when several groups[183-187] obtained crystals that were large enough to be used in X-ray diffraction. In 1971, crystals of yeast tRNA^{Phe}, yielding an X-ray diffraction pattern with a resolution of 2.3 Å, were obtained.[188] This allowed extensive studies which have recently been reviewed[18,19,399] and which led to the model shown in Figure 6.

The yeast tRNA^{Phe} molecule has an L shape: one limb is formed by the amino acid (or acceptor) stem and the TΨC stem, the other by the D stem and the anticodon stem. These stem regions are essentially normal, double helical RNA. In the loop, there are two types of tertiary bonding interactions. The first is base-base interactions involving

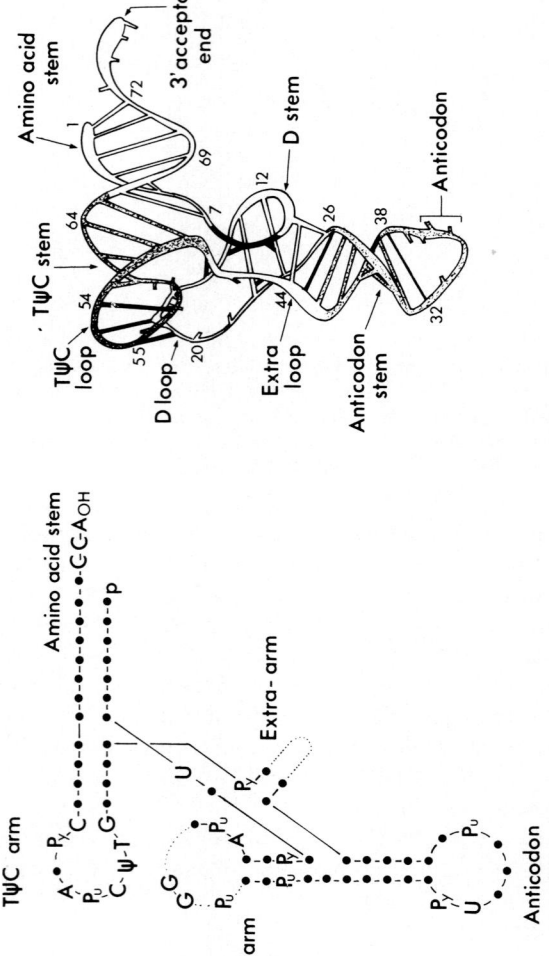

FIGURE 6.   Three-dimensional structure of yeast tRNA^Phe. Left: a diagram showing the L-form, using the same symbols as in Figure 4. (Reproduced from Kim, S. H., *Prog. Nucleic Acid Res. Mol. Biol.*, 17, 181, 1976, Academic Press, New York. With permission.) Right: a diagam where the ribose-phosphate backbone is depicted as a coiled tube, and the numbers refer to nucleotide residues in the sequence. Hydrogen-bonding interactions between bases are shown as cross-rungs; tertiary interactions between bases are shown as solid blackrungs which indicate either one, two, or three hydrogen bonds between them. Those bases that are not involved in hydrogen bonding to other bases are shown as shortened rods attached to the coiled backbone. (Reproduced, with permission, from Rich, A., and Rajbhandary, U. L., Transfer RNA: molecular structure, sequence and properties, *Annu. Rev. Biochem.*, 45, 805, 1976. Copyright © 1976 by Annual Reviews Inc., All rights reserved.)

one, two, or three hydrogen bonds. There are ten base-base interactions involving (only one of them is of the Watson-Crick type) forming a network which maintains almost all thhe bases of this tRNA in two stacking regions (the two limbs of the L-shaped molecule). The second is base-backbone interactions formed between an N atom of the base and an O atom of either a ribrose or a phosphate. There are nine base-backbone interactions which apparently stabalize the sharp bends in the polynucleotides present in the extra-loop (loop III) and in the D loop (loop I).

It should be pointed out that this three-dimentional conformation appeats to agree with the requirements for the biological functions of tRNA in protein biosynthesis. The 3′ CCA$_{OH}$-end remains single-stranded, easily accessible, and flexible, which may be important for the aminoacylation of tRNA and for the amino acid transfer involved in peptide bond formation. The anticodon loop has a confotmation allowing hydrogen bonds to be formed between the bases of the anticodon and those of the corresponding codon in the mRNA. The interactions between the D loop and the TΨC loop are not too extensive, and this may explain the opening up which seems to occur when the anticodon combines with the codon, resulting in the binding of the TΨC sequence to the complementary GAA sequence present in the 5S ribosomal RNA, as suggested by several authors.[191,192]

In spite of all information obtained during the last years on the primary, secondary, and three-dimentional structures of tRNAs, the relationships between structural features and biological functions remain obscure in many instances. One of the most intriguing mysteries is that of the role of modified nucleotides present in different amounts depending on the tRNA considered, and so varied in their structures (over 50 different minor nucleotides are presently known.)[171,200] In one well-documented case, that of *Salmonella* tRNA^{His}, two Ψ residues seem necessary for this tRNA to participate in repression of the histidine operon[193,194] (See Section VII. B. 3).

It is still not clear whether minor nucleotides are important for tRNAs to function in protein biosynthesis, although it has been suggested that they may be involved in specificity and rate of tRNA aminoacylation by the cognate aminoacyl-tRNA synthetase,[195-197,384] or in codon recognition.[189,198,199]

It has recently been observed that in vitro methylation, by an *E. Coli* uridine methylase, of a mammalian liver tRNA^{Phe} which had only a partial U → T conversion, results in an increased V_{max} for polyphenylalanine synthesis, thus suggesting that T might be involved in the regulation of protein synthesis at the level of translation.[385] The significance of variations in the number of nucleotides in the D loop, and especially in the extra loop (which can have up to 21 nucleotides and account for almost 1/4 of the whole tRNA molecule) is also unclear. It is not known whether these structural features are related to the functions of tRNAs in protein biosynthesis.

V. tRNA CISTRONS IN NUCLEAR AND ORGANELLAR GENOMES

The characterization of tRNA species located in only one cellular compartment, for instance the chloroplast (see Table 1) and specifically aminoacylated by the enzyme found in the same compartment (see Section VII,A,1,b) raises the problem of their origin. Are cytoplasmic tRNAs coded for by nuclear genes and organellar tRNAs by the DNA of the corresponding organelle?

This problem can be approached by DNA-tRNA hybridization techniques. For this purpose, radioactive tRNA can be obtained by in vivo labeling with ^{32}P, although this is easier with unicellular algae such as *Euglena*[201] than with higher plants. Alternatively, tRNA may be labeled in vitro with radioactive iodine (^{125}I)[202] This is a procedure which has been used for *Euglena* tRNAs,[203,204] but can be applied to any tRNA after extraction. Hybridization is usually performed essentially as described by Gillespie and Spiegelman[205] with denatured DNA immobilized on nitrocellulose filters. Sometimes high temperatures (57°C[203] or 70°C[204]) for 18 to 20 hr are used, but much lower temperatures can be used if the hybridization is performed in the presence of 30 to 50% formamide or of 8 *M* urea.[206,207] If the hybridization of an individual tRNA species is to be studied, there are two possibilities: (1) when the tRNA species can be fractionated and completely purified the above mentioned techniques (either [^{32}P] tRNA labeled in vivo or [^{125}I] tRNA labeled in vitro) can be used; (2) when the tRNA species is not completely pure, but has been separated from the other isoaccepting species, it can be specifically labeled by aminoacylation with the cognate (radioactive) amino acid. However, in the latter case, to prevent deacylation (and loss of the label), hybridization of the ^{3}H- or ^{14}C-aminoacyl-tRNA to DNA must be performed at low temperature in the presence of formamide or urea.[208]

Using both green and chloroplast-free strains of *Euglena*, with a flotation method to purify the chloroplasts and magnesium-cesium sulfate density gradients to fractionate large amounts of DNA, Gruol and Haselkorn[201] found that a *Euglena* cell contains 800 nuclear genes for each cytoplasmic rRNA and approximately 740 nuclear genes for sRNA (tRNAs). These figures differ from those obtained in other organisms *Escherichia coli*,[209] *Bacillus subtilis*,[209,210] *Saccharomyces cerevisiae*,[211] *Drosophila melanogaster*,[212] HeLa cells,[213] and *Xenopus laevis*,[214] where the tRNA genes to rRNA genes ratio is higher than two. If these 740 *Euglena* nuclear genes code for 50 different tRNA species (a reasonable assumption) the average reiteration frequency is only about 15, a very low figure compared with an average reiteration frequency of 200 in *X. laevis* which has about 7800 tRNA genes.[214] Gruol and Haselkorn[201] found a large number of chloroplast genes for 4S RNA in the *Euglena* cell: approximately 10,000, a figure which corresponds (taking into account that there are 400 chloroplast chromosomes per *Euglena* cell) to 25 4S RNA cistrons on each chloroplast chromosome. In these studies, they found 2 genes for each rRNA on each chloroplast chromosome of 92,000,000 mol wt.

The genes coding for *Euglena* tRNAs were also investigated by McCrea and Hershberger,[203] who showed that total cell tRNA hybridized to an equivalent of 25 cistrons on chloroplast DNA. They separated total cell tRNA into two fractions by chromatography on DBAE cellulose (see Section II. B). Fraction I hybridized to both nuclear and chloroplastic DNAs, and hybridization to chloroplast DNA indicated 18 cistrons. Fraction II hybridized only to chloroplast DNA, indicating 7 cistrons. The tRNA obtained from isolated chloroplasts hybridized to both chloroplastic and nuclear DNAs. The presence of nuclear coded tRNAs in the chloroplasts is also suggested in studies performed by other authors — for instance in comparisons of chromatographic profiles of tRNAs from cytoplasm and from chloroplasts.[108-109,111] The possibility that they represent a contamination of chloroplast tRNAs with cytoplasmic components is not yet excluded, although McCrea and Hershberger[203] note that their chloroplasts do not contain detectable levels of contamination by cytoplasmic rRNA. In a more recent article, these authors have reported that chloroplast DNA codes for tRNAs found on cytoplasmic polyribosomes.[412] That tRNAs transcribed from nuclear DNA can be transported into the organelles has been suggested in the case of *Tetrahymena* mitochondria.[335]

Martin et al. after comparing the two-dimensional polyacrylamide gel electrophoresis patterns of tRNAs obtained from clean yeast mitochondria and of yeast mitochondrial tRNAs purified by hybridization to mitochondrial DNA, have come to the conclusion that all mitochondrial tRNAs (except one tRNA^{LYS}) are coded for by mitochondrial DNA.[413] However, Schwartzbach et al.[204] (who also found that *Eu-*

glena chloroplast genome contains approximately 25 tRNA cistrons) observed no hybridization of chloroplast tRNA with nuclear DNA. They found that neither *E. coli* tRNA, nor tRNA from a blue-green alga *(Agmenellum quadraduplicatum)* was able to compete with ^{125}I-labeled *Euglena* chloroplast tRNA in the hybridization to *Euglena* chloroplast DNA. Hybridizing purified chloroplast tRNA[Phe] and tRNA[Asp], they found only one cistron for each of these two tRNA species on the chloroplast genome. Furthermore, these two tRNAs hybridize with DNA fractions of different buoyant densities, suggesting that the two cistrons are not closely linked.

In addition to studies performed on *Euglena* by these three groups, the work of Tewari et al.,[215,216] performed on tea, indicates that the chloroplast genome contains about 30 to 40 tRNA cistrons, apparently enough to code for a full complement of tRNAs. Haff and Bogorad[217] have hybridized ^{125}I-tRNA to maize chloroplast DNA and have also concluded that there are 20 to 26 tRNA cistrons on chloroplast DNA. Furthermore, studying the hybridization of tRNA aminoacylated with all 20 amino acids, they found that tRNAs charging at least 16 different amino acids can hybridize to chloroplast DNA. This also suggests that a complete (or nearly complete*) set of tRNAs can be transcribed from chloroplast DNA.

In contrast to observations of Williams et al.[218] reporting that all seven isoaccepting tRNAs[Leu] found in bean leaves hybridize with both chloroplastic and nuclear DNAs, Steinmetz and Weil[208] have shown that only chloroplast-specific tRNAs[Leu] and tRNAs[Phe] hybridize with chloroplast DNA. Cytoplasmic tRNAs[Leu] and tRNAs[Phe] do not. Three isoaccepting leucyl-tRNAs and two isoaccepting phenylalanyl-tRNAs were found in bean chloroplasts (see Table 1) by fractionation with RPC-5[108] or BD cellulose[156] chromatography, respectively. If there are approximately 25 tRNA cistrons on chloroplast DNA, this accounts for one chloroplastic tRNA per amino acid, but not for two or three. To determine whether the three chloroplast specific isoaccepting tRNAs[Leu] are coded for by the same or by different gene(s), Steinmetz and Weil[208] looked for additivity and competition in the hybridization of these isoacceptors to chloroplast DNA. All three combinations of two chloroplast-specific [3H] leucyl-tRNAs showed no additivity. Control experiments, in which one chloroplastic [3H] leucyl-RNA and one chloroplastic [3H] phenylalanyl-tRNA were mixed, showed the expected additivity. In the competition experiments, the hybridization of each of the three chloroplastic [3H] leucyl-tRNAs was similarly decreased upon addition of the same (uncharged) isoacceptor or of any of the other two chloroplastic tRNA[Leu] isoacceptors, but not upon addition of a chloroplastic tRNA[Phe]. These results suggest that the chloroplast isoacceptors are coded for either by the same gene(s), thus differing only in the extent of posttranscriptional modification, or by very similar genes which may have evolved from one gene by undergoing different mutations during evolution. Preliminary results of structural studies performed on the three bean chloroplast tRNAs [Leu] are in favor of the second hypothesis, as these three chloroplast isoacceptors have been found to differ in their nucleotide sequences.[416]

Recent reports have described the hybridization of 4S RNA to *Euglena*[417] and *Chlamydomonas*[418] chloroplast DNA fragments (obtained after action of restriction endonucleases) and have shown that a tRNA gene is located in the spacer region between the 16S and the 23S ribosomal RNA genes. Similar studies[419,428] have been performed with spinach chloroplast DNA fragments, but using purified and identified individual tRNAs fractionated by two-dimensional gel electrophoresis: 21 RNA genes have been localized on the circular map of spinach chloroplast DNA.[428] A tRNA[Ile] gene has been

---

* In fact, to read unambiguously all 61 sense condons, at least 31 different tRNAs would be necessary according to the wobble hypothesis of Crick.[372]

situation in yeast[219,220,398] and HeLa cell[221] mitochondria, nothing is presently known about the location of tRNA genes in plant mitochondria.

In bacteria, tRNA gene amplification has been detected. In the case of *E. coli* tRNAs^{Gly}, one gene copy was found for each of the first two isoacceptors, and three copies for tRNA$_3$^{Gly}.[222] Gene clustering has been revealed using genetic and physicochemical methods. A tRNA gene triplet, which can be carried by a transducing phage,[223] has been mapped by coupling tRNA to the electron opaque compound ferritin, allowing tRNA-DNA hybrids to be vizualised by electron microscopy. It was found to have the following sequence:[224] tRNA$_3$^{Thr} — 260 nucleotides spacer — tRNA$_2$^{Tyr} — 140 nucleotides spacer — tRNA$_2$^{Gly}. This technique will perhaps help localize tRNA cistrons on plant nuclear and organellar DNAS.

It is not known presently whether plant tRNA genes contain intervening sequences (or introns) such as those found in yeast tRNA genes for instance.[407,408]

VI. BIOSYNTHESIS OF tRNAs

The biosynthesis of tRNAs involves two main steps. The first is transcription of the corresponding genes into tRNA precursors by a DNA-dependent RNA polymerase. The second is maturation, during which trimming of the precursors (by action of nucleases) and posttranscriptional modifications (by action of specific enzymes) occur. These processes have been studied in bacteria, in phage-infected bacteria, and in animal cells. Except for a few studies on modifying enzymes, there are as yet no such studies in plants; however, the basic mechanisms are probably similar. Therefore, research on tRNA biosynthesis in plants should be inspired by what is known in other organisms.[318] For a review, see Reference 381.

A. Transcription

Transcription of bacterial[225] and mammalian cell[226] tRNA genes in vivo yields precursor molecules that are longer than mature tRNAs. In the case of *E. coli* su $_3$$^+$ tRNA^{Tyr} the isolated precursor has 41 additional nucleotides terminating in pppGp at the 5'-end, suggesting that it is probably the primary transcript and not a cleavage product. There are three more nucleotides at the 3'-end beyond the CCA sequence (which is present and therefore appears, at least in this case, to be coded for by the tRNA gene and not added after transcription).[227] Of the seven modified nucleotides present in the mature tRNA, i^{6}A (next to the anticodon) and T and ψ (in the TψC loop) were found in the precursor.[228]

Eight different tRNAs are coded for by phage T$_4$, and their genes are clustered on the DNA of this phage.[229,230] On infection of *E. coli* cells, in addition to mature tRNAs, three long precursors accumulate. One of them is 175 nucleotides long and has the following (5' → 3') sequence: nine nucleotides — tRNA^{Pro} — CU — tRNA^{Ser} — UAA$_{OH}$. In contrast to what was observed in the case of *E. coli* su $_3$$^+$ tRNA^{Tyr} monomeric precursor, this phage-coded dimeric precursor lacks both of the 3'-terminal CCA sequences. These are added, by a tRNA nucleotidyltransferase,[231] after exonucleolytic removal of U in the case of tRNA^{Pro} and of UAA in the case of tRNA^{Ser}. All modified nucleotides are present in this dimeric precursor (except for a 2′ *0*-methyl G in tRNA^{Ser}), and there is no 5' triphosphate group indicating that this precursor is probably not the primary transcript.[232] Recent studies on *E. coli* cells deficient in tRNA nucleotidyl transferase suggest that this enzyme is not required for the biosynthesis of tRNAs.[401]

In vitro transcription of tRNA cistrons by purified DNA-dependant RNA polymerase has been attempted to obtain unmodified tRNA precursors. The following approaches have been used to obtain tRNA genes (tDNA) to serve as templates in transcription experiments:

1. Isolation of tRNA-DNA hybrids trimmed with a single-strand specific DNase.[233,234] If a mature tRNA is used in the hybridization, the promoter and terminator regions will not be present in the isolated DNA fragment. Hence, a tRNA precursor should be used in the hybridization.
2. Isolation of DNA fragments containing tRNA genes, by hybridization of tRNA to DNA fragments obtained by shearing.[235]
3. Isolation of tRNA cistrons by differential melting, when these cistrons have a higher G + C content than the rest of the genome.[236]
4. Hybridization of the heavy strands of two specialized transducing phages carrying the tRNA^{Tyr} gene of *E. coli* inserted in their DNA (in opposite orientation), followed by enzymatic trimming of the single-stranded regions.[237]
5. Chemical-enzymatic synthesis of a tRNA gene, that corresponding to yeast tRNA^{Ala},[238, 239] or of a tRNA precursor gene (including the promoter and terminator regions), that corresponding to *E. coli* suppressor tRNA^{Tyr}.[240]

A major example of success with in vitro transcription is that of *E. coli* su $_3^+$ tRNA^{Tyr},[241,242] the transcript was found to be active in suppression of an amber mutation.[243] Other examples of in vitro transcription are eight tRNAs coded by phage T$_4$ DNA,[244] and two *E. coli* tRNA gene clusters.[245] In vitro transcription by purified *E. coli* RNA polymerase yields high molecular weight tRNA precursors[389,390] which can be processed by *E. coli* extracts to mature size tRNA molecules.[245] The ability of the in vitro synthesized tRNAs to undergo nucleotide modifications and aminoacylations has been demonstrated.[391]

B. Maturation

The conversion of tRNA precursors into mature tRNAs has been followed by kinetic studies in various organisms including animal cells.[246,251,393] This maturation involves trimming of the precursor by nucleases and enzymatic modifications of some nucleotides to form the so-called "minor" ("rare" or "modified") nucleotides. It is not established in which order exactly these two processes take place (e.g., whether any modification is required for proper trimming or vice-versa). However, as the number and nature of modifications differ from one tRNA to the other, it is possible that maturation processes vary depending on the organism and the tRNA considered.

1. Trimming of the Precursors

The 41 extra nucleotides present at the 5'-end of *E. coli* su $_3^+$ tRNA precursor (see above) are removed as a single fragment (exposing the 5'-terminus of the mature tRNA) by a specific endonuclease called RNase P. This RNase was detected in crude *E. coli* extracts and purified,[252] but similar enzymatic activities have been found in other organisms. Among *E. coli* mutants defective in tRNA biosynthesis, some have a temperature-sensitive ability to process tRNA precursors, as a result of a temperature-sensitive RNaseP.[253] In such mutants, processing of most, if not all, tRNA precursors is defective, and different types of precursors accumulate. In addition to RNase P, at least two more enzymes participate in the processing of tRNA precursors.[254] One enzyme removes the extra nucleotides from the 3'-end, and the other (RNase P$_2$) endonucleolytically cleaves tRNA precursors at a site different from that attacked by RNAse P. For instance, in polycistronic precursors, RNase P$_2$ cleaves in the spacer sequences separating the different tRNAs in the precursor.

2. Nucleotide Modification

Over fifty modified nucleotides have so far been characterized in tRNAs. Several reviews have been devoted to their structure, formation, and possible function.[12,200,255-257,392,397] They all seem to result from enzymatic modifications of the four major nucleotides after synthesis of the polynucleotide chain, and are hence termed "posttranscriptional modifications." The modifying enzymes and the reactions they catalyze are difficult to study because of problems encountered in obtaining the proper substrates. Even the tRNA precursors that have been isolated are often already modified, as in the case of the T_4-coded dimeric precursor (see above). In some instances, however, it has been possible to isolate appropriate substrates from mutant strains having altered modifying enzymes.[199,258]

a. Methylation

One of the simplest modifications is methylation, a reaction catalyzed by various tRNA methyltransferases (or tRNA methylases). Extensive studies[259] of these enzymes have shown them to be species-specific and generally to use S-adenosyl-methionine as a donor of methyl groups.[374] However, it has recently been shown that in some bacteria ribothymidine is formed by a folate-dependent methylation.[375,376] Methylation can occur either on the 2'OH of a ribose residue or on a purine or pyrimidine base, and each tRNA methylase has a specificity with regard to: (1) the nucleotide (A, G, U, or C); (2) the position on the base (one enzyme will, for example, methylate adenine only on carbon 1); (3) the position of the nucleotide in the molecule (out of 15 or 20 A residues, one enzyme will only methylate one of them).

Many studies on the role of methylated nucleotides in tRNAs have been performed using methyl-deficient tRNAs synthesized by relaxed methionine requiring mutants of *E. coli* when starved for methionine. The tRNAs isolated are, in fact, mixtures of methylated and methyl deficient tRNAs. No significant difference was observed in the amino acid accepting capacity of tRNAs. Even after separation of some methyl deficient species from their normal counterparts, little or no difference was found in the rate and extent of aminoacylation.[12] Some differences were noted in the codon recognition pattern,[12] but as several methylated nucleotides are missing in each methyl deficient tRNA molecule, it is impossible to ascribe a specific role to any single methylated nucleotide. A more promising approach is the use of mutants lacking a specific tRNA methylase, such as uracil tRNA methylase deficient *E. coli* mutants which produce ribothymidine-free tRNAs.[190,199] An *E. coli* mutant producing m^7G-deficient tRNA[190] and a yeast mutant producing $m_2^2 G$-free tRNA[290] have also been isolated and characterized.

Some tRNA methylases from plants have been studied.[260-265] Yellow lupin seed tRNA methylases have been purified;[266] they do not methylate homologous tRNA, but methylate *E. coli* tRNA and tRNA from two other lupin species, and m^1A, m^5C, and T are formed. In two tRNAs^{Gly} (including tRNA $_1{}^{Gly}$ which has been sequenced[158]) and three tRNAs^{Thr} from wheat embryo, T_{54} (in the TψC loop) is absent, but can be formed upon action of *E. coli* uracil tRNA methylase.[267]

In plant chloroplasts and mitochondria, specific tRNA methylases (not present in the plant cytoplasm) have been characterized using pure heterologous tRNA species as substrates.[268,269] Because the sequence of these pure tRNA species (from yeast) was known, it was possible to determine the nature and position of the nucleotides methylated by the organellar enzymes. In *Phaseolus vulgaris* chloroplasts and mitochondria there is an m^1A-tRNA methylase which modifies A_7 in yeast tRNA^{Asp} [268] and an m^7G-tRNA methylase which modifies G_{46} in yeast tRNA^{Ala}.[269] It is not known whether these organelle-specific tRNA methylases function in the maturation of tRNAs coded for by organellar or by nuclear DNA.

Changes in the levels of tRNA methylases have been observed in animals under various physiological conditions — for instance after hormone administration or upon tumor formation (see Reference[270,272] for a review and the whole issue of cancer research devoted to these problems.)[271] In plants, methylation of purine residues has been shown to be enhanced after gibberellic acid treatment.[272] As far as plant tumors are concerned, an m^7G tRNA methylase, absent in normal *Parthenocissus tricuspidata* cells, has been characterized in crown-gall cells.[273] The methylation of tRNAs has been recently reviewed,[274] and a mechanism of tRNA methylase-tRNA recognition has been proposed.[275]

b. Other Modifications

Other tRNA modifying enzymes also act on the polynucleotide chain, resulting in ψ formation,[276] thiolation,[277,278] esterification of 5-carboxymethyluridine,[279] isopentenylation,[280,281] formation of threonylcarbamoyladenosine,[282,283] and some other "exotic" modifications.

The formation of ψ, which differs from U in that the C_1 of the ribose is linked to the C_5 (instead of the N_1) of uracil, has been particularly well studied in *Salmonella typhimurium* in which two pseudouridine synthetases have been characterized. Pseudouridine synthetase I is coded by the His T gene.[284] Using a His T mutant, where repression of the histidine operon is altered, sequence analyses of $tRNA^{His}$ [193] (and of $tRNA^{Leu}$ [285]) showed that two ψ, present in the anticodon region of the wild type tRNAs, were absent in the mutant. Unmodified U was present in these two positions. Pseudouridine synthetase I is responsible for the formation of these two ψ in the anticodon region while pseudouridine synthetase II is involved in the formation of another ψ, that present (even in the His T mutant tRNAs) next to T in the $T\psi C$ loop.

The formation of the hypermodified nucleosides isopentenyladenosine (i^6A) and 2-methylthio-isopentenyladenosine ($ms^2 i^6A$) is of special interest because of their possible involvement in codon recognition,[198] and their cytokinin activity. Using *Mycoplasma* sp. Kid tRNA (which lacks i^6A) or undermodified *E. coli* $tRNA^{Tyr}$ as substrates, an enzyme was purified from *E. coli* which transfers the isopentenyl group from isopentenyl pyrophosphate to the proper adenosine residue (next to the anticodon, on the 3' side);[281] i^6A can be further modified into s^2i^6A and finally into ms^2i^6A, a nucleoside found in *E. coli* tRNAs, and also in *Phaseolus vulgaris* chloroplast $tRNA.^{Phe}$ [155,156] In fact, both i^6A and ms^2i^6A have been found in tRNAs from a large variety of organisms, including bacteria, fungi, higher animals, and plants. In yeast, they are located next to the 3'-end of the anticodon in tRNA species which can translate codons starting with U[286] (especially in tRNAs corresponding to the aminoacids, Ser, Tyr, and Cys). It is generally considered that these hypermodified nucleosides increase the specificity of the codon-anticodon recognition and facilitate the binding of tRNA to the ribosome-mRNA complex. In plant tRNAs, cytokinins also appear to be restricted to some species which recognize codons beginning with U.[431,432]

Cytokinins are plant hormones which exert regulatory functions on plant growth and development. They are essentially N^6-substituted adenosine derivatives and among the naturally occurring derivatives, the most active are 6-isopentenyladenosine derivatives such as i^6A, ms^2i^6A and ribosyl *trans-* (or *cis*) zeatin, or the corresponding free bases, respectively i^6Ade, ms^2i^6Ade and *trans-* (or *cis*) zeatin. The structure of these compounds can be seen on Table 2, but a more complete list of cytokinins and their structures can be found in a review by Skoog.[287]

Free cytokinins (ribonucleotides, ribonucleosides, and bases) are certainly released as tRNAs are degraded. Although it has been suggested that this is how the hormones originate in the cell,[288] some authors think that there is an independent pathway for

TABLE 2.

Structure of the Main Cytokinins

Compound name	Abbreviation	R_1	R_2	R_3
N^6-Δ^2-isopentenyladenosine	i^6 A or 2iPA		H	Ribose
N^6-Δ^2-isopentenyladenine	i^6 Ade or 2iP		H	H
2 methylthio N^6-Δ^2-isopentenyl-adenosine	ms^{2}i^{6}A or ms^{2}iPA		H_3CS	Ribose
2 methylthio N^6-Δ^2-isopentenyl-adenine	ms^{2}i^6 Ade or ms^2 iP		H_3CS	H
Ribosyl (trans) zeatin	t-io^6 A		H	Ribose
(trans) zeatin	t-io^6 Ade		H	H

the biosynthesis of cytokinins. The mode of action of cytokinins, in relation to nucleic acid and protein metabolism, is still a matter of discussion (for a review, see Reference 287). Recently, a specific binding site for cytokinins has been found on higher plant ribosomes.[289]

VII. FUNCTIONS OF tRNAs

A. Role of tRNAs in Protein Biosynthesis
1. Aminoacylation of tRNA
a. Mechanism

According to the generally accepted scheme, the aminoacylation of a tRNA occurs in two steps, as illustrated Figure 1. Amino acid activation results in the formation of an enzyme-bound aminoacyl-adenylate complex [E(AA-AMP)] which can be isolated by gel filtration, nitrocellulose membrane filtration, or electrophoresis.[291] However, some aminoacyl-tRNA synthetases are unable to form this complex in the absence of tRNA. This is the case for the arginyl-tRNA synthetases from *E. coli*,[292,293] *B. stearothermophilus*[294] and *Phaseolus vulgaris*,[41] and the glutamyl- and glutaminyl-tRNA synthetases of *E. coli*,[295-297] rat liver,[298] and *P. aureus*[299] (mungbean). These observations represent one of the arguments against the two-step mechanism, and Loftfield has proposed an alternative mechanism in which all the substrates (amino acid, ATP, and tRNA) are bound to the enzyme and react to form aminoacyl-tRNA, pyrophosphate, and AMP in a single concerted reaction having no discrete intermediates.[11,300]

Amino acid activation can be conveniently assayed by following either the amino acid-dependent ATP-PP exchange reaction and measuring the incorporation of ^{32}P from ^{32}PP into ATP, or the formation of amino acid hydroxamate (upon addition of hydroxylamine) by a colorimetric technique or by a sensitive radioactive method using a labeled amino acid.[11]

Attachment (or transfer) of the amino acid to the corresponding tRNA results in the formation of an ester bond between the carboxyl group of the amino acid and the 2' or 3' hydroxyl group[301,302,383] of the ribose moiety of the terminal adenosine residue of the cognate tRNA. The formation of an aminoacyl-tRNA can be conveniently assayed by measuring the amount of acid-insoluble radioactive material after incubation of the tRNA in the presence of the corresponding aminoacyl-tRNA synthetase, the corresponding (^{3}H or ^{14}C) amino acid and ATP, under optimal conditions of temperature and pH and with optimal monovalent ion concentrations. These optimal conditions vary depending upon the system considered and upon the amino acid studied. The rate of aminoacyl-tRNA synthesis may appear lower in vitro than in vivo (in bean leaves), mainly because of the dilution of the labeled amino acid by amino acid present in the enzyme and/or tRNA preparation.[303]

In fact, the aminoacylation of tRNAs is a rather complex process involving three substrates (amino acid, ATP, tRNA) and yielding three products (aminoacyl-tRNA, PP, AMP). Many studies have been performed, mainly with *E. coli* and yeast tRNA and enzymes to derive a detailed reaction mechanism. Those interested in these problems should consult two recent reviews devoted to the aminoacyl-tRNA synthetases and to the reactions catalyzed by these enzymes[14,15] as well as other reviews on aminoacyl-tRNA synthetases and on tRNA-aminocyl-tRNA synthetases interactions which have appeared during the past few years.[11-13,16,17,394]

In the review by Söll and Schimmel,[14] the molecular and catalytic properties of these enzymes are discussed in detail. The very intriguing problem of the sites on the tRNA molecule recognized by the cognate aminoacyl-tRNA synthetase, and responsible for the specificity of the tRNA aminoacylation reaction, is also discussed by these authors who have reviewed the major approaches used to study this recognition problem, namely: (1) the determination and comparison of tRNA sequences, (2) the study of the aminoacylation properties of mutant tRNAs (differing from the wild-type tRNA by a single base change) and of chemically modified tRNAs, (3) the study of the aminoacylation properties of tRNA fragments and of products of recombined fragments (in which parts of the tRNA molecules are missing), (4) the study of tRNA mischarging in heterologous and homologous systems, and (5) the study of the topology of tRNA-aminoacyl-tRNA synthetase complexes (by determination of the parts of the tRNA molecule which are protected by the enzyme against nuclease attack, or by comparing the hybridization of complementary oligonucleotides to free and enzyme-bound tRNA, or by cross linking the tRNA and the enzyme in the complex to determine which parts of the two molecules are in close proximity). These studies suggest that certain structural features in a given tRNA are responsible for the recognition by the cognate aminoacyl-tRNA synthetase, but these structural features are not located at similar sites of the cloverleaf model in all of the tRNA enzyme systems studied. The conclusion, therefore, is that either the recognition mechanisms are different depending on the tRNA enzyme system considered, or — if they are identical — that they involve structural features which only appear similar when one considers the three-dimensional structure of tRNAs. In the latter case, the problem will only be solved when more tertiary structures of tRNAs and possibly of tRNA-aminoacyl-tRNA synthetase complexes are known. The results of cross linking[3] and tritium exchange[379] experiments suggest that the acceptor stem and the D stem are part of the recognition site.

b. Plant Aminoacyl-tRNA Synthetases

Plant aminoacyl-tRNA synthetases have been thoroughly reviewed by Lea and Norris[20] so that it is not necessary to discuss them here in detail, and we shall only stress a few important points.

Purification — Plant aminoacyl-tRNA synthetases can be isolated and purified using the general methods available for enzyme purification and especially those described to purify aminoacyl-tRNA synthetases from other sources.[14,15,130a] Although a rather large number of aminoacyl-tRNA synthetases (above 30) have been purified to homogeneity from various organisms (especially from *E. coli* and yeast), few plant aminoacyl-tRNA synthetases have so far been obtained pure. This is because most of them appear to be labile even at low temperature and in the presence of their substrates, of sulfhydryl reducing agents (such as glutathione, mercaptoethanol, or dithiothreitol), and of polyols (such as glycerol or propylene glycol). Another difficulty in purification of plant synthetases is the presence of proteases which are often not inhibited by phenylmethylsulfonylfluoride (PMSF), an inhibitor of proteases having a serine in their active site. However, effective purification procedures are now available that have been applied to plant aminoacyl-tRNA synthetases. Hydrophobic chromatography on aminohexyl Sepharose[304] and affinity chromatography on Sepharose columns containing either the corresponding amino acid[305] or the cognate tRNA[306] appear especially useful. When tRNA is used to purify the cognate aminoacyl-tRNA synthetase by affinity chromatography, other proteins such as tRNA nucleotidyltransferase and tRNA methylases, may bind to the tRNA.[307] In this case, purification is improved when two columns are used successively, the first containing all tRNAs minus the cognate tRNA, the second containing only the cognate tRNA.

Intracellular localization and specificity — It has recently been shown that in plant cells, aminoacyl-tRNA synthetases are present not only in the free form but also (as in animal cells) in the form of enzymatic complexes. In quiescent wheat germ, arginyl-, leucyl, methonyl- , and phenylalanyl-tRNA synthetases participate to various high molecular weight complexes (MW = 3×10^5 to 3×10^6), which have a sedimentation constant of 18 to 20S (or even higher).[427] It is not known at present whether these complexes contain a limited number or the majority of the cell (cytoplasmic) aminoacyl-tRNA synthetases and whether these enzymes are associated (as in animal cells) with other molecules participating to the translation apparatus (tRNAs, ribosomes or ribosomal RNAs, elongation factors, etc.); also unknown is the nature of the bonds involved in these complexes the advantage (if any) resulting from such a macrostructure organization, and whether this organization varies depending upon the physiological conditions in the plant cells.

For several plants, and especially in the case of *Euglena,* chloroplast-specific and mitochondria-specific synthetases have been characterized which differ from their cytoplasmic counterparts.[20] In addition to their subcellular localization, the organellar and cytoplasmic synthetases differ in regard to their chromatographic mobilities (e.g., on hydroxylapatite or DEAE cellulose). This is fortunate since it facilitates their separation. Also, they have different substrate (tRNA) specificities. This was shown by homologous and heterologous aminoacylation reactions (also reviewed by Lea and Norris[20]) performed using tRNAs obtained from the same or from another compartment of that organism, or a prokaryotic tRNA. The overall picture emerging from these studies is that aminoacyl-tRNA synthetases and tRNAs from chloroplasts and prokaryotes (*E. coli, Anacystis nidulans*) have a certain degree of similarity because cross-aminoacylation reactions are possible.[108,109,111,308-310] As far as plant mitochondrial aminoacyl-tRNA synthetases are concerned, in some cases, these enzymes resemble their chloroplastic and bacterial counterparts,[109,111] but in other cases, they appear closer to their cytoplasmic counterparts.[100,108,136]

Origin — Chloroplast-specific tRNAs are coded for by chloroplast DNA (Section V) and it was of interest to determine if chloroplast-specific aminoacyl-tRNA synthetases are coded for by nuclear or by chloroplast genes, and whether they are synthesized on cytoplasmic or on chloroplastic ribosomes. Studies using *Euglena* have shown that a mutant lacking detectable plastid DNA and plastid-coded structure (W₃BU1) did contain low levels of chloroplastic synthetases. Also, the light-induced synthesis of chloroplastic phenylalanyl-, valyl-, lysyl- and leucyl-tRNA synthetases was completely inhibited by cycloheximide which inhibits cytoplasmic protein synthesis, but was not affected by streptomycin which blocks protein synthesis on organellar ribosomes. These studies on the origin of *Euglena* chloroplast aminoacyl-tRNA synthetases suggest that they are coded for by nuclear genes, synthesized on cytoplasmic ribosomes, and transported into the chloroplasts.[135,142,311-313,396] Nevertheless, it is not known at present how similar (or how different) a chloroplastic aminoacyl-tRNA synthetase is compared with its cytoplasmic counterpart. To answer this question, one will probably have to wait for the complete purification of two enzymes specific for the same amino acid but located in two different cellular compartments. This will allow comparative studies on their structural and catalytic properties. Progress in this direction has recently been made by Locy and Cherry who have purified soybean cytoplasmic and chloroplastic tyrosyl-tRNA synthetases,[387] and by Sarantoglou et al. who have purified *Euglena* cytoplasmic and chloroplastic valyl-tRNA synthetases.[410] As the same question can be raised concerning the origin of mitochondrial aminoacyl-tRNA synthetases, it is interesting to note that distinct nuclear genes have been recently shown to exist for yeast cytoplasmic and mitochondrial methionyl-tRNA synthetases.[426]

Variations during Development — Studies on the changes in the levels of aminoacyl-tRNA synthetases during plant growth and development have been reviewed by Lea and Norris.[20] Variations have been reported following a variety of changes in physiological conditions, such as cold treatment of pear embryos to break dormancy, germination, hormone action, illumination of dark-grown *Euglena* cells, root development, seed maturation, and senescence. It should, however, be noted that in several cases: (1) the pyrophosphate exchange reaction (rather than the aminoacylation reaction) was used and not always under optimal conditions; and (2) even when the aminoacylation reaction was used, the existence or the appearance (upon greening, for instance) of organellar aminoacyl-tRNA synthetases which may not necessarily recognize the tRNA used as a substrate in the reaction (see above) was not considered. Thus, many studies, especially the early ones, should probably be reconsidered. Furthermore, in most of these studies, only changes in enzymatic activities have been reported. This is unfortunate, since an increase in enzymatic activity can be due to an increased rate of enzyme synthesis, a decreased rate of degradation, or the activation of preexisting enzyme molecules. In one case, deuterium labeling of chloroplastic leucyl-tRNA synthetase in greening cells of *Euglena* and subsequent enzyme fractionation by CsCl gradient centrifugation have been performed to demonstrate *de novo* synthesis of the enzyme.[314]

2. Transfer of the Amino Acid from the Aminoacyl-tRNA to the Growing Polypeptide Chain

a. Elongation

Transfer of the bound amino acid involves formation of a complex between the aminoacyl-tRNA and the elongation factor EF-Tu in prokaryotes (or EF 1 in eukaryotes) in the presence of GTP, and the aminoacyl-tRNA (selected because its anticodon is complementary to the codon being read) is then bound to the A site on the ribosome. The following steps of the translation process (peptide bond formation, translocation, and release of the tRNA which can then either accept a new molecule of amino acid or be degraded) are similar in all organisms and will not be discussed here in detail. Those interested in protein synthesis in plants should consult recent review articles dealing with various aspects of this problem.[8,131,315-321,373]

It should be remembered that in plants, three protein synthesizing systems coexist, as already mentioned in the introduction. In many respects, protein biosynthesis in the organelles (chloroplasts and mitochondria) resembles prokaryotic protein synthesis rather than that taking place in the surrounding cytoplasm. This can be illustrated by several examples, but as this chapter deals with tRNAs, the author will concentrate on the process of initiation because, as far as tRNAs are concerned, initiation of protein synthesis in the organelles resembles that occurring in prokaryotes and differs from that occurring in the cytoplasm of eukaryotic cells.

b. Initiation

After a report showing that formyl-methionine is found as the N-terminal amino acid of polypeptides synthesized in a cell-free system on *Euglena* chloroplast ribosomes,[322] N-formyl-methionyl-tRNA was first characterized in the chloroplasts of *Phaseolus vulgaris*[147] and then in etioplasts and mitochondria of the same plant.[148] Further studies showed that mitochondrial $tRNA_f^{Met}$ was different from chloroplastic $tRNA_f^{Met}$.[110] Presence of a formylatable $tRNA_f^{Met}$ was also shown in the chloroplasts of other plants, such as wheat,[85] *Acetabularia*,[323] spinach,[324] and cotton.[325] In plant cytoplasm, on the contrary, the initiator tRNA species is a $tRNA^{Met}$ which cannot be formylated even when a bacterial transformylase is used.[84,86,326-328] This contrasts to yeast or animal cytoplasmic initiator $tRNA^{Met}$ (although partial formylation of wheat germ $tRNA_i^{Met}$ was observed using *Anacystis nidulans* transformylase.)[329] Those interested in the initiation of protein synthesis in plants should read two review articles by Marcus et al.[330,331]

c. Coding Properties of tRNAs

The coding properties of wheat germ tRNAs are similar to those of *E. coli* tRNAs,[332] thus confirming that the genetic code is universal. Wheat embryo methionyl-tRNAs have been fractionated by reversed-phase chromatography into two peaks, both binding to AUG and (to a smaller extent) to GUG with wheat embryo ribosomes.[315] Black-eyed peas $tRNAs^{Lys}$ can be fractionated into two fractions: one recognizes AAG, and the other recognizes AAA and AAG.[145] "In lupin seeds, where five isoaccepting $tRNAs^{Lys}$ have been detected, $tRNA_1^{Lys}$ recognizes AAG, $tRNA_2^{Lys}$ AAA and $tRNA_4^{Lys}$ both AAA and AAG;[369] two $tRNA^{Ile}$ fractions have also been studied, using synthetic polynucleotides, one fraction recognizes AAU, the other AUU and AUC.[67] Patterns of codon recognition by isoacceptor aminoacyl- tRNAs from wheat germ have recently been studied in detail by Hatfield and Rice.[411]

Chloroplast-specific tRNAs different from their cytoplasmic counterparts have been characterized. They are recognized specifically by the corresponding chloroplastic aminoacyl-tRNA synthetase (and usually not by the cytoplasmic enzyme), and are coded for by chloroplast DNA. To study the function of these chloroplast-specific tRNAs, it was of interest to compare the codon recognition pattern of the various isoaccepting tRNAs from the two cellular compartments. Ramiasa et al.[333] have recently compared the coding properties of *Phaseolus vulgaris* cytoplasmic and chloroplastic $tRNAs^{Phe}$, $tRNAs^{Lys}$, and $tRNAs^{Leu}$ and found that among the six leucine codons, only UUG is recognized by the chloroplast-specific $tRNAs^{Leu}$. This result agrees with the fact that these three tRNAs seem to be coded for by the same or similar gene(s) on chloroplast DNA, as shown by absence of additivity and existence of competition in hybridization experiments.[208] There are three cytoplasmic tRNA[Leu] species, the first responds to UUG, the second to CUU, and the third to CUG. These results show that, in some cases at least, chloroplast-specific tRNAs recognize code words different from those recognized by their cytoplasmic counterparts. Similar results have been obtained in the case of mitochondria-specific tRNAs from *Tetrahymena pyriformis*.[334,335] Organelle-specfic tRNAs could, therefore, translate codons present in organellar mRNAs which would not be recognized by cytoplasmic tRNAs. Whether this actually occurs in the organelles and whether it plays a role in the control of protein biosynthesis in the organelles, remains to be demonstrated. Nevertheless, experimental evidence (see Section VII,B,4) has recently been reported[365,366,377] in favor of a regulation of protein synthesis at the translation level by the availability of one (or several) isoaccepting tRNA(s).

B. Other Biological Functions of tRNAs

1. Transfer of Amino Acids in Processes Different from mRNA Translation

In this type of transfer, which takes place in the absence of ribosomes and mRNA and which is catalyzed by aminoacyl-tRNA transferases, the aminoacyl moiety of an aminoacyl-tRNA is transferred to a specific acceptor such as:

1. a protein (the amino acid is transferred to the N-terminus of the protein)
2. a phosphatidyl-glycerol (the aminoacylesters of phosphatidyl-glycerol which are formed, are components of cellular membranes).
3. an *N*-acetylmuramide peptide (an intermediate in the formation of interpeptide bridges in bacterial cell walls).

The tRNAs involved in these transfers, especially the staphylococcal tRNA[Gly] species which have been particularly studied, lack several of the invariant and semi-invariant nucleotides present in most tRNAs (see Section IV,B,2).

The aminoacyl-tRNA transferases have been recently reviewed.[336]

2. Involvement of tRNAs in Polynucleotide Synthesis

It has been shown that in some cases, a tRNA species is used as a primer by reverse transcriptase, an enzyme responsible for synthesizing DNA using the RNA of an oncogenic virus as a template.[337,337a]

A number of plant viral RNAs such as turnip yellow mosaic virus,[338] brome mosaic virus,[339] tobacco mosaic virus[340,341] (and even an animal viral RNA[342]) can be enzymatically aminoacylated in vitro (see Reference 343 for a review). Sequence studies have shown that at least in some cases, these RNAs have a "tRNA-like" structure at their 3'-end[344] (for a review, see Reference 345). It is not known whether this aminoacylation occurs in vivo in the infected cell and what the biological function of this aminoacylation could be. If, as it has been suggested, it plays a role in the replication of viral RNA, the "tRNA-like" structure of the 3'-end of viral RNA could be involved in polynucleotide synthesis, thus justifying the description in this section of this puzzling aminoacylation phenomenon.

3. Role of tRNAs in Regulatory Processes

Uncharged or charged tRNAs have been shown to be involved in control mechanisms. For instance, tRNAs appear to participate in the control of gene expression in bacteria, and it has been shown that tRNA, together with guanosine 5'-diphosphate-2'(3') diphosphate, can stimulate transcription of the lactose operon.[353] The involvement of tRNA in the control of enzyme synthesis is particularly well documented in the case of the histidine operon of *Salmonella typhimurium*. Many constitutive mutants have been isolated in which the enzymes responsible for histidine biosynthesis are no longer repressed, and the mutations have been found to map at six different loci. For instance, one of them corresponds to the structural gene of tRNA^{His}, another to the gene of histidyl-tRNA synthetase, and yet another one to the gene of a tRNA modifying enzyme which converts two U residues located near the anticodon of tRNA^{His} into two ψ residues. Repression of the histidine operon is directly related to the in vivo concentration of hystidyl-tRNA^{His}, but in addition to aminoacylation, the presence of the two ψ residues close to the anticodon is necessary for tRNA^{His} to function in repression.[193,194] There is evidence in bacteria favoring a role for tRNAs in the control of the biosynthesis not only of histidine, but of isoleucine-valine and of leucine; whereas, results obtained with other pathways, such as those leading to the synthesis of arginine or methionine, are more difficult to interpret (for a review, see References 17 and 346).

That tRNA methylation is involved in transcriptional control and particularly in repression of amino acid biosynthetic enzymes is suggested by recent observations[400] showing that in a methionine auxotroph of *Salmonella typhimurium* deprived of its essential amino acid, there are incomplete modifications of tRNAs (evidenced by the ability of total tRNA to serve as substrate for homologous tRNA methylases), changes in the chromatographic elution profiles of tRNAs^{Leu}, tRNA^{Ile}, and tRNA^{Val}, and a concomitant five-1 tenfold increase in the isoleucine-valine biosynthetic enzymes even in the presence of an excess of these amino acids. In yeast, there is also evidence suggesting that aminoacyl-tRNAs might be involved in the specific regulation of the biosynthesis of the corresponding amino acid[347-349] or in the "general derepression" of amino acid biosynthetic enzymes.[350]

The synthesis of some aminoacyl-tRNA synthetases appears also to be controlled by mechanisms involving the corresponding tRNA, at least in bacteria (for a review, see Reference 16) and in yeast.[351,352] In plants, no clear demonstration of tRNA involvement in a regulatory process has yet been made.

4. Changes in the Levels of tRNAs Under Different Physiological Conditions

A great number of published reports describe changes in the tRNA complement usually observed by comparing chromatographic fractionation profiles under a variety of different physiological or pathological conditions. These changes in the tRNA population of microorganisms, animals, and plants have been reviewed by Littauer and Inouye,[12] and those observed more specifically in plants have been recently reviewed by Lea and Norris.[20] Changes in the tRNA population have been observed: (1) when the growth conditions are modified, especially in the case of microorganisms, (2) depending upon the tissue considered (in one given organism), (3) during embryogenesis and differentiation, (4) upon action of hormones, (5) upon adaptation to the production of specific proteins, and (6) in tumors and upon viral infection.

As far as variations during embryogenesis and differentiation in plants are concerned, changes have been observed in the relative levels of the six isoaccepting species of tRNA^{Leu} and the three isoaccepting species of tRNA^{Tyr} found in soybean cotyledons following germination and during development and senescence.[40,103,104,137,138,354] Changes in isoaccepting tRNA^{Leu} species have also been shown to occur in germinating pea seedlings and developing pods.[140] Changes have also been observed in the relative levels of the five isoaccepting species of tRNA^{Tyr} found in pea roots when dividing cells (in the apical part) were compared to nondividing (but maturing) cells.[141] Variations also occur in the relative levels of the isoaccepting tRNA^{Phe} species found in barley following germination.[152] During the differentiation of wheat seedlings, significant changes in the chromatographic profiles of seryl-, lysyl-, and prolyl-tRNAs were observed.[133] However, no difference in the composition of the tRNA pool was found in cotton, when the young embryo, the cotyledons, and the roots were compared, except for a change in the cotyledons upon germination as a result of a 7- to 8-fold increase in chloroplast tRNA.[130] A light-stimulated synthesis of chloroplastic tRNA species has been observed in *Euglena*,[99] and an increase in the levels of chloroplast-specific tRNAs occurs upon conversion of etioplasts into chloroplasts in *Phaseolus vulgaris*.[355] It has recently been reported that three minor (methylated) nuckeotides are missing in etiolated wheat germ tRNA^{Phe}, perhaps as a result of the lack of light activation (via the phytochrome system) of the synthesis of some modifying enzymes.[415] Another environmental factor, chilling stress, causes a shift in the ratio of the two glycyl-tRNAs found in soybean seedlings hypocotyls.[134] The level of aminoacylation of the two methionine tRNAs (tRNA$_1^{Met}$ and tRNA$_2^{Met}$) has been shown to increase from 0 to 81% in dry and 2 day-old germinating lupin cotyledons.[425]

A heat-labile factor present in pea roots can specifically degrade tRNA$_6^{Leu}$ into a new species called tRNA$_L^{Leu}$ which appears to consist of at least half the original tRNA$_6^{Leu}$ molecule (from the 3'-end to the anticodon).[356] This is reminiscent of what happens in *E. coli* following infection by phage T$_2$ where host tRNA$_1^{Leu}$ disappears to yield (probably by action of a phage-induced nuclease) tRNA$_F^{Leu}$. This phenomenon plays a role in the inhibition of host protein synthesis.[357-359] In soybean cotyledons, a ribonuclease apparently associated with the chloroplasts, specifically degrades chloroplastic tRNA$_3^{Tyr}$ and tRNA$_4^{Tyr}$ to yield fragments approximately the size of a half-molecule.[360]

Plant hormones, especially cytokinins, have been found to cause modificaions of tRNAs. For instance, they change the relative levels of isoaccepting tRNAs^{Leu} in soybean cotyledons[103,138,139,431] and in lupin embryos.[371] On genetically male *Mercurialis annua* plants (a dioecious species), cytokinin spraying induced female flowers that contained a tRNA population different from that found in male flowers.[143] Quantitative changes in the relative proportions of tRNAs^{Leu}, tRNAs^{Lys} and tRNAs^{Met} have been observed during ethylene-induced ripening of tomato fruits.[144]

In some tissues specialized in synthesis of specific proteins, a correlation has been observed between the tRNA population and the amino acid composition of these proteins. This has been well documented in the case of the silk gland of *Bombyx mori* in which fibroin, a protein with a very unusual amino acid composition (four amino acids — Gly, Ala, Ser, Tyr — represent over 90% of the fibroin amino acids), is synthesized. When this gland becomes functional, there is a very large increase in the amounts of the corresponding tRNAs, a phenomenon which has been called "functional adaptation" of tRNAs.[361,362] Similar adaptation phenomena have been described in other animal tissues (for a review, see Reference 12). In plants, during the later stages of the maturation of cereal grains, storage protein rich in proline and glutamine, but low in lysine and tryptophane is laid down in discrete protein bodies in the endosperm and/ or aleurone layer. This prompted Norris et al.[146] to measure the levels of lysyl- and prolyl-tRNAs at various stages of the maturating wheat grain, to fractionate the isoaccepting lysyl- and prolyl-tRNAs from the embryo and the endosperm and to study the levels of aminoacylation of the individual isoaccepting tRNAs^{Lys} and tRNAs^{Pro} during seed maturation. The small changes observed were insufficient to account for the changes in lysine and proline content in storage protein. In maize seed, the major storage protein, zein, is synthesized in the endosperm where it represents up to 60% of the total proteins. In maize endosperm tRNA, Viotti et al.[380] have recently observed an increase in glutamine (also in alanine and leucine) accepting activity as compared to that of maize embryo tRNA. This increase is accompanied by a change in the relative amounts of the two isoaccepting tRNAs^{Gln} as shown by studying the RPC-5 chromatographic profiles. In the embryo, tRNA$_2^{Gln}$ represents approximately 25%, while in the endosperm it represents approximately 40%. These changes appear to be correlated with the high glutamine content of zein (approximately 20%) as compared to that of embryo proteins (approximately 1 or 2%).

Many tRNA modifications have been observed upon viral infection and in animal tumor cells when compared with the corresponding normal cells (for a review, see References 12 and 271). In plants, tobacco normal tissues and crown-gall tumor tissues have been compared.[132] No difference was found in the RPC-5 chromatographic profiles of tRNAs corresponding to ten amino acids; however, an extra peak of tRNA^{Phe} was seen in tumor tissues which differed from the major tRNA^{Phe} species present in both normal and tumor tissues in that it lacked the Y base. This phenomenon has also been observed in some animal tumors.[363,364]

It has, however, been recently pointed out that if complete aminoacylation of all the isoacceptors present in a tRNA mixture is not accomplished, then the observed differences may be due not to true variations in the relative proportions of isoaccepting species present, but to the possibility that different isoacceptors may be aminoacylated at different rates, and thus to different extents under conditions of incomplete acylation.[386] Since a majority of studies reporting changes in tRNA complements have not considered this possibility, the interpretation of some of these studies must be questioned.

But, assuming that the observed changes in the tRNA population are real, what is their biological significance?

It should first be stressed that in many studies, especially the earlier ones, the existence of organellar tRNAs was not taken into account. This makes it difficult to interpret the variations observed. It is possible, at least in some cases, that these variations reflect the participation of tRNAs in control mechanisms such as those mentioned in the preceding section. It has been suggested that protein synthesis can be regulated at the translational level by the availability of one (or several) isoaccepting tRNA(s). A minor tRNA species could be a limiting factor, for instance, if it is the only tRNA whose anticodon is able to read a codon present in mRNA molecules which are to be translated in a given tissue or at a given time in differentiation. Recent studies have furnished experimental evidence in favor of this hypothesis. The translation of egg white proteins and globin mRNAs in a tRNA-dependent cell-free system derived from Krebs ascites cells has been shown to be optimal in the presence of tRNA from the homologous tissue.[377] Ascites cell tRNAs and rabbit liver tRNAs promote efficient translation of globin mRNA, oviduct mRNA, and encephalomyocarditis viral RNA, while reticulocyte tRNAs participate efficiently only in globin mRNA translation and probably lack some isoaccepting tRNA species required for the translation of the other two RNAs.[365] In extracts from interferon-treated mouse cell, different minor tRNA^{Leu} species are required for the translation of globin mRNA and mengovirus RNA.[366] These studies have been performed in animal systems, and it is clear that more work is needed to determine the relationship between the observed variations in the tRNA population and the events occurring during plant growth and development.

REFERENCES

1. Hoagland, M. R., Zamecnik, P. C., and Stephenson, M. L., Intermediate reactions in protein biosynthesis, *Biochim. Biophys. Acta,* 24, 215, 1957.

2. Ogata, K., Nohara, H., and Morita, T., The effect of ribonuclease on the amino acid-dependent exchange between labeled inorganic pyrophosphate and adenosine triphosphate by the pH 5 enzyme, *Biochim. Biophys. Acta,* 26, 656, 1957.

3. Cohn, W. E. and Volkin, E., Nucleoside 5′ phosphates from ribonucleic acid, *Nature (London),* 167, 483, 1951.

4. Zamecnik, P. C. and Keller, E., Relation between phosphate energy donors and incorporation of labeled amino acids into proteins, *J. Biol. Chem.,* 209, 337, 1954.

5. Chapeville, F., Lipmann, F., Von Ehrenstein, G., Weisblum, B., Ray, W., and Benzer, S., On the role of soluble ribonucleic acid in coding for amino acids, *Proc. Natl. Acad. Sci. U.S.A.,* 48, 1086, 1962.

6. Barnett, W. E. and Brown, D., Mitochondrial transfer ribonucleic acids, *Proc. Natl. Acad. Sci. USA.,* 57, 452, 1967.

7. Barnett, W. E., Brown, D., and Epler, J., Mitochondrial-specific aminoacyl-tRNA synthetases, *Proc. Natl. Acad. Sci. USA.,* 57, 1775, 1967.

8. Ellis, R. J., Blair, G. E., and Hartley, M., The nature and function of chloroplast protein synthesis, *Biochem. Soc. Symp.,* 38, 137, 1973.

9. Borst, P., Mitochondrial nucleic acids, *Annu. Rev. Biochem.,* 41, 333, 1972.

10. Cramer, F., Three-dimensional structure of tRNA, *Prog. Nucleic Acid Res. Mol. Biol.,* 11, 391, 1971.

11. Loftfield, R. B., The mechanisms of aminoacylation of transfer RNA, *Prog. Nucleic Acid Res. Mol. Biol.,* 12, 87, 1972.

12. Littauer, U. Z. and Inouye, H., Regulation of tRNA, *Annu. Rev. Biochem.,* 42, 439, 1973.

13. Kisselev, L. L. and Farova, O. O., Aminoacyl-tRNA synthetases: some recent results and achievements, in *Advances in Enzymology,* Interscience, New York, 40, 1974, 141.

14. Söll, D. and Schimmel, P. R., Aminoacyl-tRNA synthetases, in *The Enzymes,* Academic Press, New York, 10, 1974, 489.

15. Kalousek, F. and Konigsberg, W., Aminoacyl-tRNA synthetases, in *Synthesis of aminoacids and proteins,* Arnstein, H. R. V., Ed., University Park Press, Baltimore, 1975, 58.

16. Neidhardt, F. C., Parker, J., and Mc Keever, G., Function and regulation of aminoacyl-tRNA synthetases in prokaryotic and eukaryotic cells, *Annu. Rev. Microbiol.,* 29, 215, 1975.

17. Brenchley, J. E. and Williams, L. S., Transfer RNA involvement in the regulation of enzyme synthesis, *Annu. Rev. Microbiol.,* 29, 251, 1975.

18. Rich, A. and Rajbhandary, U. L., Transfer RNA: molecular structure, sequence and properties, *Annu. Rev. Biochem.,* 45, 805, 1976.

19. Kim, S. H., Three-dimensional structure of transfer RNA, *Prog. Nucleic Acid Res. Mol. Biol.,* 17, 181, 1976.

20. Lea, P. J. and Norris, R. D., tRNA and aminoacyl-tRNA synthetases from higher plants, in *Prog. Phytochem.,* 4, 121, 1977.

21. Lea, P. J. and Norris, R. D., tRNA and aminoacyl-tRNA synthetases from plants, *Phytochemistry,* 11, 2897, 1972.

22. Schatz, G. and Mason, T. L., The biosynthesis of mitochondrial proteins, *Annu. Rev. Biochem.,* 43, 51, 1974.

23. Von Ehrenstein, G., Isolation of sRNA from intact *Escherichia coli* cells, in *Methods in Enzymology,* Academic Press, New York, 12(A), 1967, 588.

24. Holley, R. W., Isolation of sRNA from intact yeast cells, in *Methods in Enzymology,* Academic Press, New York, 12(A), 1967, 596.

25. Kirby, K. S., Isolation of nucleic acids with phenolic solvents, in *Methods in Enzymology,* Academic Press, New York, 12(B), 1968, 87.

26. Fraenkel-Conrat H., Singer, B., and Tsugita, A., Purification of viral RNA by means of bentonite, *Virology,* 14, 54, 1961.

27. Dingman, W. and Sporn, M. B., The isolation and physical characterization of nuclear and microsomal RNA from rat brain and liver, *Biochim. Biophys. Acta,* 61, 164, 1962.

28. Solymosy, F., Fedorcsak, I., Gulyas, A., Farkas, G. L., and Ehrenberg, L., A new method based on the use of diethylpyrocarbonate as a nuclease inhibitor for the extraction of undegraded nucleic acid from plant tissues, *Eur. J. Biochem.,* 5, 520, 1968.

29. Leonard, N. J., Mc Donald, J. J. and Reichmann, M. E., Reaction of diethylpyrocarbonate with nucleic acid components. I. Adenine, *Proc. Natl. Acad. Sci. USA.,* 67, 93, 1970.

30. Ortwerth, B. J., The effect of diethylpyrocarbonate on the transfer RNA of bovine muscle, *Biochim. Biophys. Acta,* 246, 344, 1971.

31. Abadom, P. N. and Elson, D., A procedure for isolating transfer ribonucleic acid from human placenta, *Biochim. Biophys. Acta,* 199, 528, 1970.

32. **Fedorcsak, I., Ehrenberg, L., and Solymosy, F.,** Diethylpyrocarbonate does not degrade RNA, *Biochem. Biophys. Res. Commun.*, 65, 490, 1975.

33. **Kirby, K. S.,** A new method for the isolation of ribonucleic acids from mammalian tissues, *Biochem. J.*, 64, 405, 1956.

34. **Aubel-Sadron, G., Beck, G., Ebel, J. P., and Sadron, C.,** Etude de la précipitation des acides nucléiques par les sels d'ammonium quaternaire, *Biochim. Biophys. Acta,* 42, 542, 1960.

35. **Burkard, G., Weil, J. H., and Ebel, J. P.,** Mise en évidence d'un effet inhibiteur des polyphosphates sur l'activité acceptrice de l'acide ribonucléique soluble de levure, *Bull. Soc. Chim. Biol.*, 47, 561, 1965.

36. **Weil, J. H., Ebel, J. P., and Monier, R.,** Conservation of the biological activity of a soluble ribonucleic acid, *Nature (London),* 192, 169, 1961.

37. **Weil, J. H., and Ebel, J. P.,** Action des sels d'ammonium quaternaire sur les acides nucléiques. II. Application à l'acide ribonucléique soluble, *Biochim. Biophys. Acta,* 55, 836, 1962.

38. **Zubay, G.,** The isolation and fractionation of soluble ribonucleic acid, *J. Mol. Biol.*, 4, 347, 1962.

39. **Augustyniak, H., Barciszewski, J., Rafalski, A., Zawielak, J., and Szyfter, K.,** Phenylalanine tRNA of *Lupinus luteus* seeds, *Phytochemistry*, 13, 2679, 1974.

40. **Anderson, M. B. and Cherry, J. H.,** Differences in leucyl-transfer RNAs and synthetases in soybean seedlings, *Proc. Natl. Acad. Sci. USA.*, 62, 202, 1969.

41. **Burkard, G., Guillemaut, P., and Weil, J. H.,** Comparative studies of the tRNAs and the aminoacyl-tRNA synthetases from the cytoplasm and chloroplasts of *Phaseolus vulgaris, Biochim. Biophys. Acta,* 224, 184, 1970.

42. **Kirby, K. S.,** Isolation and characterization of ribosomal ribonucleic acid, *Biochem. J.*, 96, 266, 1965.

43. **Vanderhoef, L. N., Bohannon, R. F., and Key, J. L.,** Purification of transfer RNA and studies on aminoacyl-tRNA synthetases from higher plants, *Phytochemistry,* 9, 2291, 1970.

44. **Cornelis, P.,** Use of high salt Sepharose 4B chromatography for the extraction of tRNA from plant tissues, *Plant Sci. Lett.*, 11, 3, 1978.

45. **Dirheimer, G., Weil, J. H., and Ebel, J. P.,** Séparation des acides ribonucléiques ribosomique et soluble de levure par filtration sur gel de dextrane, *C. R. Acad. Sci.,* 255, 2312, 1962.

46. **McCoy, T. A. and Carter, E. A.,** The separation of ribonucleic acids on Sephadex columns, *J. Chromatogr.*, 37, 458, 1968.

47. **Ingle, J., Key, J. L., and Holm, R. E.,** Demonstration and characterization of a DNA-like RNA in excised plant tissue, *J. Mol. Biol.*, 11, 730, 1965.

48. **Cherry, J. H. and Chroboczek, H.,** Factors affecting nucleic acid extractability in plants, *Phytochemistry,* 5, 411, 1966.

49. **Glitz, D. G. and Dekker, C.,** The purification and properties of ribonucleic acid from wheat germ, *Biochemistry,* 2, 1185, 1963.

50. **Kothari, R. M. and Taylor, M. W.,** RNA fractionation on modified celluloses. II. DEAE-cellulose, *J. Chromatogr.,* 73, 463, 1972.

51. **Gilbert, W.,** Polypeptide synthesis in *E. coli.* II. The polypeptide chain and sRNA, *J. Mol. Biol.*, 6, 389, 1963.

52. **Loening, U. E.,** The fractionation of high-molecular-weight ribonucleic acid by polyacrylamide-gel electrophoresis, *Biochem. J.*, 102, 251, 1967.

53. **Sarin, P. S. and Zamecnik, P. C.,** On the stability of aminoacyl-sRNA to nucleophilic catalysis, *Biochim. Biophys. Acta,* 91, 653, 1964.

54. **Tao, K. L. and Hall, T. C.,** Extensive charging of transfer ribonucleic acid by bean leaf extracts in vitro, *Biochem. J.*, 125, 975, 1971.

55. **Ehresmann, B., Imbault, P., and Weil, J. H.,** Determination of the degree of *in vivo* tRNA aminoacylation in yeast cells, *Anal. Biochem.*, 61, 548, 1974.

56. **Griffiths, E. and Bayley, S. T.,** Properties of transfer ribonucleic acid and aminoacyl-transfer ribonucleic acid synthetases from an extremely halophilic bacterium, *Biochemistry,* 8, 541, 1969.

57. **Ritter, P. O., Kull, F., and Jacobson, K. B.,** Effect of Tris and dimethylsulfoxide on the aminoacylation of *E. coli* valine tRNA by *N. crassa* phenylalanyl-tRNA synthetase, *Biochim. Biophys. Acta,* 179, 524, 1969.

58. **Giege, R., Kern, D., and Ebel, J. P.,** Incorrect aminoacylations catalyzed by *E. coli* valyl-tRNA synthetase, *Biochimie,* 54, 1245, 1972.

59. **Vold, B. S. and Sypherd, P. S.,** Changes in soluble RNA and ribonuclease activity during germination of wheat, *Plant Physiol.*, 43, 1221, 1968.

60. **Novelli, G. D.,** Amino acid activation for protein synthesis, *Annu. Rev. Biochem.*, 36, 449, 1967.

61. Lea, P. J. and Fowden, L., Amino acid substrate specificity of asparaginyl-, aspartyl-, and glutaminyl-tRNA synthetase isolated from higher plants, *Phytochemistry*, 12, 1903, 1973.

62. Doctor, B. P., Apgar, J., and Holley, R. W., Fractionation of yeast aminoacid acceptor ribonucleic acids by countercurrent distribution, *J. Biol. Chem.*, 236, 1117, 1961.

63. Morris, C. J. and Morris, P., Counter-current distribution, in *Separation Methods in Biochemistry*, Sir Isaac Pitman and Sons, London, 1963, 595.

64. Ayad, S. R., Partition chromatography and countercurrent distribution, in *Techniques of Nucleic Acid Fractionation*, Wiley-Interscience, Chichester, 1972, 85.

65. Dirheimer, G., and Ebel, J. P., Fractionnement des tRNA de levure de bière par distribution en contre-courant, *Bull. Soc. Chim. Biol.*, 49, 1679, 1967.

66. Doctor, B. P., Fractionation of RNAs by countercurrent distribution, *Methods Enzymol.*, 12(A), 644, 1967.

67. Legocki, A. B., Szymkowiak, A., Hierowski, M., and Pawelkiewicz, J., Heterogeneity of transfer ribonucleic acids from yellow lupin seeds *Acta Biochim. Pol.*, 15, 197, 1968.

68. Legocki, A. B. and Pawelkiewicz, J., Fractionation of plant transfer ribonucleic acid by hydroxyapatite column, *Bull. Acad. Pol. Sci. Sev. Sci. Biol.*, 15, 517, 1967.

69. Hartmann, G. and Coy, U., Fraktionierung der Aminosaure spezifischer loslichen Ribonukleinsauren, *Biochim. Biophys. Acta*, 47, 612, 1961.

70. Muench, K. and Berg, P., Resolution of aminoacyl-transfer ribonucleic acids by hydroxyapatite chromatography, *Biochemistry*, 5, 982, 1966.

71. Pearson, R. L. and Kelmers, A. D., Separation of transfer ribonucleic acids by hydroxyapatite columns, *J. Biol. Chem.*, 241, 767, 1966.

72. Dirheimer, G., Chromatographie des acides ribonucléiques de transfer de levure de bière sur colonne d'hydroxyapatite, *Bull. Soc. Chim. Biol.*, 50, 1221, 1968.

73. Bernardi, G., Chromatography of nucleic acids on hydroxyapatite columns, *Methods Enzymol.*, 21, 95, 1971.

74. Sueoka, N. and Yamane, T., Fractionation of aminoacyl acceptor RNA on a methylated albumin column, *Proc. Natl. Acad. Sci. USA.*, 48, 1454, 1962.

75. Okamoto, T. and Kawade, Y., Fractionation of soluble RNA by a methylated albumin column of increased capacity, *Biochem. Biophys. Res. Commun.*, 13, 324, 1963.

76. Melchers, F. and Zachau, H. G., Chromatographic von seryl-RNA and Cysteinyl-RNA und Säulen aus methylierten Albumin auf Kieselguhr, *Biochim. Biophys. Acta*, 95, 380, 1965.

77. Stern, R. and Littauer, U. Z., Fractionation of transfer ribonucleic acid on a methylated albumin-silicic acid column, *Biochemistry*, 7, 3469, 1968.

78. Vanderhoef, L. N. and Key, J. L., RNA ratios in dividing and nondividing cells of pea root, *Biochim. Biophys. Acta*, 240, 62, 1971.

79. Gillam, I., Millward, S., Blew, D., Von Tigerstrom, M., Wimmer, E., and Tener, G. M., The separation of soluble ribonucleic acids on benzoylated diethylaminoethyl-cellulose, *Biochemistry*, 6, 3043, 1967.

80. Wimmer, E., Maxwell, I. H., and Tener, G. M., A simple method for isolating highly purified yeast phenylalanine transfer ribonucleic acid, *Biochemistry*, 7, 2623, 1968.

81. Maxwell, I. H., Wimmer, E., and Tener, G. M., The isolation of yeast tyrosin and tryptophan transfer acids, *Biochemistry*, 7, 2629, 1968.

82. Gillam, I., Blew, D., Warrington, R. C., Von Tigerstrom, M., and Tener, G. M., A general procedure for the isolation of specific transfer ribonucleic acids, *Biochemistry*, 7, 3459, 1968.

83. Kothari, R. M. and Taylor, M. W., RNA fractionation on modified celluloses. III. BD-cellulose, *J. Chromatogr.*, 73, 479, 1972.

84. Leis, J. P. and Keller, E. B., Protein chain initiation by methionyl-tRNA, *Biochem. Biophys. Res. Commun.*, 40, 416, 1970.

85. Leis, J. P. and Keller, E. B., Protein chain-initiating methionine tRNAs in chloroplasts and cytoplasm of wheat leaves, *Proc. Natl. Acad. Sci. USA*, 67, 1593, 1970.

86. Yarwood, A., Boulter, D., and Yarwood, J. N., Methionyl-tRNAs and initiation of protein synthesis in *Vicia faba*, *Biochem. Biophys. Res. Commun.*, 44, 353, 1971.

87. Dudock, B. S., Katz, G., Taylor, E. K., and Holley, R. W., Primary structure of wheat germ phenylalanine transfer RNA, *Proc. Natl. Acad. Sci. USA.*, 62, 941, 1969.

88. Fairfield, S. A. and Barnett, W. E., On the similarity between the tRNAs of organelles and prokaryotes, *Proc. Natl. Acad. Sci. USA.*, 68, 2972, 1971.

89. Tanaka, K., Richards, H. H., and Cantoni, G. L., Studies on soluble ribonucleic acid. IX. Partition chromatography of yeast soluble ribonucleic acid on Sephadex, *Biochem. Biophys. Acta*, 61, 846, 1962.

90. **Nathenson, S. G., Dohan, F. C., Richards, H. H., and Cantoni, G. L.,** Partition chromatography of yeast and *E. coli* soluble ribonucleic acid. Relation of coding properties to fractionation, *Biochemistry,* 4, 2412, 1965.

91. **Bergquist, P. L., Baguley, B. C., Roberston, J. M., and Ralph, K. K.,** Fractionation of amino acid acceptor ribonucleic acid, *Biochim. Biophys. Acta,* 108, 531, 1965.

92. **Muench, K. H. and Berg, P.,** Fractionation of transfer ribonucleic acid by gradient partition chromatography on Sephadex columns, *Biochemistry,* 5, 970, 1966.

93. **Everett, G. A., Merrill, S. H., and Holley, R. W.,** Separation of amino acid-specific soluble ribonucleic acids by partition chromatography, *J. Am. Chem. Soc.,* 82, 757, 1960.

94. **Schweiger, M. and Zachau, H. G.,** Oligo- und Polynucleotidtrennungen Verteilungschromatographie von löslicher Ribonucleinsäure, *Hoppe Seylers Z. Physiol. Chem.,* 342, 93, 1965.

95. **Kelmers, A. D., Novelli, G. D., and Stulberg, M. D.,** Separation of transfer ribonucleic acids by reverse phase chromatography, *J. Biol. Chem.,* 240, 3979, 1965.

96. **Kothari, R. M. and Taylor, M. W.,** RNA fractionation on reversed phase columns, *J. Chromatogr.,* 86, 289, 1973.

97. **Weiss, J. F. and Kelmers, A. D.,** A new chromatographic system for increased resolution of transfer ribonucleic acids, *Biochemistry,* 6, 2507, 1967.

98. **Yoshikami, D., Katz, G., Keller, E. B., and Dudock, B. S.,** A fluorescence assay for phenylalanine transfer RNA, *Biochim. Biophys. Acta,* 166, 714, 1968.

99. **Barnett, W. E., Pennington, C. J., and Fairfield, S. A.,** Induction of *Euglena* transfer RNAs by light, *Proc. Natl. Acad. Sci. USA.,* 63, 1261, 1969.

100. **Guderian, R. H., Pulliam, R. L., and Gordon, M. P.,** Characterization and fractionation of tobacco leaf transfer RNA, *Biochim. Biophys. Acta,* 262, 50, 1972.

101. **Jones, D. S. and Jay, F. T.,** Transfer ribonucleic acid from *Scenedesmus obliquus* D3, *Biochem. Soc. Trans.,* 3, 660, 1975.

102. **Cherry, J. H. and Osborne, D. J.,** Specificity of leucyl-tRNA and synthetase in plants, *Biochem. Biophys. Res. Commun.,* 40, 763, 1970.

103. **Bick, M. D., Liebke, H., Cherry, J. H., and Strehler, B.,** Changes in leucyl- and tyrosyl-tRNA of soybean cotyledons during plant growth, *Biochim. Biophys. Acta,* 204, 175, 1970.

104. **Bick, M. D. and Strehler, B.,** Leucyl-transfer RNA synthetase changes during soybean cotyledon senescence, *Proc. Natl. Acad. Sci. USA.,* 68, 224, 1971.

105. **Kanabus, J. and Cherry, J. H.,** Isolation of an organ-specific leucyl-tRNA synthetase from soybean seedlings, *Proc. Natl. Acad. Sci. USA.,* 68, 873, 1971.

106. **Pearson, R. L., Weiss, J. F., and Kelmers, A. D.,** Improved separation of transfer RNAs on polychlorotrifluoroethylene-supported reversed phase chromatography columns, *Biochim. Biophys. Acta,* 228, 770, 1971.

107. **Kelmers, A. D. and Heatherly, D. E.,** Columns for rapid chromatographic separation of small amounts of tracer-labeled transfer ribonucleic acids, *Anal. Biochem.,* 4, 486, 1971.

108. **Guillemaut, P., Steinmetz, A., Burkard, G., and Weil, J. H.,** Aminoacylation of tRNA^{Leu} species from *E. coli* and from the cytoplasm, chloroplasts, and mitochondria of *Phaseolus vulgaris* by homologous and heterologous enzymes, *Biochim. Biophys. Acta,* 378, 64, 1975.

109. **Jeannin, G., Burkard, G., and Weil, J. H.,** Aminoacylation of *Phaseolus vulgaris* cytoplasmic, chloroplastic and mitochondrial tRNAs^{Pro} and tRNAs^{Lys} by homologous and heterologous enzymes, *Biochim. Biophys. Acta,* 442, 24, 1976.

110. **Guillemaut, P., Burkard, G., Steinmetz, A., and Weil, J. H.,** Comparative studies on the tRNAs^{Met} from the cytoplasm, chloroplasts and mitochondria of *Phaseolus vulgaris,* *Plant Sci. Lett.,* 1, 141, 1973.

111. **Guillemaut, P. and Weil, J. H.,** Aminoacylation of *Phaseolus vulgari* cytoplasmic, chloroplastic and mitochondrial tRNAs^{Met} and of *E. coli* tRNAs^{Met} by homologous and heterologous enzymes, *Biochim. Biophys. Acta,* 407, 240, 1975.

112. **Holmes, W. M., Hurd, R. E., Reid, B. R., Rimerman, R. A., and Hatfield, G. W.,** Separation of transfer ribonucleic acid by Sepharose chromatography using reverse salt gradient, *Proc. Natl. Acad. Sci. USA.,* 72, 1068, 1975.

113. **Singhal, R. P., Griffin, G. D., and Novelli, G. D.,** Separation of transfer ribonucleic acids on polystyrene anion exchangers, *Biochemistry,* 15, 5083, 1976.

114. **Salomon, R., Fuchs, S., Aharonov, A., Giveon, D., and Littauer, U. Z.** Detection and purification of isoaccepting tRNA^{Phe} species containing Y base by affinity chromatography on columns of anti-Y antibodies, *Biochemistry,* 14, 4046, 1975.

115. **Grosjean, H., Takada, C., and Petre, J.,** Complex formation between transfer RNAs with complementary anticodons: use of matrix-bound tRNA, *Biochem. Biophys. Res. Commun.,* 53, 882, 1973.

116. **Vögeli, G., Grosjean, H., and Söll, D.,** A method for the isolation of specific tRNA precursors, *Proc. Natl. Acad. Sci. USA.,* 72, 4790, 1975.

117. Littauer, U. Z., The synthesis of peptidyl-aminoacyl-tRNA and its use as a method for the purification of tRNA, in *Methods in Enzymology,* Academic Press, New York, 20, 70, 1971.

118. Sokoloff, L. and Rappaport, H. P., Isolation of isoaccepting tRNAs, *Arch. Biochem. Biophys.,* 153, 788, 1972.

119. Goss, D. J. and Parkhurst, L. J., Ultra-rapid quantitative isolation of specific transfer ribonucleic acids, a solid-phase method, *Biochem. Biophys. Res. Commun.,* 59, 181, 1974.

120. Zamecnik, P. C., Stephenson, M. L., and Scott, J. F., Partial purification of soluble RNA, *Proc. Natl. Acad. Sci. USA.,* 46, 811, 1960.

121. Zachau, H. G., Tada, M., Lawson, W. B., and Schweiger, M., Fraktionierung der löslichen Ribonucleinsäuren, *Biochim. Biophys. Acta,* 53, 221, 1961.

122. Duncan, R. E. and Gilham, P. T., Isolation of transfer RNA isoacceptors by chromatography on dihydroxyboryl-substituted cellulose, polyacrylamide, and glass, *Anal. Biochem.,* 66, 532, 1975.

123. Mc Cutchan, T. F., Gilham, P. T., and Söll, D., An improved method for the purification of tRNA by chromatography on dihydroxyborylsubstituted cellulose, *Nucleic Acid Res.,* 2, 853, 1975.

124. Klyde, B. J. and Bernfield, M. R., Purification of chicken liver seryl-transfer ribonucleic acid by complex formation with elongation factor EF-Tu : GTP. A general micromethod of aminoacyl-transfer ribonucleic acid purification, *Biochemistry,* 12, 3753,1973.

125. Bartkowiak, S., Radlowski, M., and Augustyniak, J., Covalent coupling of aminoacyl-tRNA to modified cellulose as a method of purification of specific tRNAs, *FEBS Lett.,* 43, 112, 1974.

126. Ikemura, T. and Dahlberg, J. E., Small ribonucleic acids of *E. coli,* I. Characterization by polyacrylamide gel electrophoresis and fingerlrint analysis, *J. Biol. Chem.,* 248, 5024, 1973.

127. Stein, M. and Varricchio, F., Separation of tRNAs by two-dimensional polyacrylamide gel electrophoresis, *Anal. Biochem.,* 61, 112, 1974.

128. Fradin, A., Gruhl, H., and Feldmann, H., Mapping of yeast tRNAs by two-dimensional electrophoresis on polyacrylamide gels, *FEBS Lett.,* 50, 185, 1975.

129. Ikemura, T., Shimura, Y., Sakano, H., and Ozeki, H., Precursor molecules of *E. coli* transfer RNAs accumulated in a temperature-sensitive mutant, *J. Mol. Biol.,* 96, 69, 1975.

129a. *Methods Enzymol.,* 20, 3—232, 1971.

130. Merrick, W. C. and Dure, L. S., The developmental biochemistry of cotton seed embryogenesis and germination. IV. Levels of cytoplasmic and chloroplastic transfer ribonucleic acid species, *J. Biol. Chem.,* 247, 7988, 1972.

130a. *Methods Enzymol.,* 29, 469, 1974.

131. Dure, L. S., Regulation of protein synthesis in cotton seed embryogenesis and germination, *Biochem. Soc. Symp.,* 38, 217, 1973.

132. Cornelis, P., Claessen, E., and Claessen, J., Reversed phase chromatography in isoaccepting tRNAs from healthy and crown-gall tissues from *Nicotiana tabacum, Nucleic Acid Res.,* 2, 1153, 1975.

133. Vold, B. S. and Sypherd, P. S., Modification in transfer RNA during the differentiation of wheat seedlings, *Proc. Natl. Acad. Sci. USA.,* 59, 453, 1968.

134. Yang, J. S. and Brown, G. N., Isoaccepting transfer ribonucleic acids during chilling stress in soybean seedling hypocotyls, *Plant Physiol.,* 53, 694, 1974.

135. Reger, B. J., Fairfield, S. A., Epler, J. L., and Barnett, W. E., Identification and origin of some chloroplast aminoacyl-tRNA synthetases and tRNAs, *Proc. Natl. Acad. Sci. USA.,* 67, 1207, 1970.

136. Kislev, N., Selsky, M. I., Norton, C., and Eisenstadt, J. M., tRNA and tRNA aminoacyl synthetases of chloroplasts, mitochondria, and cytoplasm from *Euglena gracilis, Biochim. Biophys. Acta,* 287, 256, 1972.

137. Venkataraman, R. and Deleo, P., Changes in leucyl-tRNA species during aging of detached soybean cotyledons, *Phytochemistry,* 11, 923, 1972.

138. Pillay, D. T. N. and Cherry, J. H., Changes in leucyl-, seryl-, and tyrosyl-tRNAs in aging soybean cotyledons, *Can. J. Bot.,* 52, 2499, 1974.

139. Wright, R. D., Pillay, D. T. N., and Cherry, J. H., Changes in leucyl-tRNA species of pea leaves during senescence and after zeatin treatment, *Mech. Ageing Dev.,* 1, 403, 1972/73.

140. Patel, H. V. and Pillay, D. T. N., Leucine specific transfer ribonucleic acids and synthetases in the cotyledons of mature and germinating pea seeds, *Phytochemistry,* 15, 401, 1976.

141. Vanderhoef, L. N. and Key, J. L., The fractionation of transfer ribonucleic acid from roots of pea seedlings, *Plant Physiol.,* 46, 294, 1970.

141a. Parthier, B.,personal communication.

142. Parthier, B., Light induced chloroplast differentiation in *Euglena gracilis,* in *Cell Differentiation in Microorganisms, Higher Plants and Animals,* Nover, L. and Mothes, K., Eds., Fisher Jena and Elsevier, Amsterdam, 1977, 602.

143. Bazin, M., Chabin, A., and Durand, R., Comparison between four isoaccepting transfer ribonucleic acids and corresponding synthetases in male and female flowers of the dioecious species *Mercurialis annua, Dev. Biol.,* 44, 288, 1975.

144. **Mettler, I. J. and Romani, R. J.,** Quantitative changes in tRNA during ethylene induced ripening (ageing) of tomato fruits, *Phytochemistry,* 15, 25, 1976.

145. **Hague, D. R. and Kofoid, E. C.,** The coding properties of lysine accepting transfer ribonucleic acids from black-eyed peas, *Plant Physiol.,* 48, 305, 1971.

146. **Norris, R. D., Lea, P. J., and Fowden, L.,** tRNA species in the developing grain of *Triticum aestivum, Phytochemistry,* 14, 1683, 1975.

147. **Burkard, G., Eclancher, B., and Weil, J. H.,** Presence of N-formyl-methionyl-transfer RNA in bean chloroplasts, *FEBS Lett.,* 4, 285, 1969.

148. **Guillemaut, P., Burkard, G., and Weil, J. H.,** Characterization of N-formyl-methionyl-tRNA in bean mitochondria and etioplasts, *Phytochemistry,* 11, 2217, 1972.

149. **Goins, D. J., Reynolds, R. J., Schiff, J. A., and Barnett, W. E.,** A cytoplasmic regulatory mutant of *Euglena*: Constitutivity for the light inducible chloroplast transfer RNAs, *Proc. Natl. Acad. Sci. USA.,* 70, 1749, 1973.

150. **Chang, S. H., Hecker, L. I., Silberklang, M., Brum, C. K., Rajbhandary, U. L., and Barnett, W. E.,** Nucleotide sequence of phenylalanine transfer RNA from the chloroplasts of *Euglena gracilis, Cell,* 9, 717, 1976.

151. **Everett, G. A. and Madison, J. T.,** Nucleotide sequence of phenylalanine transfer ribonucleic acid from pea (*Pisum sativum,* Alaska), *Biochemistry,* 15, 1016, 1976.

152. **Hiatt, V. S. and Snyder, L. A.,** Phenylalanine transfer RNA species in early development of barley, *Biochim. Biophys. Acta,* 324, 57, 1973.

153. **Labuda, D., Janowicz, Z., Haertle, T., and Augustyniak, J.,** Isolation and chromatographic behaviour of phenylalanine tRNA from barley embryos, *Nucleic Acid Res.,* 1, 1703, 1974.

154. **Rafalski, A. J., Barciszewski, J., Guzewicz, K., Twardowski, T., and Keith, G.,** Nucleotide sequence of tRNA[Phe] from the seeds of lupin. Comparison of the major species with wheat germ tRNA[Phe], *Acta Biochem. Pol.,* 24, 301, 1977.

155. **Guillemaut, P. and Keith, G.,** Primary structure of bean chloroplast tRNA[Phe]: comparison with *Euglena* chloroplast tRNA[Phe], *FEBS. Lett.,* 84, 351, 1977.

156. **Guillemaut, P., Martin, R., and Weil, J. H.,** Purification and base composition of a chloroplastic tRNA[Phe] from *Phaseolus vulgaris, FEBS Lett.,* 63, 273, 1976.

157. **Yoshikami, D. and Keller, E. B.,** Chemical modification of the fluorescent base in phenylalanine transfer ribonucleic acid, *Biochemistry,* 10, 2969, 1971.

158. **Marcu, K. B., Mignery, R. E., and Dudock, B. S.,** Complete nucleotide sequence and properties of the major species of glycine tRNA from wheat germ, *Biochemistry,* 16, 797, 1977.

159. **Preddie, D. L., Preddie, E. C., Guerrini, A. M., and Cremona, T.,** Two isoaccepting species of tryptophan tRNA from *Chlamydomonas reinhardii, Can. J. Bot.,* 51, 951, 1973.

160. **Brownlee, G. G.,** Determination of sequences in RNA, *Laboratory Techniques in Biochemistry and Molecular Biology,* Work, T. S. and Work, E., Eds., North-Holland, Amsterdam, 1976.

161. **Sanger, F., Brownlee, G. G., and Barrel, B. G.,** A two-dimensional fractionation procedure for radioactive nucleotides, *J. Mol. Biol.,* 13, 373, 1965.

162. **Simsek, M., Ziegenmeyer, J., Heckman, J., and Rajbhandary, U. L.,** Absence of the sequence G-T-ψ-C-G(A) in several eukaryotic cytoplasmic initiator transfer RNAs, *Proc. Natl. Acad. Sci. USA.,* 70, 1041, 1973.

163. **Simsek, M., Petrissant, G., and Rajbhandary, U. L.,** Replacement of the sequence G-T-ψ-C-G(A) by G-A-U-C-G in initiator transfer RNA of rabbit liver cytoplasm, *Proc. Natl. Acad. Sci. USA.,* 70, 2600, 1973.

164. **Gillum, A. M., Urquhart, N., Smith, M., and Rajbhandary, U. L.,** Nucleotide sequence of salmon testes and salmon liver cytoplasmic initiator tRNA, *Cell,* 6, 395, 1975.

165. **Szeto, K. S. and Söll, D.,** Fingerprinting non radioactive ribonucleic acid with the aid of polynucleotide phosphorylase, *Nucleic Acid Res.,* 1, 171, 1974.

166. **Szeto, K. S. and Söll, D.,** Sequence studies of nonradioactive Mycoplasma tRNA[Phe] with the aid of polynucleotide phosphorylase and polynucleotide kinase, *Nucleic Acid Res.,* 1, 1733, 1974.

167. **Randerath, K., Randerath, E., Chia, L. S. Y., Gupta, R. C., and Sivarajan, M.,** Sequence analysis of unradioactive RNA fragments by periodate phosphatase digestion and chemical tritium labeling: characterization of large oligonucleotides and oligonucleotides containing modified nucleosides, *Nucleic Acid Res.,* 1, 1121, 1974.

168. **Sivarajan, M., Gupta, R. C., Chia, L. S. Y., Randerath, E., and Randerath, K.,** Tritium sequence analysis of oligoribonucleotides: a combination of postlabeling and thin-layer chromatographic techniques for the analysis of partial snake venom phosphodiesterase digests, *Nucleic Acid Res.,* 1, 1329, 1974.

169. **Holley, R., Apgar, J., Everett, G. A., Madison, J. T., Marquisee, M., Merrill, S. H., Penswick, J. R., and Zamir, A.,** Structure of a ribonucleic acid, *Science,* 147, 1462, 1965.

170. Barrel, B. G. and Clark, B. F. C., *Handbook of Nucleic Acid Sequences,* Joynson Bruvvers, Oxford, 1974.

171. Sodd, M. A., Analysis of the primary and secondary structure of tRNA, in *Handbook of Biochemistry and Molecular Biology,* Vol. 2, 3rd ed., Fasman, G. D., Ed., CRC Press, Cleveland, 1976, 423.

172. Dirheimer, G., Keith, G., Martin, R., and Weissenbach, J., Primary sequences of tRNAs, in *Synthesis, Structure and Chemistry of tRNA and Their Components,* Proc. Int. Conf., Dymaczewo, Poznan, 1976, 273.

173. Dube, S. K., Marcker, K. A., Clark, B. F. C., and Cory, S., Nucleotide sequence of N-formylmethionyl-transfer RNA, *Nature (London),* 218, 232, 1968.

174. Ecarot, B. and Cedergren, R. J., Structure-function correlations in formylmethionine transfer RNA, *Biochem. Biophys. Res. Commun.,* 59, 400, 1974.

175. Delk, A. S. and Rabinowitz, J. C., Partial nucleotide sequence of a prokaryote initiator tRNA that functions in its non-formylated form, *Nature (London),* 252, 106, 1974.

176. Walker, R. T. and Rajbhandary, U. L., Formylatable methionine transfer RNA from Mycoplasma: purification and comparison of partial nucleotide sequence with those of other procaryotic initiator tRNAs, *Nucleic Acids Res.,* 2, 61, 1975.

177. Simsek, M. and Rajbhandary, U. L., The primary structure of yeast initiator transfer ribonucleic acid, *Biochem. Biophys. Res. Commun.,* 49, 508, 1972.

178. Piper, P. W. and Clark, B. F. C., Primary structure of a mouse myeloma cell initiator transfer RNA, *Nature (London),* 247, 516, 1974.

179. Simsek, M., Rajbhandary, U. L., Boisnard, M., and Petrissant, G., Nucleotide sequence of rabbit liver and sheep mammary gland cytoplasmic initiator transfer RNAs, *Nature (London),* 247, 518, 1974.

180. Ghosh, K., Ghosh, H. P., Simsek, M., and Rajbhandary, U. L., Initiator methionine transfer ribonucleic acid from wheat embryo. Purification, properties and partial nucleotide sequences, *J. Biol. Chem.,* 249, 4720, 1974.

181. Hecker, L. I., Uziel, M., and Barnett, W. E., Comparative base composition of chloroplast and cytoplasmic tRNAs^{Phe} from *Euglena gracilis, Nucleic Acids Res.,* 3, 371, 1976.

182. Weil, J. H., Burkard, G., Guillemaut, P., Jeannin, G., Martin, R., and Steinmetz, A., tRNAs and aminoacyl-tRNA synthetases in plant cytoplasm, chloroplasts and mitochondria, in *Nucleic Acids and Protein Synthesis in Plants,* Bogorad, L. and Weil, J. H., Eds., Plenum Press, New York, 1977, 97.

183. Clark, B. F. C., Doctor, B. P., Holmes, K. C., Klug, A., Marcker, K. A., Morris, S. J., and Paradies, H. H., Crystallization of transfer RNA, *Nature (London),* 219, 1222, 1968.

184. Kim, S. H. and Rich, A., Single crystals of transfer RNA: An X-ray diffraction study, *Science,* 162, 1381, 1968.

185. Hampel, A., Labanauskas, M., Connors, P. G., Kirkegard, L., Rajbhandary, U. L., Sigler, P., and Bock, R. M., Single crystals of transfer RNA from forylmethionine and phenylalanine transfer RNAs, *Science,* 162, 1384, 1968.

186. Cramer, F., Von der Haar, F., Saenger, W., and Schlimme, E., Single crystals of phenylalanine-specific transfer ribonucleic acid, *Angew. Chem. Int. Ed. Engl.,* 7, 895, 1968.

187. Fresco, J. R., Blake, R. D., and Langridge, R., Crystallization of transfer ribonucleic acids from unfractionated mixtures, *Nature (London),* 220, 1285, 1968.

188. Kim, S. H., Quigley, G., Suddath, F. L., and Rich, A., High resolution X-ray diffraction patterns of crystalline transfer RNA that show helical regions, *Proc. Natl. Acad. Sci. USA.,* 68, 841, 1971.

189. Colby, D. S., Schedl, P., and Guthrie, C., A functional requirement for the wobble nucleotide in the anticodon of a T$_4$ suppressor tRNA, *Cell,* 9, 449, 1976.

190. Marinus, M. G., Morris, N. R., Söll, D., and Kwong, T. C., Isolation and partial characterization of three *E. coli* mutants with altered transfer ribonucleic acid methylases, *J. Bacteriol.,* 122, 257, 1975.

191. Erdmann, V. A., Sprinzl, M., and Pongs, O., The involvement of 5S RNA in the binding of tRNA to ribosomes, *Biochem. Biophys. Res. Commun.,* 54, 942, 1973.

192. Schwarz, U., Lührmann, R., and Gassen, H. G., On the mRNA induced conformational change of AA-tRNA exposing the TψCG sequence for the binding to the 50S ribosomal subunit, *Biochem. Biophys. Res. Commun.,* 56, 807, 1974.

193. Singer, C. E., Smith, G. R., Cortese, R., and Ames, B. N., Mutant tRNA^{His} ineffective in repression and lacking two pseudouridine modifications *Nature (London), New Biol.,* 238, 72, 1972.

194. Goldberger, R. F., Autogeneous regulation of gene expression, *Science,* 183, 810, 1974.

195. Shugart, L., Chastain, B. H., Novelli, G. D., and Stulberg, M. D., Restoration of aminoacylation activity of undermethylated transfer RNA by *in vitro* methylation, *Biochem. Biophys. Res. Commun.,* 31, 404, 1968.

196. **Yoshida, M., Takeishi, K., Ukita, T.,** Anticodon structure of GAA-specific glutamic acid tRNA from yeast, *Biochem. Biophys. Res. Commun.,* 39, 852, 1970.

197. **Roe, B., Sirover, M., and Dudock, B.,** Kinetics of homologous and heterologous aminoacylation with yeast phenylalanyl-transfer ribonucleic acid synthetase, *Biochemistry,* 12, 4146, 1973.

198. **Gefter, M. and Russel, R. L.,** Role of modifications in tyrosine transfer RNA: A modified base affecting ribosome binding, *J. Mol. Biol.,* 39, 145, 1969.

199. **Björk, G. R. and Isaksson, L. A.,** Isolation of mutants of *E. coli* lacking 5-methyluracil in transfer ribonucleic acid of 1-methylguanine in ribosomal RNA, *J. Mol. Biol.,* 51, 83, 1970.

200. **Dunn, M. B. and Hall, R. H.,** Purines, pyrimidines, nucleosides and nucleotides: Physical constants and spectral properties, in *Handbook of Biochemistry and Molecular Biology,* Vol. 1, Fasman, G. D., Ed., CRC Press, 1975, 65.

201. **Gruol, D. J. and Haselkorn, R.,** Counting the genes for stable RNA in the nucleus and chloroplasts of *Euglena, Biochim. Biophys. Acta,* 447, 82, 1976.

202. **Commerford, S. L.,** Iodination of nucleic acids *in vitro, Biochemistry,* 10, 1993, 1971.

203. **Mc Crea, J. M. and Hershberger, C. L.,** Chloroplast DNA codes for transfer RNA, *Nucleic Acids Res.,* 3, 2005, 1976.

204. **Schwartzbach, S. D., Hecker, L. I., and Barnett, W. E.,** Transcriptional origin of *Euglena* chloroplast tRNA, *Proc. Natl. Acad. Sci. USA.,* 73, 1984, 1976.

205. **Gillespie, D. and Spiegelman, S.,** A quantitative assay for DNA-RNA hybrids with DNA immobilized on a membrane, *J. Mol. Biol.,* 12, 829, 1965.

206. **Bonner, J., Kung, G., and Bekhor, I.,** A method for the hybridization of nucleic acid molecules at low temperature, *Biochemistry,* 6, 3650, 1967.

207. **Kourilsky, P., Manteuil, S., Zamansky, M. H., and Gros, F.,** DNA-RNA hybridization at low temperature in the presence of urea, *Biochem. Biophys. Res. Commun.,* 41, 1080, 1970.

208. **Steinmetz, A. and Weil, J. H.,** Hybridization of bean chloroplast transfer RNAs to chloroplast DNA, *Biochim. Biophys. Acta,* 454, 429, 1976.

209. **Morell, P., Smith, I., Dubnau, D., and Marmur, J.,** Isolation and characterization of low molecular weight ribonucleic acid species from *Bacillus subtilis, Biochemistry,* 6, 259, 1967.

210. **Smith, I., Dubnau, D., Morell, P., and Marmur, J.,** Chromosomal location of DNA base sequences complementary to transfer RNA and to 5S, 16S and 23S ribosomal RNA in *Bacillus subtilis, J. Mol. Biol.,* 33, 123, 1968.

211. **Schweizer, E., Mac Kechnie, C., and Halvorson, H. O.,** The redundancy of ribosomal and transfer RNA genes in *Saccharomyces cerevisiae, J. Mol. Biol.,* 40, 261, 1969.

212. **Tartof, K. D. and Perry, R. P.,** The 5S RNA genes of *Drosophila melanogaster, J. Mol. Biol.,* 51, 171, 1970.

213. **Hatzen, L. and Attardi, G.,** Proportion of the HeLa cell genome complementary to transfer RNA and 5S RNA, *J. Mol. Biol.,* 56, 535, 1971.

214. **Clarkson, S. G., Birnstiel, M. L., and Serra, V.,** Reiterated transfer RNA genes of *Xenopus laevis, J. Mol. Biol.,* 79, 391, 1973.

215. **Tewari, K. K. and Wildman, S. C.,** Control of organelle development, in *Symposia of the Society for Experimental Biology,* Vol. 24, Miller, P. L., Ed., Cambridge University Press, London, 1970, 147.

216. **Tewari, K. K., Kolodner, R., Chu, N. M., and Meeker, R. M.,** Structure of chloroplast DNA, in *Nucleic Acids and Protein Synthesis in Plants,* Bogorad, L. and Weil, J. H., Eds., Plenum Press, New York, 1977, 15.

217. **Haff, L. A. and Bogorad, L.,** Hybridization of maize chloroplast DNA with transfer ribonucleic acids, *Biochemistry,* 15, 4105, 1976.

218. **Williams, G. R., Williams, A., and George, S. A.,** Hybridization of leucyl-transfer ribonucleic acid isoacceptors from green leaves with nuclear and chloroplastic deoxyribonucleic acid, *Proc. Natl. Acad. Sci. USA.,* 70, 3498, 1973.

219. **Fukuhara, H., Bolotin-Fukuhara, M., Hsu, H. J., and Rabinowitz, M.,** Deletion mapping of mitochondrial transfer RNA genes in *Saccharomyces cerevisiae* by means of cytoplasmic petite mutants, *Molec. Gen. Genet.,* 145, 7, 1976.

220. **Martin, N. and Rabinowitz, M.,** Transfer RNAs of yeast mitochondria, in *Genetics and Biogenesis of Chloroplasts and Mitochondira,* Bücher, Th., Neufert, W., Sebald, W., and Werner, S., Eds., Elsevier, Amsterdam, 1976, 749.

221. **Attardi, G., Albring, M., Amalric, F., Gelfand, R., Griffith, J., Lynch, D., Merkel, C., Murphy, W., and Ojala, D.,** Organization and expression of the mitochondrial genome in HeLa cells, in *Genetics and Biogenesis of Chloroplasts and Mitochondria,* Bücher, Th. Neufert, W., Sebald, W., and Werner, S., Eds., Elsevier, Amsterdam, 1976, 573.

222. **Squires, C. and Carbon, J.,** Normal and mutant glycine transfer RNAs, *Nature (London), New Biol.,* 233, 274, 1971.

223. Squires, C., Konrad, B., Kirschbaum, J., and Carbon, J., Three adjacent transfer RNA genes in *E. coli*, *Proc. Natl. Acad. Sci. USA.*, 70, 438, 1973.

224. Wu, M., Davisson, N., and Carbon, J., Physical mapping of the transfer RNA genes on λ h 80 d gly Tsu₃₆, *J. Mol. Biol.*, 78, 23, 1976.

225. Altman, S., Isolation of tyrosine tRNA precursor molecules, *Nature, London, New Biol.*, 229, 19, 1971.

226. Burdon, R. H., Ribonucleic acid maturation in animal cells, *Prog. Nucleic Acid Res. Mol. Biol.*, 11, 33, 1971.

227. Altman, S. and Smith, J. D. Tyrosine tRNA precursor molecule polynucleotide sequence, *Nature (London), New Biol.*, 233, 35, 1971.

228. Schaeffer, K. P., Altman, S., and Söll, D., Nucleotide modification *in vitro* of the precursor of transfer RNA¹ʸʳ of *E. coli*, *Proc. Natl. Acad. Sci. USA.*, 70, 3626, 1973.

229. Wilson, J. H., Kim, J. S., and Abelson, J. N., Bacteriophage T₄ transfer RNA. III. Clustering of the genes for the T₄ transfer RNAs, *J. Mol. Biol.*, 71, 547, 1972.

230. Mc Clain, W. H., Guthrie, C., and Barrell, B. G., Eight transfer RNAs induced by infection of *E. coli* with bacteriophage T₄, *Proc. Natl. Acad. Sci. USA.*, 69, 3703, 1972.

231. Deutscher, M. D., Synthesis and functions of the CCA terminus of transfer RNA, *Prog. Nucleic Acid Res. Mol. Biol.*, 13, 51, 1973.

232. Seidman, J. G., Barrell, B. G., and Mc Clain, W. A., Five steps in the conversion of a large precursor RNA into bacteriophage proline and serine transfer RNAs, *J. Mol. Biol.*, 99, 733, 1975.

233. Marks, A. and Spencer, J. H., Isolation of *E. coli* transfer RNA-gene hybrids, *J. Mol. Biol.*, 51, 115, 1970.

234. Marks, A., Isolation of an aminoacyl tyrosine tRNA-tsDNA cistron hybrid, *Biochimie*, 55, 443, 1973.

235. Brenner, D. J., Fournier, M. J., and Doctor, B. P., Isolation and partial characterization of the transfer ribonucleic acid cistrons from *E. coli*, *Nature (London)*, 227, 448, 1970.

236. Ryhn, J. L. and Morowitz, H. J., Partial purification of native rRNA and tRNA cistrons from Mycoplasma sp. (Kid), *Proc. Natl. Acad. Sci. USA;* 63, 1282, 1969.

237. Daniel, V., Beckmann, J. S., Sarid, S., Grimberg, J. I., Herzberg, M., and Littauer, U. Z., Purification and *in vitro* transcription of a transfer RNA gene, *Proc. Natl. Acad. Sci. USA.*, 68, 2268, 1971.

238. Agarwal, K. L., et al., Total synthesis of the gene for an alanine transfer ribonucleic acid from yeast, *Nature (London)*, 227, 27, 1970.

239. Khorana, H. G., et al., Studies on polynucleotides. III. Total synthesis of the structural gene for an alanine transfer ribonucleic acid from yeast, *J. Mol. Biol.*, 72, 209, 1962.

240. Khorana, H. G., et al., Total synthesis of the structural gene for the precursor of a tyrosine suppressor transfer RNA from *E. coli*, *J. Biol. Chem.*, 251, 565, 1976.

241. Daniel, V., Sarid, S., Beckmann, J., and Littauer, U. Z., *In vitro* transcription of a transfer RNA gene, *Proc. Natl. Acad. Sci. USA.*, 66, 1260, 1970.

242. Bikoff, E. and Gefter, M., *In vitro* synthesis of transfer RNA, *J. Biol. Chem.*, 250, 6240, 6248, 1975.

243. Zubay, G., Cheong, L., and Gefter, M., DNA-directed cell-free synthesis of biologically active transfer RNA: su⁺ₗₗₗ tyrosyl-tRNA, *Proc. Natl. Acad. Sci. USA.*, 68, 2195, 1971.

244. Nierlich, D. P., Lamfrom, H., Sarabhai, A., and Abelson, J., Transfer RNA synthesis *in vitro*, *Proc. Natl. Acad. Sci. USA.*, 70, 179, 1973.

245. Daniel, V., Grimberg, J. I., and Zeevi, M., *In vitro* synthesis of tRNA precursors and their conversion to mature size tRNA, *Nature (London)*, 257, 193, 1975.

246. Burdon, R. H. and Clason, A. E., Intracellular location and molecular characteristics of tumor cell transfer RNA precursors, *J. Mol. Biol.*, 39, 113, 1969.

247. Bernhardt, D. and Darnell, J. E., tRNA synthesis in HeLa cells: a precursor to tRNA and the effects of methionine starvation on tRNA synthesis, *J. Mol. Biol.*, 42, 43, 1969.

248. Kay, J. E. and Cooper, H. L., Rapidly labeled cytoplasmic RNA in normal and phytohaemagglutinin-stimulated human lymphocytes, *Biochim. Biophys. Acta*, 186, 62, 1969.

249. Mowshowitz, D. B., Transfer RNA synthesis in HeLa cells, II: Formation of tRNA from a precursor *in vitro* and formation of pseudo-uridine, *J. Mol. Biol.*, 50, 143, 1970.

250. Friedlander, A. and Buonassisi, V., Kinetics of synthesis of cytoplasmic RNA with transfer properties in cultures of adrenal tumor cells, *Biochim. Biophys. Acta*, 213, 101, 1970.

251. Choe, B. K. and Taylor, W. M., Kinetics of synthesis and characterization of transfer RNA precursors in mammalian cells, *Biochim. Biophys. Acta*, 272, 275, 1972.

252. Robertson, H. D., Altman, S., and Smith, J. D., Purification and properties of a specific *E. coli* ribonuclease which cleaves a tyrosine transfer ribonucleic acid precursor, *J. Biol. Chem.*, 247, 5243, 1972.

253. **Schedl, P. and Primakoff, P.**, Mutants of *E. coli* thermosensitive for the synthesis of transfer RNA, *Proc. Natl. Acad. Sci. USA.*, 70, 2091, 1973.

254. **Schedl, P., Roberts, J., and Primakoff, P.**, *In vitro* processing of *E. coli* tRNA precursors, *Cell*, 8, 581, 1976.

255. **Hall, R. H.**, *The modified nucleosides in nucleic acids*, Columbia University Press, New York, 1971.

256. **Söll, D.**, Enzymatic modification of transfer RNA, *Science*, 173, 293, 1971.

257. **Nishimura, S.**, Minor components in transfer RNA: Their characterization, location and function, *Prog. Nucleic Acid Res. Mol. Biol.*, 12, 49, 1972.

258. **Wilson, J. H. and Abelson, J. N.**, Bacteriophage T₄ transfer RNA. II. Mutants of T₄ defective in the formation of functional suppressor transfer RNA, *J. Mol. Biol.*, 69, 57, 1972.

259. **Borek, E. and Srinivasan, P. R.**, The methylation of nucleic acids, *Annu. Rev. Biochem.*, 35, 275, 1966.

260. **Srinivasan, P. R. and Borek, E.**, The species variation of RNA methylase, *Proc. Natl. Acad. Sci. USA.*, 49, 529, 1963.

261. **Stone, B. P., Whitty, C. D., and Cherry, J. H.**, Effect of ethionine on invertase development and methylation of ribonucleic acid, *Plant Physiol.*, 45, 636, 1970.

262. **Streeter, D. G. and Lane, B. G.**, Studies of the biogenesis of N^2-dimethylguanylate. I. Generation of N^2-dimethylguanylate when bulk *E. coli* transfer RNA is used as a substrate for wheat embryo methyltransferases, *Biochim. Biophys. Acta*, 199, 394, 1970.

263. **Stone, B. P. and Cherry, J. H.**, Induced production of invertase in sugar beet root by γ-irradiation: role of RNA, *Planta*, 102, 179, 1972.

264. **Abeels, M., Digneffe, C., and Dubois, E.**, Methylation of nucleic acids by higher plants, *Symp. Biol. Hung.*, 13, 69, 1972.

265. **King, B. and Chapman, J. M.**, Non-coordinate synthesis and methylation of tRNA following excision of plant tissue, *Planta*, 114, 227, 1973.

266. **Wierzbicka, H., Jakubowski, H., and Pawelkiewicz, J.**, Transfer RNA methyltransferases from yellow lupin seeds: Purification and properties, *Nucleic Acids Res.*, 2, 101, 1975.

267. **Marcu, K., Mignery, M., Reszelbach, R., Roe, B., Sirover, M., and Dudock, B.**, The absence of ribothymidine in specific eukaryotic transfer RNAs: I. Glycine and threonine tRNAs of wheat embryo, *Biochem. Biophys. Res. Commun.* 55, 477, 1973.

268. **Dubois, E. G., Dirheimer, G., and Weil, J. H.**, Methylation of yeast tRNA^{Asp} by enzymes from cytoplasm, chloroplasts and mitochondria of *Phaseolus vulgaris*, *Biochim. Biophys. Acta*, 374, 332, 1974.

269. **Dubois, E. G., Dirheimer, G., and Weil, J. H.**, Characterization of an organelle-specific 7-methylguanine tRNA methylase in the chloroplasts and mitochondria of *Phaseolus vulgaris*, *Plant Sci. Lett.*, 5, 17, 1975.

270. **Borek, E.**, Modification of nucleic acids in relation to differentiation, *Trends Biochem. Sci.*, 2, 3, 1977.

271. Transfer RNA and transfer RNA modification in differentiation and neoplasia, *Cancer Res.*, 31, 596, 1971.

272. **Chandra, G. R. and Duynstee, E. E.**, Methylation of ribonucleic acids and hormone-induced α-amylase synthesis in the aleurone cells, *Biochim. Biophys. Acta*, 232, 514, 1971.

273. **Dubois, E. G. and Weil, J. H.**, A 7-methylguanine tRNA methylase present in crown-gall cells but absent in normal *Parthenocissus tricuspidata* L cells, *Plant Sci. Lett.*, 8, 385, 1977.

274. **Nau, F.**, The methylation of tRNA, *Biochimie*, 58, 629, 1976.

275. **Gambaryan, A. S., Venkstern, T. V., and Bayev, A. A.**, On the mechanism of tRNA methylase-tRNA recognition, *Nucleic Acid Res.*, 3, 2079, 1976.

276. **Johnson, L. and Söll, D.**, *In vitro* biosynthesis of pseudouridine at the polynucleotide level by an enzyme extract from *E. coli*, *Proc. Natl. Acad. Sci. USA.*, 67, 943, 1970.

277. **Wong, T. W., Weiss, S. B., Eliceiri, G., and Bryant, J.**, Ribonucleic acid sulfurtransferase from *B. subtilis* W168. Sulfuration with β-mercaptopyruvate and properties of the enzyme system, *Biochemistry*, 9, 2376, 1970.

278. **Lipsett, M. N.**, Biosynthesis of 4-thiouridylate, *J. Biol. Chem.*, 247, 1458, 1972.

279. **Bronskill, P., Kennedy, T. D., and Lane, B. G.**, Cell-free enzymic esterification of 5-carboxymethyluridine residues in bulk yeast transfer RNA, *Biochim. Biophys. Acta*, 262, 275, 1972.

280. **Bartz, J. K. and Söll, D.**, N^6 (Δ^2-isopentenyl)adenosine : Biosynthesis *in vitro* in transfer RNA by an enzyme purified from *E. coli*, *Biochimie*, 54, 31, 1972.

281. **Rosenbaum, N. and Gefter, M. L.**, Δ^2-isopentenylpyrophosphate: transfer ribonucleic acid Δ^2-isopentenyltransferase from *E. coli*, *J. Biol. Chem.*, 247, 5675, 1972.

282. **Powers, D. M. and Peterkofsky, A.**, Biosynthesis and specific labeling of N-(purin-6-ylcarbamoyl) threonine of *E. coli* transfer RNA, *Biochim. Biophys. Res. Commun.*, 46, 831, 1972.

283. Körner, A. and Söll, D., N-(purin-6-ylcarbamoyl)threonine: Biosynthesis *in vitro* in transfer RNA by an enzyme purified from *E. coli*, *FEBS Lett.*, 39, 301, 1974.

284. Cortese, R., Kammen, H. O., Spengler, S. J., and Ames, B. N., Biosynthesis of pseudouridine in transfer ribonucleic acid, *J. Biol. Chem.*, 249, 1103, 1974.

285. Allauden, H. S., Yang, S. K., and Söll, D., Leucine tRNA₁ from His T mutant of *Salmonella typhimurium* lacks two pseudouridines, *FEBS Lett.*, 28, 205, 1972.

286. Armstrong, D. J., Skoog, F., Kirkegaard, L. H., Hampel, A. E., Bock, R. M., Gillam, I., and Tener, G. M., Cytokinins: Distribution in species of yeast transfer RNA, *Proc. Natl. Acad. Sci. USA.*, 63, 504, 1969.

287. Skoog, F., Survey of cytokinins and cytokinin antagonists with reference to nucleic acid and protein metabolism, *Biochem. Soc. Symp.*, 38, 195, 1973.

288. Hall, R. H., N⁶-(Δ²-isopentenyl)adenosine: Chemical reactions, biosynthesis, metabolism, and significance to the structure and function of tRNA, *Prog. Nucleic Acid Res. Mol. Biol.*, 10, 57, 1970.

289. Fox, J. E. and Erion, J. L., A cytokinin binding protein from higher plant ribosomes, *Biochem. Biophys. Res. Commun.*, 64, 964, 1975.

290. Phillips, J. H. and Kjellin-Straby, K., Studies on microbial ribonucleic acid. IV. Two mutants of *Saccharomyces cerevisiae* lacking N²-dimethylguanine in soluble ribonucleic acid, *J. Mol. Biol.*, 26, 509, 1967.

291. Allende, J. E. and Allende, C. C., Detection and isolation of complexes between aminoacyl-tRNA synthetases and their substrates, in *Methods in Enzymology*, Academic Press, New York, 20, 1971, 210.

292. Mitra, S. K. and Mehler, A., The arginyl transfer ribonucleic acid synthetase of *E. coli*, *J. Biol. Chem.*, 242, 5490, 1967.

293. Mitra, S. K. and Smith, C. J., Absolute requirement for transfer RNA in the activation of arginine by arginyl transfer RNA synthetase of yeast, *Biochim. Biophys. Acta*, 190, 222, 1969.

294. Parfait, R. and Grosjean, H., Arginyl transfer ribonucleic acid synthetase from *Bacillus stearothermophilus*, *Eur. J. Biochem.*, 30, 242, 1972.

295. Ravel, J. M., Wang, S. F., Heinemeyer, C., and Shive, W., Glutamyl and glutaminyl ribonucleic acid synthetases of *E. coli* W, *J. Biol. Chem.*, 240, 432, 1965.

296. Lee, L. W., Ravel, J. M., and Shive, W., A general involvement of acceptor ribonucleic acid in the initial activation step of glutamic acid and glutamine, *Arch. Biochem. Biophys.*, 121, 614, 1967.

297. Lapointe, J. and Söll, D., Glutamyl transfer ribonucleic acid synthetase of *E. coli*, *J. Biol. Chem.*, 247, 4966, 1972.

298. Deutscher, M. D., Rat liver glutamyl ribonucleic acid synthetase II. Further properties and anomalous pyrophosphate exchange, *J. Biol. Chem.*, 242, 1132, 1967.

299. Lea, P. J. and Fowden, L., Amino acid substrate specificity of asparaginyl-, aspartyl-, and glutaminyl- tRNA synthetase isolated from higher plants, *Phytochemistry*, 12, 1903, 1973.

300. Loftfield, R. B. and Eigner, E. A., Mechanism of action of amino acid transfer ribonucleic acid ligases, *J. Biol. Chem.*, 244, 1746, 1969.

301. Fraser, T. H. and Rich, A., Amino acids are not all initially attached to the same position on transfer RNA molecules, *Proc. Natl. Acad. Sci. USA.*, 72, 3044, 1975.

302. Sprinzl, M. and Cramer, F., Site of aminoacylation of tRNAs from *E. coli* with respect to the 2′- or 3′-hydroxyl group of the terminal adenosine, *Proc. Natl. Acad. Sci. USA.*, 72, 3049, 1975.

303. Hall, T. C. and Tao, K. L., Rates of aminoacyl-transfer ribonucleic acid synthesis *in vivo* and *in vitro* by bean leaves, *Biochem. J.*, 117, 853, 1970.

304. Jakubowski, H. and Pawelkiewicz, J., Chromatography of plant aminoacyl-tRNA synthetases on ω-aminoalkyl Sepharose columns, *FEBS Lett.*, 34, 150, 1973.

305. Forrester, P. J. and Hancock, R. L., Affinity chromatography of phenylalanine-tRNA ligase, *Can. J. Biochem.*, 51, 231, 1973.

306. Bartkowiak, S. and Pawelkiewicz, J., The purification of aminoacyl-tRNA synthetases by affinity chromatography, *Biochim. Biophys. Acta*, 272, 137, 1972.

307. Remy, P., Birmele, C., and Ebel, J. P., Purification of yeast phenylalanyl-tRNA synthetase by affinity chromatography on a tRNA^Phe-Sepharose column, *FEBS Lett.*, 27, 134, 1972.

308. Beauchemin, N., Larue, B., and Cedergren, R. J., The characterization of the tRNAs and aminoacyl-tRNA synthetases of the blue-green alga, *Anacystis nidulans*, *Arch. Biochem. Biophys.*, 156, 17, 1973.

309. Merrick, W. C. and Dure, L. S., Developmental biochemistry of cottonseed embryogenesis and germination. Preferential charging of cotton chloroplastic transfer ribonucleic acid by *E. coli* enzymes, *Biochemistry*, 12, 629, 1973.

310. Parthier, B. and Krauspe, R., Chloroplast and cytoplasmic transfer RNA of *Euglena gracilis*. Transfer RNA^Leu of blue-green algae as a substitute for chloroplast tRNA^Leu, *Biochem. Physiol. Pflanz.*, 165, 1, 1974.

311. **Parthier, B.,** Cytoplasmic site of synthesis of chloroplast aminoacyl-tRNA synthetases in *Euglena gracilis, FEBS Lett.,* 38, 70, 1973.

312. **Hecker, L. I., Egan, J., Reynolds, R. J., Nix, C. E., Schiff, J. A., and Barnett, W. E.,** The sites of transcription and translation for *Euglena* chloroplastic aminoacyl-tRNA synthetases, *Proc. Natl. Acad. Sci. USA.,* 71, 1910, 1974.

313. **Barnett, W. E., Schwartzbach, S. D., Farelly, J. G., Schiff, J. A., and Hecker, L. I.,** Comments on the translational and transcriptional origin of *Euglena* chloroplastic aminoacyl-tRNA synthetases, *Arch. Microbiol.,* 109, 201, 1976.

314. **Nover, L.,** Density labeling of chloroplast-specific leucyl-tRNA synthetase in greening cells of *Euglena gracilis, Plant Sci. Lett.,* 7, 403, 1976.

315. **Allende, J. E.,** Protein synthesis in plant systems, in *Techniques in Protein Biosynthesis,* Campbell, P. N. and Sargent, J. R., Eds., Academic Press, New York, 1969, 2, 55.

316. **Legocki, A. B.,** Function of elongation factors in peptide synthesis, *Biochem. Soc. Symp.,* 38, 57, 1973.

317. **Allende, J. E., Tarrago, A., Monasterio, O., Litvak, S., Gatica, M., Ojeda, J. M., and Matamala, M.,** The binding of aminoacyl-transfer ribonucleic acid to wheat ribosomes, *Biochem. Soc. Symp.,* 38, 77, 1973.

318. **Chua, N. H., Blobel, G., and Siekevitz, P.,** Isolation and characterization of chloroplast 70S and of 80S ribosomes of *Chlamydomonas reinhardii:* Protein synthesis *in vitro, Biochem. Soc. Symp.,* 38, 163, 1973.

319. **Ciferri, O., Tiboni, O., Munoz-Calvo, M. L., and Camerino, G.,** Protein synthesis in plants: Specificity and role of the cytoplasmic and organellar systems, in *Nucleic Acids and Protein Synthesis in Plants,* Bogorad, L. and Weil, J. H., Eds., Plenum Press, New York, 1977, 155.

320. **Ellis, R. J.,** The synthesis of chloroplast proteins, in *Nucleic Acids and Protein Synthesis in Plants,* Bogorad, L. and Weil, J. H., Eds., Plenum Press, New York, 1977, 195.

321. **Leaver, C. J. and Pope, P. K.,** Biosynthesis of plant mitochondria proteins, in *Nucleic Acids and Protein Synthesis in Plants,* Bogorad, L. and Weil, J. H., Eds., Plenum Press, New York, 1977, 213.

322. **Schwartz, J., Meyer, R., Eisenstadt, J., and Brawerman, G.,** Involvement of N-formylmethionine in initiation of protein synthesis in cell-free extracts of *Euglena gracilis, J. Mol. Biol.,* 25, 571, 1967.

323. **Bachmayer, H.,** Initiation of protein synthesis in intact cells and in isolated chloroplasts of *Acetabularia mediterranea, Biochim. Biophys. Acta,* 209, 584, 1970.

324. **Bianchetti, R., Lucchini, G., and Sartirana, M. L.,** Endogenous synthesis of formyl-methionine peptides in isolated mitochondria and chloroplasts, *Biochem. Biophys. Res. Commun.,* 42, 97, 1971.

325. **Merrick, W. C. and Dure, L. S.,** Specific transformylation of one methionyl-tRNA from cotton seedling chloroplasts by endogenous and *E. coli* transformylases, *Proc. Natl. Acad. Sci. USA.,* 68, 641, 1971.

326. **Marcus, A., Week, D. P., Leis, J. P., and Keller, E. B.,** Protein chain initiation by methionyl-tRNA in wheat embryo, *Proc. Natl. Acad. Sci. USA.,* 67, 1681, 1970.

327. **Tarrago, A., Monasterio, O., and Allende, J. E.,** Initiator-like properties of a methionyl-tRNA from wheat embryos, *Biochem. Biophys. Res. Commun.,* 41, 765, 1970.

328. **Ghosh, K., Grishko, A., and Ghosh, H. P.,** Initiation of protein synthesis in eukaryotes, *Biochem. Biophys. Res. Commun.,* 42, 462, 1971.

329. **Ecarot, B. and Cedergren, R. J.,** Methionine transfer RNAs from the blue-green alga *Anacystis nidulans, Biochim. Biophys. Acta,* 340, 130, 1974.

330. **Marcus, A., Weeks, D. P., and Seal, S. N.,** Protein chain initiation in wheat embryo, *Biochem. Soc. Symp.,* 38, 97, 1973.

331. **Seal, S. N., Giesen, M., Roman, R., and Marcus, A.,** Functional characterization of the initiation factors of wheat germ, in *Nucleic Acids and Protein Synthesis in Plants,* Bogorad, L. and Weil, J. H., Eds., Plenum Press, New York, 1977, 167.

332. **Basilio, C., Bravo, M., and Allende, J. E.,** Ribonucleic acid code words in wheat germ, *J. Biol. Chem.,* 241, 1917, 1966.

333. **Ramiasa, J., Guillemaut, P., and Weil, J. H.,** Codon recognition pattern of *Phaseolus vulgaris* cytoplasmic and chloroplastic transfer RNAs, *FEBS Lett.,* 75, 128, 1977.

334. **Chiu, N., Chiu, A. O., and Suyama, Y.,** Three isoaccepting forms of leucyl-transfer RNA in mitochondria, *J. Mol. Biol.,* 82, 441, 1974.

335. **Chiu, N., Chiu, A. O., and Suyama, Y.,** Native and imported transfer RNA in mitochondria, *J. Mol. Biol.,* 99, 37, 1975.

336. **Soffer, R. L.,** Aminoacyl-tRNA transferases, *Adv. Enzymol.,* 40, 91, 1974.

337. **Sawyer, R. C., Harada, F., and Dahlberg, J. E.,** Virion-associated RNA primer for Rous Sarcoma Virus DNA synthesis: isolation from uninfected cells, *J. Virol.,* 13, 1302, 1974.

337a. **Taylor, J. M.**, An analysis of the role of tRNA species as primers for the transcription into DNA of RNA tumor virus genomes, *Biochim. Biophys. Acta*, 473, 57, 1977.

338. **Pinck, M., Yot, P., Chapeville, F., and Duranton, H.**, Enzymatic binding of valine to the 3′ end of TYMV-RNA, *Nature (London)*, 226, 954, 1970.

339. **Hall, T. C., Shih, D. S., and Kaesberg, P.**, Enzyme mediated binding of tyrosine to Brome Mosaic Virus ribonucleic acid, *Biochem. J.*, 129, 969, 1972.

340. **Oberg, B. and Philipson, L.**, Binding of histidine to Tobacco Mosaic Virus RNA, *Biochem. Biophys. Res. Commun.*, 48, 927, 1972.

341. **Litvak, S., Tarrago, A., Tarrago-Litvak, L., and Allende, J. E.**, Elongation factor-viral genome interaction dependent on the aminoacylation of TYMV and TMV RNAs, *Nature (London), New Biology*, 241, 88, 1973.

342. **Salomon, R. and Littauer, U. Z.**, Enzymatic aminoacylation of histidine to mengovirus RNA, *Nature (London)*, 249, 32, 1974.

343. **Pinck, M., Genevaux, M., Lestienne, P., and Duranton, H.**, Aminoacylation of viral RNAs, in *Nucleic Acids and Protein Synthesis in Plants*, Bogorad, L. and Weil, J. H., Eds., Plenum Press, New York, 1977, 377.

344. **Briand, J. P., Jonard, J., Guilley, H., Richards, K. E., and Hirth, L.**, Nucleotide sequence (n = 159) of the amino acid accepting 3′OH extremity of Turnip Yellow Mosaic Virus RNA and the last portion of its coat protein cistron, *Eur. J. Biochem.*, 72, 453, 1977.

345. **Briand, J. P., Fritsch, C., Guilley, H., Jonard, G., Klein, C., Lamy, D., Richards, K., and Hirth, L.**, Nucleotide sequences of plant viral RNAs, in *Nucleic Acids and Protein Synthesis in Plants*, Bogorad, L. and Weil, J. H., Eds., Plenum Press, New York, 1977, 343.

346. **Calhoun, D. H. and Hatfield, G. W.**, Autoregulation of gene expression, *Annu. Rev. Microbiol.*, 29, 275, 1975.

347. **McLaughlin, C. S., Magee, P. T., and Hartwell, L. H.**, Role of isoleucyl-transfer ribonucleic acid synthetase in ribonucleic acid synthesis and enzyme repression in yeast, *J. Bacteriol.*, 100, 579, 1969.

348. **Bolon, A. and Magee, P. T.**, Involvement of threonine deaminase in repression of the isoleucine-valine and leucine pathways in *Saccharomyces cerevisiae*, *J. Bacteriol.*, 113, 1333, 1973.

349. **Cherest, H., Surdin-Kerjan, Y., and De Robichon-Szulmajster, H.**, Methionine-mediated repression in *Saccharomyces cerevisiae*: A pleitropic regulatory system involving methionyl-transfer ribonucleic acid and the product of gene *eth*2, *J. Bacteriol.*, 106, 758, 1971.

350. **Messenguy, F. and Delforge, J.**, Role of transfer ribonucleic acids in the regulation of several biosyntheses in *Saccharomyces cerevisiae*, *Eur. J. Biochem.*, 67, 335, 1976.

351. **Ehresmann, B., Karst, F., and Weil, J. H.**, Regulation of the biosynthesis of valyl-tRNA synthetase in yeast, *Biochim. Biophys. Acta*, 254, 226, 1971.

352. **Ehresmann, B., Imbault, P., and Weil, J. H.**, Role of valyl-tRNA in the regulation of the biosynthesis of valyl-, isoleucyl-, and leucyl-tRNA synthetases, *Biochimie*, 56, 1351, 1974.

353. **Aboud, M. and Pastan, I.**, Stimulation of lac transcription by guanosine 5′-diphosphate 3′(3′)diphosphate and transfer ribo-nucleic acid, *J. Biol. Chem.* 249, 3356, 1973.

354. **Shridhar, V. and Pillay, D. T. N.**, Changes in leucyl-tRNAs and aminoacyl-tRNA synthetases in developing and aging soybean cotyledons, *Phytochemistry*, 15, 1809, 1976.

355. **Burkard, G., Vaultier, J. P., and Weil, J. H.**, Differences in the level of plastid-specific tRNAs in chloroplasts and etioplasts of *Phaseolus vulgaris*, *Phytochemistry*, 11, 1351, 1972.

356. **Babcock, D. F. and Morris, R. O.**, Specific degradation of a plant leucyl-transfer ribonucleic acid by a factor in the homologous synthetase preparation, *Plant Physiol.*, 52, 292, 1973.

357. **Kano-Sueoka, T., Nirenberg, M., and Sueoka, N.**, Effect of bacteriophage infection upon the specificity of leucine transfer RNA for RNA codewords, *J. Mol. Biol.*, 35, 1, 1968.

358. **Kano-Sueoka, T. and Sueoka, N.**, Characterization of a modified leucyl-tRNA of *E. coli* after bacteriophage T$_2$ infection, *J. Mol. Biol.*, 37, 475, 1968.

359. **Kano-Sueoka, T. and Sueoka, N.**, Leucine tRNA and cessation of *E. coli* protein synthesis upon phage T$_2$ infection, *Proc. Natl. Acad. Sci. USA.*, 62, 1229, 1969.

360. **Locy, R. D. and Cherry, J. H.**, Evidence for a chloroplast specific tyrosyl tRNA degrading activity, *Biochem. Biophys. Res. Commun.*, 72, 15, 1976.

361. **Chavancy, G., Daillie, J., and Garel, J. P.**, Adaptation fonctionelle des tRNA à la biosynthèse protéique dans un système cellulaire hautement différencié, *Biochimie*, 53, 1187, 1971.

362. **Garel, J. P.**, Functional adaptation of tRNA population, *J. Theor. Biol.*, 43, 211, 1974.

363. **Grunberger, D., Weinstein, I. B., and Mushinski, J. F.**, Deficiency of the Y base in a hepatoma phenylalanine tRNA, *Nature (London)*, 253, 66, 1975.

364. **Salomon, R., Giveon, D., Kimhi, Y., and Littauer, U. Z.**, Abundance of tRNA^{Phe} lacking the peroxy-Y base in mouse neuroblastoma, *Biochemistry*, 15, 5258, 1976.

365. **Sharma, O. K., Beezley, D. N., and Roberts, W. K.**, Limitation of reticulocyte transfer RNA in the translation of heterologous messenger RNAs, *Biochemistry*, 15, 4313, 1976.

366. **Zilberstein, A., Dudock, B., Berissi, H., and Revel, M.,** Control of messenger RNA translation by minor species of leucyl-transfer RNA in extracts from interferon-treated L cells, *J. Mol. Biol.,* 108, 43, 1976.

367. **Romani, R. J., Sprole, B. V., Mettler, I. J., and Tuskes, S. E.,** Extraction and purification of tRNA from fruit tissues, *Phytochemistry,* 14, 2563, 1975.

368. **Augustyniak, H. and Pawelkiewicz, J.,** Isolation and properties of the main isoleucine tRNAs from *Lupinus luteus* seeds, *Acta Biochem. Pol.,* 25, 81, 1978.

369. **Augustyniak, H. and Pawelkiewicz, J.,** Lysyl-tRNAs of *Lupinus luteus* seeds, *Phytochemistry,* 17, 15, 1978.

370. **Legocki, A. B., Szymkowiak, A., and Pawelkiewicz, J.,** Isolation of isoleucine-specific transfer ribonucleic acid from yellow lupin seeds, *Acta Biochim. Pol.,* 15, 183, 1968.

371. **Legocki, A. B., Wojciechowska, K., and Pech, K.,** Effect of plant growth substances on the differentiation of tRNA in embryos of yellow lupin *L. luteus, Bull. Acad. Pol. Sci.,* 18, 63, 1970.

372. **Crick, F. H. C.,** Codon-anticodon pairing: the wobble hypothesis, *J. Mol. Biol.,* 19, 548, 1966.

373. **Boulter, D., Ellis, R. J., and Yarwood, A.,** Biochemistry of protein synthesis in plants, *Biol. Rev. Cambridge Philos. Soc.,* 47, 113, 1972.

374. **Kerr, S. J. and Borek, E.,** Enzymic methylation of natural polynucleotides, in *The Enzymes,* Boyer, P., Ed., Vol. 9, 3rd ed., Academic Press, New York, 1973, 167.

375. **Delk, A. S. and Rabinowitz, J. C.,** Biosynthesis of ribosylthimine in the transfer RNA of *Streptococcus faecalis:* a folate-dependent methylation not involving S-adenosylmethionine, *Proc. Natl. Acad. Sci. USA.,* 72, 528, 1975.

376. **Schmidt, W., Arnold, H. H., and Kersten, H.,** Biosynthetic pathway of ribothymidine in *B. subtilis* and *M. lysodeikticus* involving different coenzymes for transfer RNA and ribosomal RNA, *Nucleic Acid Res.,* 2, 1043, 1975.

377. **Le Meur, M. A., Gerlinger, P., and Ebel, J. P.,** Messenger RNA translation in the presence of homologous tRNA, *Eur. J. Biochem.,* 67, 519, 1976.

378. **Shoemaker, H. J. P., Budzik, G. P., Giégé, R., and Schimmel, P. R.,** Three photocross-linked complexes of yeast phenylalanine transfer ribonucleic acid with aminoacyl-transfer ribonucleic acid synthetases, *J. Biol. Chem.,* 250, 4440, 1975.

379. **Shoemaker, H. J. P., Gamble, R. C., Budzik, G. P., and Schimmel, P. R.,** Comparison of isotope labeling patterns of purines in three specific transfer RNAs, *Biochemistry,* 15, 2800, 1976.

380. **Viotti, A., Balducci, C., and Weil, J. H.,** Adaptation of the tRNA population of maize endosperm for zein synthesis, *Biochim. Biophys. Acta,* 517, 125, 1978.

381. **Smith, J. D.,** Transcription and processing of transfer RNA precursors, *Prog. Nucleic Acid Res. Mol. Biol.,* 16, 25, 1976.

382. **Dirheimer, G., Ebel, J. P., Bonnet, J., Gangloff, J., Keith, G., Krebs, B., Kuntzel, B., Roy, A., Weissenbach, J., and Werner, C.,** Structure primaire des tRNA, *Biochimie,* 54, 127, 1972.

383. **Chinault, A. C., Tan, K. H., Hassur, S. M., and Hecht, S. M.,** Initial position of aminoacylation of individual *E. coli,* yeast and calf liver transfer RNAs, *Biochemistry,* 16, 766, 1977.

384. **Stern, L. and Schulman, L. H.,** Role of anticodon bases in aminoacylation of *E. coli* methionine tRNAs, *J. Biol. Chem.,* 252, 6403, 1977.

385. **Roe, B. A. and Tsen, H. Y.,** Role of ribothymidine in mammalian tRNA^Phe^, *Proc. Natl. Acad. Sci. USA.,* 74, 3696, 1977.

386. **Gussek, D. J.,** On dealing with anomalies in the transfer RNA aminoacylation reaction in partially purified systems, *Arch. Biochem. Biophys.* , 182, 533, 1977.

387. **Locy, R. D. and Cherry, J. H.,** Purification and characterization of two tyrosyl-tRNA synthetase activities from soybean cotyledons, *Phytochemistry,* 17, 19, 1978.

388. **Jeannin, G., Burkard, G., and Weil, J. H.,** Characterization of *Phaseolus vulgaris* cytoplasmic, chloroplastic and mitochondrial tRNAs^Phe^. Aminoacylation by homologous and heterologous enzymes, *Plant Sci. Lett.,* 13, 75, 1978.

389. **Daniel, V., Sarid, S., Beckmann, J. S., and Littauer, U. Z.,** In vitro transcription of a transfer RNA gene, *Proc. Natl. Acad. Sci. U.S.A.,* 66, 1260, 1970.

390. **Grimberg, J. I. and Daniel, V.,** In vitro transcription of three adjacent *E. coli* transfer RNA genes, *Nature (London),* 250, 320, 1974.

391. **Zeevi, M. and Daniel, V.,** Aminoacylation and nucleoside modification of in vitro synthesized transfer RNA, *Nature (London),* 260, 72, 1976.

392. **Mc Closkey, J. A. and Nishimura, S.,** Modified nucleosides in transfer RNA, *Acc. Chem. Res.,* 10, 403, 1977.

393. **Mc Clain, W. H.,** Seven terminal steps in a biosynthetic pathway leading from DNA to transfer RNA, *Acc. Chem. Res.,* 10, 418, 1977.

394. **Schimmel, P. R.,** Approaches to understanding the mechanism of specific protein-transfer RNA interactions, *Acc. Chem. Res.,* 10, 411, 1977.

395. **Goddard, J. P.**, The structures and functions of transfer RNA, *Prog. Biophys. Mol. Biol.*, 32, 233, 1977.

396. **Parthier, B. and Neumann, D.**, Structural and functional analysis of some plastid mutants of *Euglena gracilis, Biochem. Physiol. Pflanz.*, 171, 547, 1977.

397. **Feldman, M. Y.**, Minor components in transfer RNA: the location-function relationship, *Prog. Biophys. Mol. Biol.*, 32, 83, 1977.

398. **Martin, N. C., Rabinowitz, M., and Fukuhara, H.**, Yeast mitochondrial DNA specifies tRNA for 19 amino acids. Deletion mapping of the tRNA genes, *Biochemistry,* 16, 4672, 1977.

399. **Rich, A. and Kim, S. H.**, The three-dimensional structure of transfer RNA, *Sci. Amer.*, 238, 52, 1978.

400. **Rizzion, R., Mastanduno, M., and Freundlich, M.**, Partial derepression of the isoleucine-valine enzymes during methionine in *Salmonella typhimurium, Biochim. Biophys. Acata,* 475, 267, 1977.

401. **Deutscher, M. P., Lin, J. J. C., and Evans, J. A.**, Transfer RNA metabolism in *E. coli* cells deficient in tRNA nucleotidyl transferase, *J. Mol. Biol.*, 117, 1081, 1977.

402. **Broker, T. R., Angerer, L. M., Yen, P. H., Hershey, N. D., and Davidson, N.**, Electron microscopic visualization of tRNA genes with ferritin-avidin: biotin labels, *Nucleic Acid Res.*, 5, 363, 1978.

403. **Augustyniak, H. and Pawelkiewicz, J.**, Lysine tRNAs from cytoplasm, chloroplasts and mitochondria of the lupin seedling cotyledons, *Acta Biochim. Pol.*, 25, 91, 1978.

404. **Olins, P. O. and Jones, D. S.**, The sequence of the major formylatable species of methionine tRNA from *Scenedesmus obliquus*, Cold Spring Harbor Symposium on tRNA, 1979, in press.

405. **Chang, S. H., Lin, F. K., Hecker, L. I., Heckman, J. E., RajBhandary, U. L., and Barnett, W. E.**, Nucleotide sequence of blue-green algae phenylalanine tRNA. Cold Spring Harbor Symposium on tRNA, 1979, in press.

406. **Chang, S. H., Brum, C. K., Schnabel, J. J., Heckman, J. I., RajBhandary, U. L., and Barnett, W. E.**, Similarities in nucleotide sequence between *Euglena gracilis* and mammalian cytoplasmic phenylalanine tRNAs, *Fed. Proc.*, 37, 1768, 1978.

407. **Valenzuela, P., Venegas, A., Weinberg, F., Bishop, R., and Rutter, W. J.**, Structure of yeast phenylalanine tRNA genes: an intervening DNA segment within the region coding for the tRNA, *Proc. Natl. Acad. Sci. USA*, 75, 190, 1978.

408. **Knapp, C., Beckmann, J. S., Johnson, P. F., Fuhrmann, S. A., and Abelson, J.**, Transcription and processing of intevening sequences in yeast tRNA genes, *Cell*, 14, 221, 1978.

409. **Gauss, D. H., Gruter, F., and Sprinzl, M.**, Compilation of tRNA sequences, *Nucleic Acid Res.*, 6 rl, 1979.

410. **Sarantoglou, V., Imbault, P., and Weil, H. H.**, Partial purifiiation and properties of *Euglena gracilis* cytoplasmic and chloroplastic valyl-tRNA synthetases in *Chloroplast Development* Akoyunoglou, G., Ed., Elsevier/North-Holland, Amsterdam, 1978, 695.

411. **Hatfield, D. and Rice, M.**, Patterns of codon recognition by isoacceptor aminoacyl-transfer RNAs from wheat germ, *Nucleic Acid Res.*, 5, 3491, 1978.

412. **McCrea, J. M. and Hershberger, C. L.**, Chloroplast DNA codes for tRNA from cytoplasmic polyribosomes, *Nature (London)*, 274, 717, 1978.

413. **Martin, R. P., Schneller, J. M., Stahl, A. J. C., and Dirheimer, G.**, Study of yeast mitochondrial tRNAs by two-dimensional gel electrophoresis: characterization of isoaccepting species and search for imported cytoplasmic tRNAs, *Nucleic Acid Res.*, 4, 3497, 1977.

414. **Goss, J. D. and Parkhurst, L. J.**, Rapid solidphase isolation of 20 specific tRNAs from *E. coli, J. Biol. Chem.*, 253, 7804, 1978.

415. **Racz, I., Juhasz, A., Kiraly, I., and Lasztity, D.**, The effect of light on the nucleotide composition of tRNA^{Phe} of wheat germ, *Plant Sci. Lett.*, in press.

416. **Osorio, L. and Guillemaut, P.**, unpublished.

417. **Hallick, R. B., Gray, P. W., Chelm, B. K., Rushlow, K. E., and Orozco, E. M.**, *Euglena gracilis* chloroplast DNA structure, gene mapping and RNA transcription in *Chloroplast Development*, Akoyunoglou, G. and Argyroudi-Akoyunoglou, J. H., Eds., Elsevier/North-Holland, Amsterdam, 1978, 619.

418. **Malnoe, P. and Rochaix, J. D.**, Localization of 4S RNA genes on chloroplast genome of *Chlamydomonas reinhardi, Mol. Gen. Genet.*, 166, 269, 1978.

419. **Steinmetz, A., Mubumbila, M., Keller, M., Burkard, G., Weil, J. H., Driesel, A. J., Crouse, E. J., Gordon, K., Bohnert, H. J., and Herrmann, R. G.**, Mapping of tRNA genes on the circular DNA molecule of *Spinacia oleracea* chloroplasts in *Chloroplast Development*, Akoyunoglou, G. and Argyroudi-Akoyunoglou, J. H., Eds., Elsevier/North-Holland, Amsterdam, 1978, 573.

420. **Selsky, M. I.**, Reverse phase chromatographic analysis of *Nostoc* and *Euglena* isoleucyl tRNAs aminoacylated *in-vitro:* in homologous and heterologous systems, *Biochem. Biophys. Acta*, 520, 555, 1978.

421. **Gozdzicka-Josefiak, A. and Augustyniak, J.,** Valine specific tRNA from barley, *Bull. Acad. Pol. Sci.,* 26, 299, 1978.
422. **Karwowska, V. and Augustyniak, J.,** Leucine transfer RNA from barley: characterization and purification of isoaccepting species, *Acta Biochim. Pol.,* 24, 319, 1977.
423. **Babcock, D. F. and Morris, R. O.,** *Plant Physiol.* 52, 292, 1973.
424. **Janowicz, Z., Wower, J. M., and Augustyniak, J.,** Primary structure of barley embryo tRNA^{Phe} and its identity with wheat germ tRNA^{Phe}, *Plant Sci. Lett.,* 14, 177, 1979.
425. **Kedzierski, W., Sulewski, T., and Pawelkiewicz, J.,** Levels of aminoacylation of methionine tRNAs in germinating lupin cotyledons, *Plant Sci. Lett.,* 14, 373, 1979.
426. **Schneller, J. M., Schneider, C., and Stahl, A. J. C.,** Distinct nuclear genes for yeast mitochondrial and cytoplasmic methionyl-tRNA synthetases, *Biochem. Biophys. Res. Comm.,* 85, 1392, 1978.
427. **Quintard, B., Mouricout, M., Carias, J. R., and Julien, R.,** Occurence of aminoacyl-tRNA synthetase complexes in quiescent wheat germ., *Biochem. Biophys. Res. Comm.,* 85, 999, 1978.
428. **Driesel, A. J., Crouse, E. J., Gordon, K., Bohnert, H. J., Herrmann, R. G., Steinmetz, A., Mubumbila, M., Keller, M., Burkard, G., and Weil, J. H.,** Fractionation and iientification of the individual spinach chloroplast transfer RNAs and mapping of their genes on the restriction endonuclease cleavage sit map of chlorplast DNA, submitted for publication.
429. **Bohnert, H. J., Driesel, A. J., Crouse, E. J., Gordon, K., Herrmann, R. G., Steinmetz, A., Mubumbila, M., Keller, M., Burkard, G., and Weil, J. H.,** Presence of a transfer RNA gene in the spacer sequence between the 16S and 23S rRNA genes of spinach chloroplast DNA, submitted for publication.
430. **Lester, B. R. and Cherry, J. H.,** Purification of leucine tRNA isoaccepting species from soybean cotyledons. I. Benzoylated diethylamino cellulose fractionation, *N*-hydroxysuccinimide modification and characterization of product, *Plant Physiol,* 63, 79, 1979.
431. **Lester, B. R., Morris, R. O., and Cherry, J. H.,** Purification of leucine tRNA isoaccepting species from soybean cotyledons. II. RPC-2 purifiiation, ribosome binding and cytokinin content, *Plant Physiol.,* 63, 87, 1979.
432. **Struxness, L. A., Amstrong, D. J., Gillam, I., Tener, G., Burrows, W. J., and Skoog, F.,** Distribution of cytokinin-active ribonucleosides in wheat germ tRNA species, *Plant Physiol.,* 63, 35, 1979.

RIBOSOMAL RNA OF PLANTS

C. J. Leaver

TABLE OF CONTENTS

I. INTRODUCTION

It is now generally accepted that all eukaryotic cells contain interdependent genetic and protein synthesizing systems in the nucleus and cytoplasm and in mitochondria. Green plants contain, in addition, a third such system in the plastids. The ribosomes which form the basis of each of these individual protein synthesizing systems are unique in their structural RNA and ribosomal proteins, but perform the same function and appear to do so by very similar biochemical reactions.

This raises questions as to why the green plant has maintained three distinct protein synthesizing systems. Is this the result of their separate evolutionary origins? Does each type of ribosome have unique features which make it especially suited to synthesize particular proteins, or does it reflect that because of differences in factors such as pH or concentration of particular ions, one type of ribosome could not function effectively in the organelles and the cytoplasm?

This chapter will review our present knowledge of the structure and synthesis of plant ribosomal RNA's (rRNA) and provide a background for studies on the regulation of synthesis and assembly of the various types of ribosomes found in the plant cell. This knowledge will be essential for eventual understanding of the regulation of protein synthesis and its involvement in the control of plant development and differentiation.(This survey was completed in June 1977.)

II. RIBOSOMES

In recent years, it has become the fashion to classify ribosomes into two groups, the so-called "70S" ribosome characteristic of prokaryotic organisms (bacteria, blue green algae) and cell organelles and the "80S" ribosome found in the cytoplasm of eukaryotic cells. These classifications were made on the basis of sedimentation coefficient, method of initiation of protein synthesis, and a selective sensitivity to certain antibiotics. This working classification, while useful, has tended to obscure the very real differences in both structure and composition of the variety of ribosomes found in living organisms. They nevertheless have the same common function of translating messenger RNA and the assembly of polypeptides. It is only now that we are beginning to understand the structure and function of the bacterial ribosome in some detail, that questions are being asked as to what was the evolutionary pressure for the accumulation of extra proteins and RNA, and consequently whether there may be functions of eukaryotic ribosomes still to be discovered.

The average plant contains approximately 10^7 ribosomes per cell; the majority are 80S cytoplasmic ribosomes, but in leaves between 25 and 40% of this number may be 70S chloroplast ribosomes. In higher plants, mitochondrial ribosomes apparently account for less than 1% of the total cellular ribosome population.[1]

Some 50 to 63% of the mass of the ribosome consists of RNA, the remainder being ribosomal proteins. This rRNA accounts for 75 to 80% of the cell RNA and plays a dominant structural role in the ribosome.

The ribosomes, when not engaged in protein synthesis, can be readily dissociated into two subunits of unequal size. In higher plants, the larger subunit, which is about twice the size of the smaller, contains one molecule of high molecular weight RNA (23 to 25S), one or two low molecular weight RNAs and between 30 and 40 ribosomal proteins. The small subunit contains a single high molecular weight RNA (16 to 18S) and 20 to 30 proteins.

III. HIGH MOLECULAR WEIGHT RIBOSOMAL RNAs

A. Cytoplasmic rRNAs

The larger 60S subunit of the cytoplasmic 80S ribosome contains a 25S rRNA which has an estimated molecular weight of 1.3×10^6 (with approximately 37,000 nucleotides) and 38 to 40 ribosomal proteins. The smaller, 40S, subunit contains an 18S rRNA with a molecular weight of approximately 0.7×10^6 (with about 2000 nucleotides) and 30 ribosomal proteins (Table 1).

These high molecular weight rRNA molecules constitute the major single structural component of each subunit, and presumably their inherent conformational properties together with the specifically associated ribosomal proteins, are the major determinants of ribosomal structure and function.

The molecular weight of the smaller rRNA (18S) component is remarkably constant in almost all eukaryotes at 0.65 to 0.7×10^6.[2] The larger rRNA (25 to 28S) has evolved with each major step of evolution from a molecular weight of about 1.3×10^6 in plants and most protozoa to 1.4×10^6 in lower animals, 1.5×10^6 in amphibians, and to 1.75

TABLE 1

Comparison of Ribosomes and Their Structural RNAs from Cytoplasm, Chloroplast, and Mitochondrion

Source of ribosome	S value ribosome	S value subunits	S value rRNAs	rRNAs (mol wt)
Cytoplasm	80S	60S	25S	1.3×10^6
			5.8S	50,400
			5S	37,500
		40S	18S	0.7×10^6
Chloroplast	70S	50S	23S	1.1×10^6
			5S	39,400
			4.5S	25,000—33,000
		30S	16S	0.56×10^6
Mitochondrion	78S	60S	24S	$1.12—1.26 \times 10^6$
			5S	38,750
		44S	18.5S	$0.69—0.78 \times 10^6$

$\times 10^6$ in mammals.[3] Thus the combined molecular weights of the rRNA varies from 2.0×10^6 in plants and protozoa to 2.45×10^6 in mammals. As suggested by Loening[3] and Maden,[4] the conservation of molecular weight throughout long periods of evolution, together with distinct increases at distinct evolutionary points, must clearly be of functional significance. Ths increase in size of rRNA with increasing evolutionary development may imply the parallel development of translational control mechanisms in protein synthesis. *Euglena* and *Acanthamoeba* are exceptions to this general observation, and they have exceptionally high molecular weight rRNA, which is also peculiarly unstable. The unusual size of *Euglena* rRNA[5,6] (1.4×10^6 and 0.9×10^6 mol wt) may relate to its unique phylogenetic position midway between plants and animals. Another parameter which has been useful in characterizing rRNA is its base composition. Despite close similarity in molecular weight of the rRNA molecules from different plants, there is considerable variation in the base composition (Table 2). The base composition of the two rRNA species tends to vary in parallel, with the 0.7×10^6 rRNA from most eukaryotic ribosomes having a characteristically lower % G + C content than its partner high molecular weight rRNA, independent of the average composition of the ribosomal RNA. An apparent evolutionary trend in this variation exists, there being a tendency for G + C content of rRNA to increase when going from the lower to the more evolved forms. Attardi[2] and Amaldi[7] have suggested that the increase in G-C pairs might be expected to lead to an increase in secondary structure, and under normal conditions it appears that up to 60 to 75% of each rRNA molecule is involved in intramolecular base pairing.

Despite changes in base composition and sequence among rRNAs of different plants and differences in size of the large rRNA between plants and animals, the overall structure of the eukaryotic ribosomal subunits seems to have been conserved. Cammarano et al.[15] have shown that hybrid 80S ribosomes prepared from the small subunit of pea ribosomes and the large subunits from either rat or mouse liver and vice versa are active in the synthesis of polyphenylalanine stimulated by poly(U).

TABLE 2

Base Composition of Cytoplasmic, Chloroplast, Mitochondrial and Procaryotic rRNA

Organism	rRNA	Moles (%)					Ref.
		C	A	G	U	G + C	
Saccharomyces cerevisiae	Cytoplasmic 25S	19.2	26.4	28.4	26	47.6	8
	18S	19.1	26.6	26.1	28.1	45.2	
	Mitochondrial 22S	11.0	40.3	14.0	34.6	25.0	
	15S	11.0	38.4	16.1	34.5	27.1	
Euglena gracilis	Cytoplasmic total	24.1	21.6	32.0	22.3	56.1	5
	Chloroplast total	20.4	25.5	31.1	23.0	51.5	
Chlorella proto-thecoides	Cytoplasmic total	23.6	23.9	30.7	21.8	54.3	9
	Chloroplast total	21.7	25.8	29.9	22.6	51.6	
Spinacia oleracea	Cytoplasmic total	23.5	21.6	31.7	23.3	55.2	10
	25S	23.4	21.3	32.5	22.8	55.9	
	18S	23.5	22.6	30.5	23.4	54.0	
	Chloroplast 23S	21.7	24.5	34.8	19.0	56.5	
	16S	24.0	23.0	30.6	22.4	54.6	
	Chloroplast total	23.4	24.2	32.4	20.0	55.8	
Phaseolus aureus	Cytoplasmic 25S	23.0	24.5	32.4	20.1	55.4	11
	18S	20.2	25.4	30.2	25.2	50.4	
Chloroplast	23S	23.4	26.3	29.9	20.4	53.3	
	16S	23.1	26.6	31.5	18.8	54.6	
Vicia faba	Cytoplasmic 5S	23.8	24.1	28.6	23.5	52.4	12
	Chloroplast 5S	22.6	26.5	28.7	22.2	51.3	
Blue-green algae *Anacystis nidulans*	23S	20.8	29.4	32.0	17.7	52.8	13
	16S	20.0	28.7	32.7	18.6	52.7	
Bacteria *Escherichia coli*	23S	21.0	25.5	32.5	21.0	53.5	14
	16S	22.3	24.2	32.1	21.3	54.4	

B. Chloroplast rRNAs

In the early 1960s, Lyttleton[16] was the first to show that chloroplasts contain a discrete class of ribosomes which sediment at approximately 70S, in contrast to the larger 80S ribosome found in the cytoplasm. This observation has since been confirmed in several algae and higher plants[17,18] leading to the conclusion that chloroplasts and blue green algae contain a 70S prokaryotic type of ribosome similar to those found in bacteria.

In common with bacterial ribosomes, the 70S chloroplast ribosomes require a higher concentration of magnesium ions (at least 5 m*M*) to maintain their structural integrity. Lowering of the magnesium ion concentration below 1 m*M* results in their dissociation into a large 50S subunit and a small 30S subunit. This is in contrast to 80S cytoplasmic ribosomes which require prolonged dialysis in the absence of magnesium ions to dissociate into subunits.

Earlier studies using sucrose density gradient techniques suggested that the chloroplast ribosome contained rRNAs of 23S and 16S which were smaller than the corresponding cytoplasmic rRNAs and similar to those described in *Escherichia coli*. Following the introduction of polyacrylamide gel electrophoresis, which affords a much higher degree of resolution of rRNA species (Figure 1) it was shown that the large subunit contains a 23S rRNA of molecular weight 1.1×10^6 containing about 3150 nucleotides. The smaller subunit contains a 16S rRNA species of molecular weight 0.56 $\times 10^6$ comprising some 1600 nucleotides.[19] (Table 1)

C. Molecular Integrity of Chloroplast rRNA

Under normal conditions of extraction and fractionation, the two chloroplast rRNAs are not found to be present in a 1:1 molecule ratio of heavy to light components, as usually observed for the cytoplasmic ribosomal RNAs. Earlier results of analysis of chloroplast rRNA from spinach, tobacco, radish, pea, and tomato leaves led to the suggestion that chloroplast ribosomes contained only one species of RNA, the light molecule. The low recovery of the heavy (1.1×10^6 mol wt) rRNA together with the presence of additional smaller components, suggested degradation of the heavy chloroplast rRNA. Studies using total RNA prepared from radish cotyledons indicated that in this tissue, the 1.1×10^6-mol wt rRNA was breaking into two molecules, with molecular weights of 0.4×10^6 and 0.7×10^6, the latter having the same mobility on electrophoresis as the cytoplasmic 0.7×10^6 rRNA.[20] After correction for such breakdown of the 1.1×10^6 rRNA, estimated from the amount of the 0.4×10^6 rRNA present, the calculated ratios of the amounts of the rRNA species were reasonably close to the ratio of their molecular weights.

Degradation is not a random process, but procedes via a number of relatively specific breaks to produce a family of RNA molecules of discrete size. The stability of the large chloroplast rRNA is a function both of the conditions employed during the extraction and fractionation of the RNA and of the leaf age and species of plant being analyzed.[21] Cleavages in three particular regions of the molecule are most frequently observed although the size and number of fragments produced varies with different plant species.

Newly synthesized radish chloroplast rRNA, as measured by the incorporation of a radioactive precursor, was found to be stable under conditions in which the older, accumulated RNA was unstable. This suggested that the newly synthesized 1.1×10^6 rRNA was stable, but that with time, something happens which results in the 1.1×10^6 rRNA being specifically broken down into smaller pieces.

Similar observations on the relative stability of newly synthesized versus older 1.1×10^6 rRNA have been made for *Phaseolus aureus*, *Spirodela*, and for the blue green alga, *Anacystis nidulans*. In *A. nidulans*, the 1.1×10^6 rRNA yields only fragments of molecular weight 0.88×10^6 and 0.17×10^6, by in vivo cleavage at a single specific site.

Leaver and Ingle[21] showed that the 1.1×10^6 rRNA from higher plant chloroplasts could be stabilized by the inclusion of divalent cations, particularly Mg^{2+} and Ca^{2+}, during both the preparation and fractionation of the RNA. In a later report,[22] it was shown that the integrity of the 1.1×10^6 rRNA could be maintained in the absence of divalent cations by extraction and fractionation of the chloroplast rRNA at temperatures between 0 and 5°C (see Figure 2). On increasing the temperatures, fragmentation to characteristic products was observed, the T_m (melting temperature) of which was markedly affected by ionic strength and the presence of divalent cations (see Figure 3).

The observations supported the conclusion that under normal conditions in the intact chloroplast 70S ribosome, the 1.1×10^6 rRNA molecule contains secondary structure in the form of helical regions or base-paired loops stabilized by hydrogen bonds. Any loops exposed at the surface of the large ribosomal subunit will be liable to en-

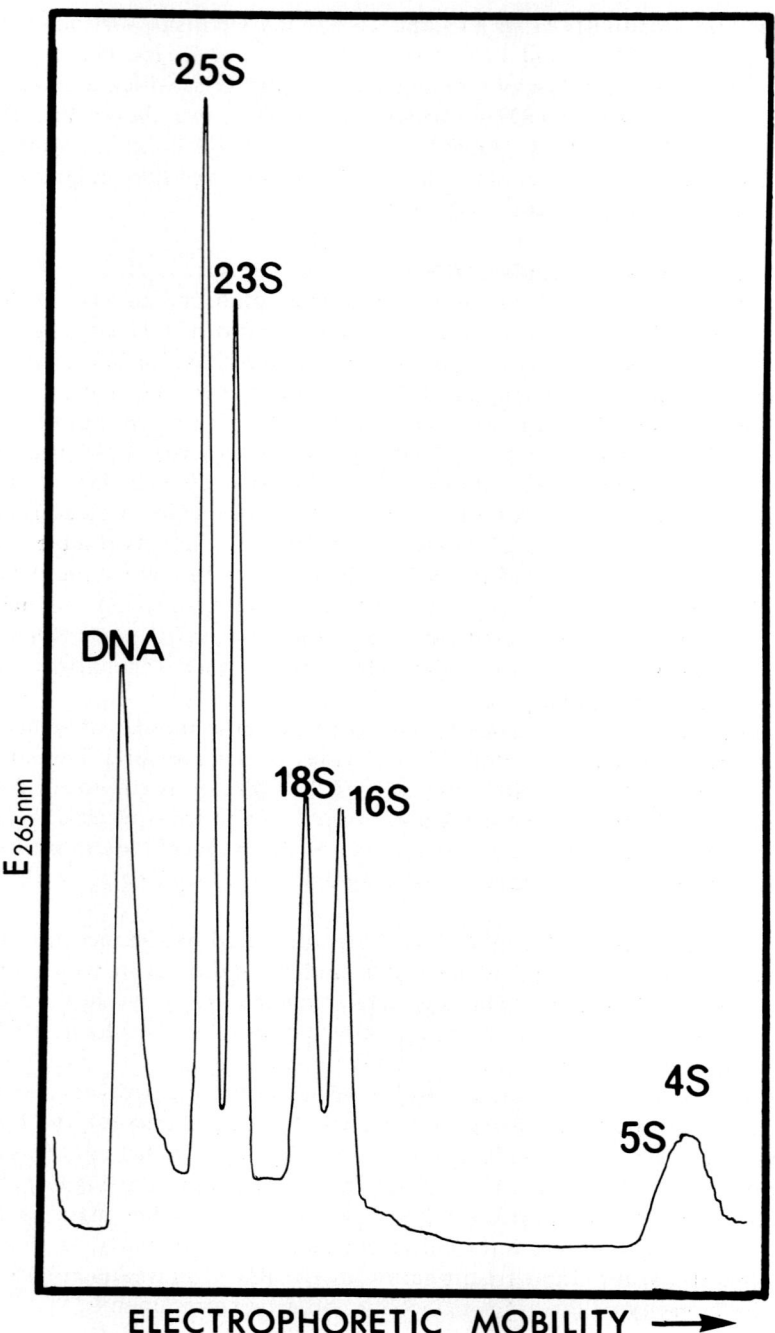

FIGURE 1.   Polyacrylamide gel electrophoresis of total nucleic acids extracted from spinach (*Spinacia oleracea*) leaf. Total nucleic acids were extracted as described[22] and fractionated by gel electrophoresis on 2.4% gels for 2 hr at 3 mA/9cm gel and at a temperature of 5°C. Sedimentation values are taken as corresponding approximately to molecular weights as follows: 25S - 1.3×10^6 mol wt, 23S - 1.1×10^6 mol wt, 18S - 0.7×10^6 mol wt, 16S - 0.56×10^6 mol wt, 5S - 37,500 mol wt, 4S - 25,000 mol wt.

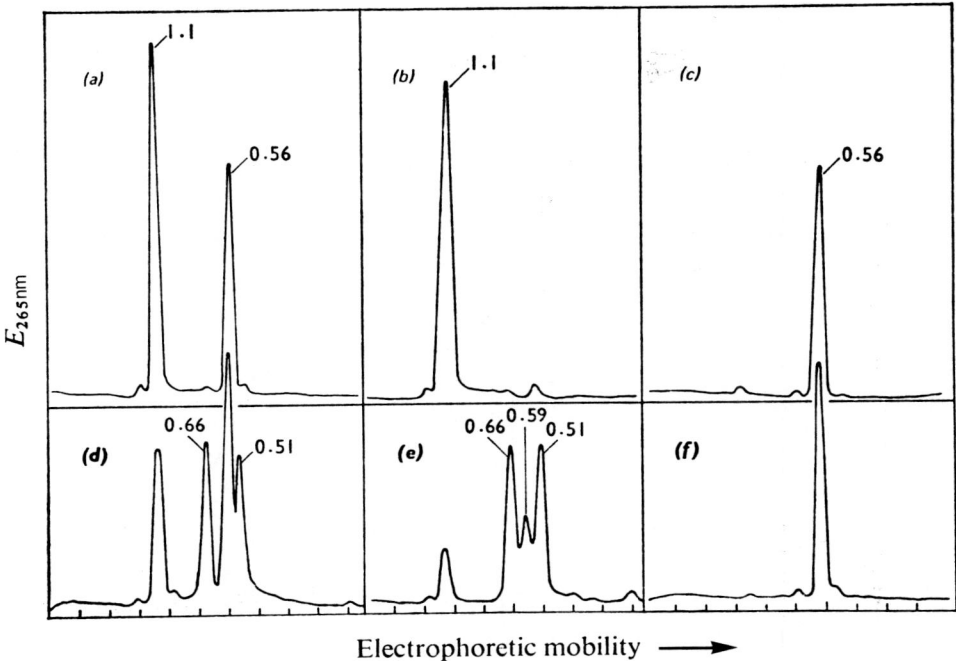

FIGURE 2.   Effect of temperature pretreatment on the stability of choloroplast rRNA species. Samples of total broad-bean chloroplast rRNA (a and d); 1.1×10^6-mol wt rRNA (b and e), and 0.56×10^6-mol wt rRNA (c and f) were exposed for 5 min to temperatures of either 5°C (a-c) or 40°C (d-f) followed by rapid cooling and electrophoresis on 2.4% polyacrylamide gels for 4 hr at 50 V (3mA/9-cm gel) and a temperature of 5°C. RNA components are referred to as $10^{-6} \times$ molecular weight.[22] (From Leaver, C. J., *Biochem. J.,* 135, 237, 1973. With permission.)

donuclease action *in situ* or during the isolation of RNA, resulting in the introduction of nicks or short deletions into the polynucleotide chain. If the base-paired region of the loop is large enough, the isolated RNA molecule in solution will remain intact, provided that the secondary structure is preserved by ionic and temperature conditions (high salt concentrations and low temperatures) that encourage hydrogen bonding and/or the stabilization of the molecule by divalent cations.

Under conditions that progressively decrease hydrogen bonding and cause unfolding of the RNA molecule (e.g., increase in the temperature, lowering of the salt concentration, or inclusion of urea in the solution), the secondary structure of the RNA molecule is lost, and the molecule dissociates into several shorter polynucleotide chains. Once the shorter polynucleotide constituents of the 1.1×10^6 rRNA have dissociated from one another, it appears that they cannot be induced to reassociate.[22] Under similar conditions, the 0.56×10^6 rRNA remains intact (Figure 2e and f). The difference in stability of the two rRNA components is probably best explained on the ground of selective nuclease action occurring either *in situ* or during the deproteinization step. The 0.56×10^6 rRNA obviously differs from the 1.1×10^6 rRNA in base sequence, configuration, or nucleic acid protein interaction in the subunit.

The biological significance of the presence of "hidden breaks" in chloroplast 1.1×10^6 rRNA is obscure, since there is no evidence to suggest that ribosomes whose RNA has such discontinuities are any less efficient in protein synthesis than those whose rRNA is continuous. Doolittle[23] suggests that the cleavage of the 1.1×10^6 rRNA in *A. nidulans* occurs in vivo during normal ribosomal maturation and ageing without loss of subunit integrity. This postmaturational cleavage in *A. nidulans* is a function of time after transcription, and is stimulated by light and retarded by inhibitors of

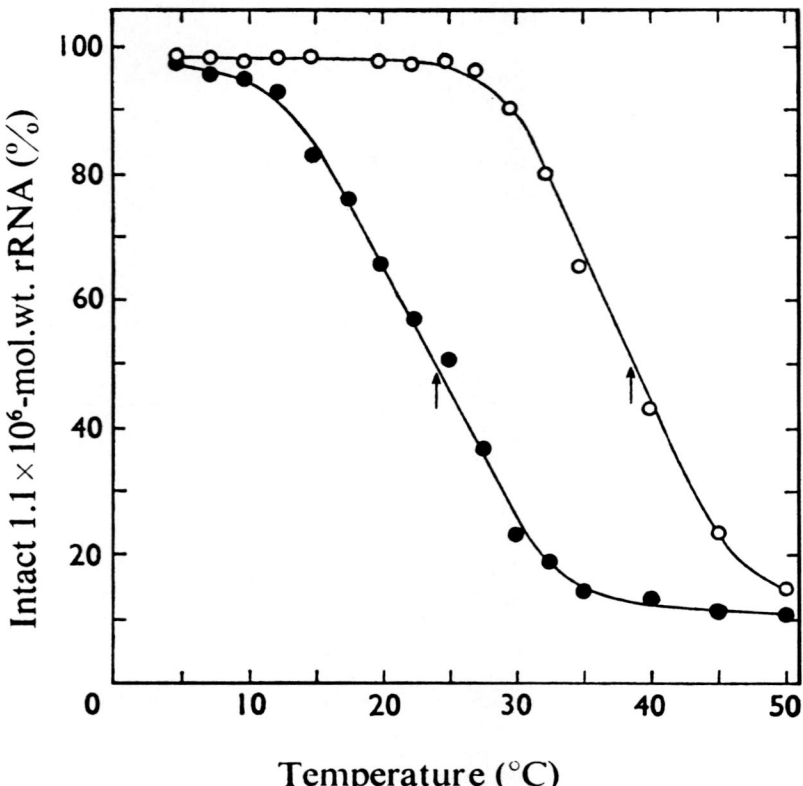

FIGURE 3.   Effect of increasing temperature on the stability of chloroplast 1.1 × 10⁶-mol wt rRNA. Chloroplast 1.1 × 10⁶-mol wt rRNA was dissolved in E buffer (●) or in E buffer containing 10 m*M*-Mg²⁺ (○), and portions were exposed for 5 min to increasing temperatures in the range 5 to 50°C, followed by rapid cooling and electrophoresis at 5°C (as described in Figure 2). The percentages of intact 1.1 × 10⁶-mol wt rRNA remaining after the different temperature treatments were calculated from the gel profiles and plotted as a function of temperature of pretreatment. The temperatures at which 50% intact 1.1 × 10⁶-mol wt rRNA remains are indicated by the arrows. (From Leaver, C. J., *Biochem. J.*, 135, 237, 1973. With permission.)

photosynthesis.[23] The relationship between rRNA cleavage and ribosome function awaits further investigation.

D. Mitochondrial Ribosomal RNAs

It has become apparent that mitochondrial ribosomes cannot be characterized as either of the 70S or the 80S type; their reported sedimentation coefficient ranges from 55 to 60S in animal mitochondria and from 70 to 80S in fungi and higher plants.[24] The available information on higher plant mitochondrial ribosomes is somewhat sparse, although the general consensus seems to be that they contain ribosomes which sediment at 77 to 78S.[24,25] These ribosomes can be dissociated into large (60S) and small (40 to 44S) ribosomal subunits which contain discrete rRNA species sedimenting at 24S to 26S and 18.5S.[24] These have estimated molecular weights of 1.12 to 1.26 × 10⁶ and 0.69 to 0.78 × 10⁶, depending on the plant species[24,25] (see Figure 4 and Table 1).

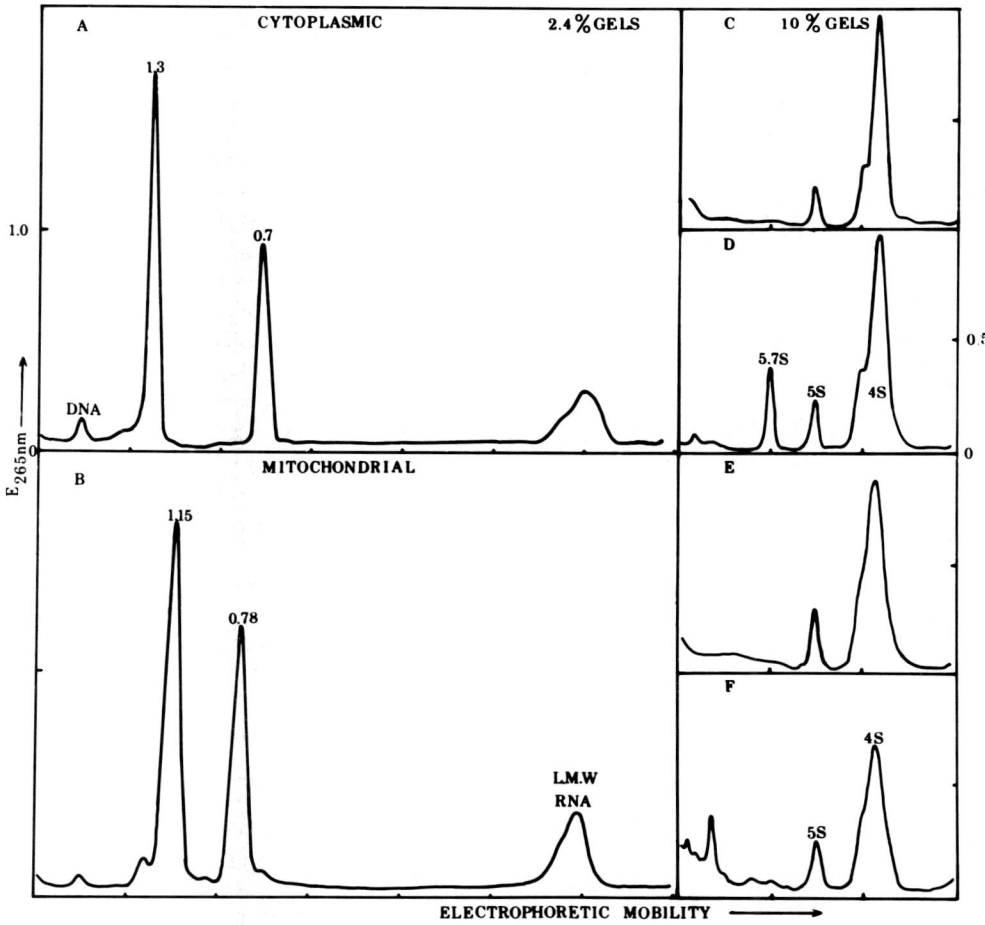

FIGURE 4. Polyacrylamide-gel electrophoresis of nucleic acids from mung bean hypocotyls. (A) Total cytoplasmic nucleic acids and (B) total mitochondrial nucleic acids were fractionated on 2.4% (w/v) polyacrylamide gels for 3.5 hr at 50 V (3mA/9cm gel) and at 5°C. Portions of the same samples were fractionated on 10% (w/v) polyacrylamide gels for 5 hr at 50 V to resolve further the low molecular weight nucleic acid components. (C) Total cytoplasmic nucleic acid; (D) as (C) but heated at 65°C for 10 min and cooled by rapid immersion in liquid N_2 before fractionation; (E) total mitochondrial nucleic acid; (F) as (E) but heated at 65°C before fractionation. RNA components are referred to as $10^{-6} \times$ molecular weight. (From Leaver, C. J. and Pope, P. K., in *Nucleic Acids and Protein Synthesis in Plants*, Bogorad, L. and Weil, J. H., Eds., Plenum Press, New York, 1976, 213. With permission.)

In the past, several unusual physical and chemical properties exhibited by mt-rRNAs from various organisms caused difficulties in characterizing the precise size of these molecules.[26] The low mole percentage of guanine plus cytosine (from as low as 26% in yeast to as high as 44% in some animals) of mt-rRNA in contrast to most bacterial and cytoplasmic rRNAs, may account for the observations that fungal mt-rRNA melts at a lower temperature than cytoplasmic-rRNA and also leads to anomalous behavior on gel electrophoresis. Several workers using animal and ascomycete mt-rRNAs have shown that the electrophoretic mobility of mt-rRNAs relative to homologous cytoplasmic rRNA and *E. coli* rRNA is strongly dependent on ionic strength and temperature.[26] As a consequence of the variable dependence of secondary structure of mt-rRNAs on physical and chemical conditions, their molecular weights cannot always be reliably inferred from their sedimentation behavior or electrophoretic mobility. However, animal mt-rRNAs sediment at 12 to 13S and 16 to 17S and have calculated molecular weights of 0.3 to 0.36 and 0.35 to 0.56×10^6. Fungal mt-rRNAs are larger and sediment at 14 to 18S and 21 to 25S corresponding to molecular weights of approximately 0.63 to 0.72×10^6 and 1.23 to 1.30×10^6.[26] Thus, the progressive decrease in mitochondrial genome size from higher plants (30 μm circle) to animals (5 μm circle) appears to be associated with or to involve the ribosomal cistrons, which have been reduced to an unprecedented small size in animal mitochondria.

Although mitochondrial ribosomes differ in size and physical properties from the cytoplasmic 80S and bacterial and chloroplast 70S ribosomes, they do have several functional similarities with the prokaryotic (70S) ribosomes. For instance, they require a formylated methionine for initiation of protein synthesis. There is an interchangeability of protein factors required for protein synthesis and a selective sensitivity to certain antibiotics.[1] Recently, this close functional similarity has been reinforced by the observation that both chloroplast and mitochondrial rRNAs are indeed prokaryotic in nature. Using the technique of T1 oligonucleotide cataloging, several workers have shown that wheat mitochondrial 18S rRNA and a range of blue-green algae and chloroplast 16S rRNAs[27] share a significantly higher number of oligonucleotide sequences with bacteria than expected by chance. These RNAs exhibit little, if any, homology with cytoplasmic 18S rRNA.[28]

The close structural and functional similarities between chloroplast, mitochondrial and prokaryotic ribosomes together with the demonstration of sequence homology between chloroplast, mitochondrial, and prokaryotic rRNAs have been used by some workers in support of the endosymbiont hypothesis for the origin of these cell organelles.

IV. RIBOSOMAL RNA SYNTHESIS

It is now generally accepted that all major classes of RNA are synthesized as larger precursor molecules which are processed (shortened) and chemically modified before acquiring their final functional form. The relative ease of purification and characterization of ribosomal RNA molecules has made it possible to study in some detail the regulation of their synthesis and maturation into functional gene products. This in turn has provided a useful model system for the study of the control of gene expression in general.

Ribosomal RNAs are a direct transcription product of the respective nuclear or organelle genome while, as far as we know, the vast majority of the ribosomal proteins are nuclear coded, synthesized in the cytoplasm, and subsequently pass through a membrane to become associated with the cognate rRNAs. This dual synthetic origin of the macromolecular components required for the assembly of active ribosomes raises many important and as yet unanswered questions. For example, how is the synthesis of rRNA and ribosomal protein coordinated, and what determines the rate of ribosome formation in different tissues in vivo and in different physiological states of the same tissue?

A. Synthesis and Processing of Cytoplasmic rRNA

In eukaryotic organisms, the cistrons for the two major rRNAs are found in tandem or clustered repeats in either one or a few sites on the chromosomes called the nucleolar organizer regions. The number of copies of the rRNA gene per nucleus (or cell) has been calculated from quantitative DNA-RNA hybridization studies and varies widely among different organisms. Fungi, such as *Neurospora crassa* and *Saccharomyces* sp., contain of the order of 200 genes, while higher animals contain several hundred to a thousand and higher plants up to 13,000 (see Ingle[29] for review).

Evidence from a wide variety of eukaryotic organisms show that rRNAs are transcribed in the nucleoli from ribosomal DNA as large precursor molecules which have been termed the rRNA transcription units (RTUs). In addition to the sequences conserved in the mature 18S, 5.8S, and 25 to 28S rRNA molecules, the precursors also contain excess RNA that is unmethylated and GC-rich. This excess RNA is lost during the processing steps of rRNA maturation and is probably degraded since components corresponding to the nonribosomal sequences do not appear, even transiently, within the nucleus. For reviews see Maden,[4] Rungger and Crippa,[30] and Perry et al.[31].

In HeLa cells, one of the best documented examples, the initial transcription product of the ribosomal genes is a single precursor molecule which sediments at 45S and has a molecular weight of 4.1×10^6. This molecule then undergoes a series of selective endonucleolytic and exonucleolytic cleavages and chemical modifications (methylation and pseudouridine formation) to produce one each of the mature 28S (1.7×10^6 mol wt) and 18S (0.7×10^6 mol wt) rRNA components.

Secondary structure mapping and a detailed investigation of the time of methylation of the precursor molecules has suggested that processing occurs by the sequence of events summarized in Figure 5. During these transformations, the rRNA molecules become associated with specific proteins, some of which are found in mature cytoplasmic ribosomes and some of which are recycled within the nucleolus. The processing of the precursor rRNA occurs within the nascent ribosomal particles; the excess RNA of the precursor is probably necessary for ensuring that rRNA adopts the correct secondary and tertiary structure required for methylation and for the addition of proteins during ribosome assembly.

The size of the initial ribosomal transcription unit varies with different organisms (Table 3). In mammals, rodents, marsupials, and birds, the transcription unit is relatively large and only 50 to 60% of the molecule is conserved. In cold-blooded animals (up to and including reptiles) and in higher plants, the size of the initial transcription unit is much less than in HeLa and up to 80% of the molecule is retained.

In higher plants, several workers have proposed schemes for the synthesis and proc-

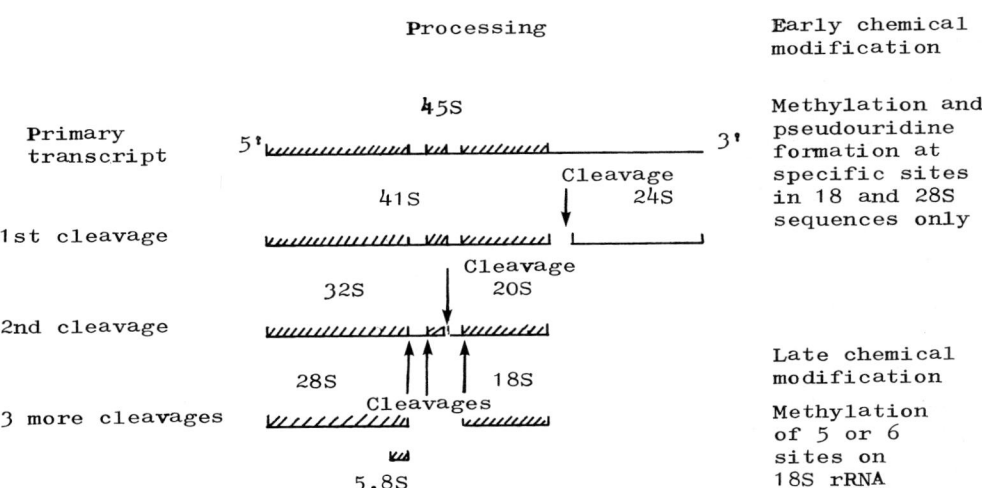

FIGURE 5. Processing and chemical modification of HeLa cell ribosomal RNA primary transcript. (From Ford, P. J., in *The Developmental Biology of Parts of Animals*, Graham and Waring, Eds., Blackwell Scientific Publications, Oxford. With permission.)

TABLE 3

Size of the Ribosomal Transcription Unit and of rRNA from Various Organisms

	Molecular weight $\times 10^{-6}$				
		rRNAs			
Organism	Transcription unit	Large	Small	% of ribosomal transcription unit conserved	Ref.
Mammal (HeLa)	4.4	1.75	0.70	54	4
Rodent (mouse)	4.19	1.70	0.65	56	31
Marsupial (potoroo)	4.19	1.70	0.65	56	31
Bird (chicken)	3.92	1.61	0.63	57	31
Reptile (iguana)	2.74	1.51	0.62	78	31
Amphibian (frog)	2.76	1.58	0.61	79	31
Fish (trout)	2.70	1.55	0.65	81	31
Insect *(Drosophila)*	2.85	1.40	0.65	72	31
Fungi *(Saccharomyces cerevisiae)*	2.6—2.3	1.30	0.72	78—86	31
Algae *(Euglena gracilis)*	4.3—2.34	1.4	0.9	53—98	50
Higher Plants					
Tobacco *(Nicotiana tabaccum)*	2.76	1.29	0.66	71	31
Pea *(Pisum sativum)*	2.3	1.3	0.70	87	34
Carrot *(Daucus carota)*	2.8—2.2	1.3	0.70	71—91	32
Artichoke *(Helianthus tuberosus)*	2.3	1.3	0.70	87	34
Sycamore *(Acer pseudoplanatanus)*	3.4—2.4	1.3	0.7	59—83	33
Mung bean *(Phaseolus aureus)*	2.5	1.3	0.7	80	11

essing of cytoplasmic rRNA based on the molecular weights and kinetics of labeling of the various components separated by gel electrophoresis,[32-34] on their base composition, and on competitive hybridization experiments.

The size of the first stable precursor appears to vary considerably in plants even between closely related species (Table 3). A further complication is the presence of several large precursor molecules whose relationship is unclear. The stages of processing are, however, analogous to those suggested for HeLa in Figure 5.

Analysis of the RNA synthesized in carrot (*Daucus carota*) root tissue discs showed, after 10 minutes incorporation of [^{32}P] orthophosphate (Figure 6), a background of polydisperse RNA plus two large precursor molecules of 2.8 and 2.2×10^6 daltons. After 20 min incubation, these molecules constituted the bulk of the newly synthesized RNA. The 1.4×10^6 dalton precursor was present after 20 min and increased in relative amount up to 1 hr. The newly synthesized 0.7×10^6-mol wt, mature rRNA first appeared at 40 min. Beyond this time, the distinction between the 1.4×10^6 and 1.3×10^6-mol wt components is lost as radioactivity accumulates in the 1.3×10^6-mol wt mature rRNA. With increasing time of incorporation, radioactivity continues to accumulate in all high molecular weight species of RNA, but the amount of label in the precursor molecules becomes proportionately less relative to the 1.3×10^6 and 0.7×10^6-mol wt stable rRNAs.[32]

Cell fractionation and kinetic analysis of the flow of radioactivity through and into various RNA species have shown that the three high molecular weight (2.8×10^6, 2.2×10^6 and 1.4×10^6) RNA components are found in the nucleus (Figure 7). These components have base compositions similar to those of rRNA and show a precursor-product relationship with the 1.3×10^6 and 0.7×10^6-mol wt cytoplasmic rRNAs.[32] When RNA extracted from purified mung bean nucleoli is examined by acrylamide gel electrophoresis, it is possible to visualize four major species of RNA in addition to the 1.3×10^6 and 0.7×10^6 mature rRNAs. These components have molecular weights of 2.54×10^6, 2.3×10^6, 1.44×10^6, and 1.02×10^6 and probably correspond to the precursor rRNAs previously described from labeling studies (Figure 8).

In the presence of inhibitors of protein synthesis (e.g., cycloheximide), an additional

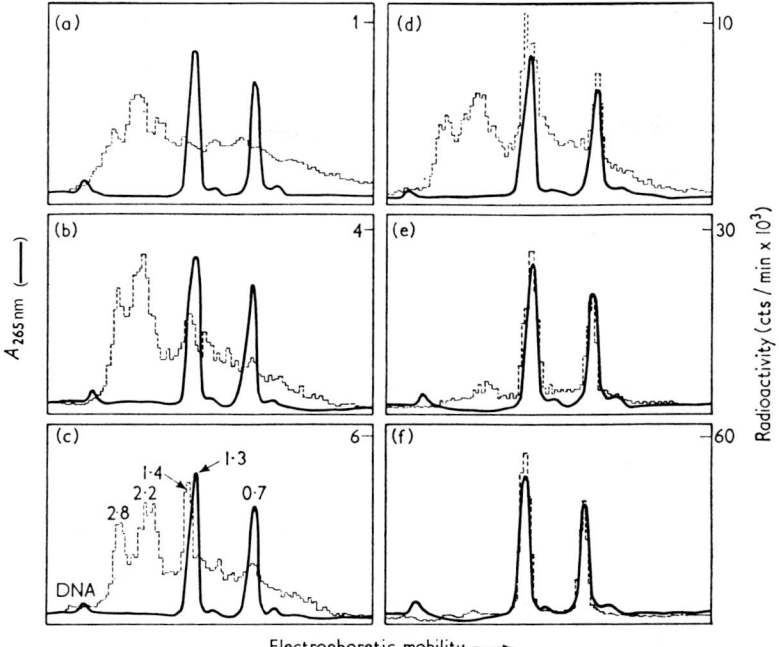

FIGURE 6. Polyacrylamide gel electrophoresis of total nucleic acid from carrot-root discs. Carrot discs were incubated with [³²P] orthophosphate (80μCi/ml) for 10, 20, 40, 60, 120, and 240 min (a to f respectively). Total nucleic acid was prepared and fractionated on 2.2% gels for 4 hr at 6 mA/8 cm gel. (—) A_{265nm} (•••) radioactivity per 0.5 mm slice. RNA components are referred to as 10^{-6} × molecular weight. (From Leaver, C. J. and Key, J. L., *J. Mol. Biol.*, 49, 671, 1970. With permission.)

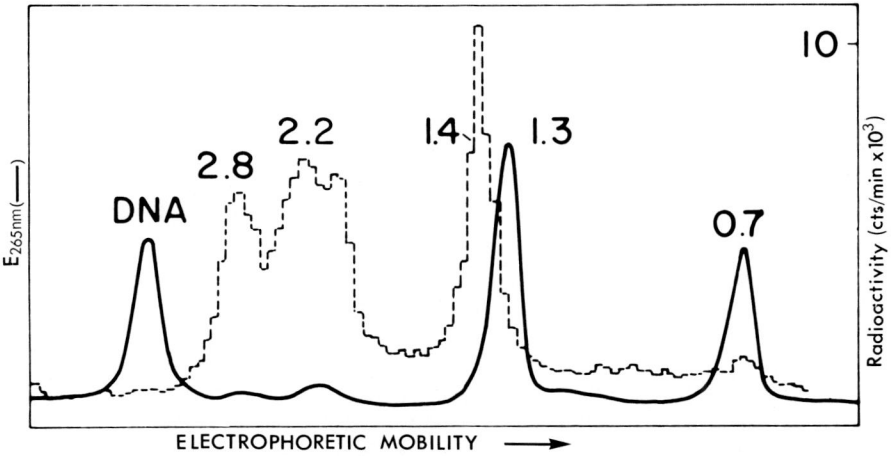

FIGURE 7. Polyacrylamide gel electrophoresis of purified high molecular weight ribosomal RNA precursor molecules. Nucleic acids were extracted from the "nuclear fraction" prepared from carrot-root discs exposed to [³²P] orthophosphate (80 μCi/ml) for 40 min and fractionated by electrophoresis on a 2.2% gel for 4 hr at 6 mA/8 cm gel. RNA components are referred to as 10^{-6} × molecular weight. (From Leaver, C. J. and Key, J. L., *J. Mol. Biol.*, 49, 671, 1970. With permission.)

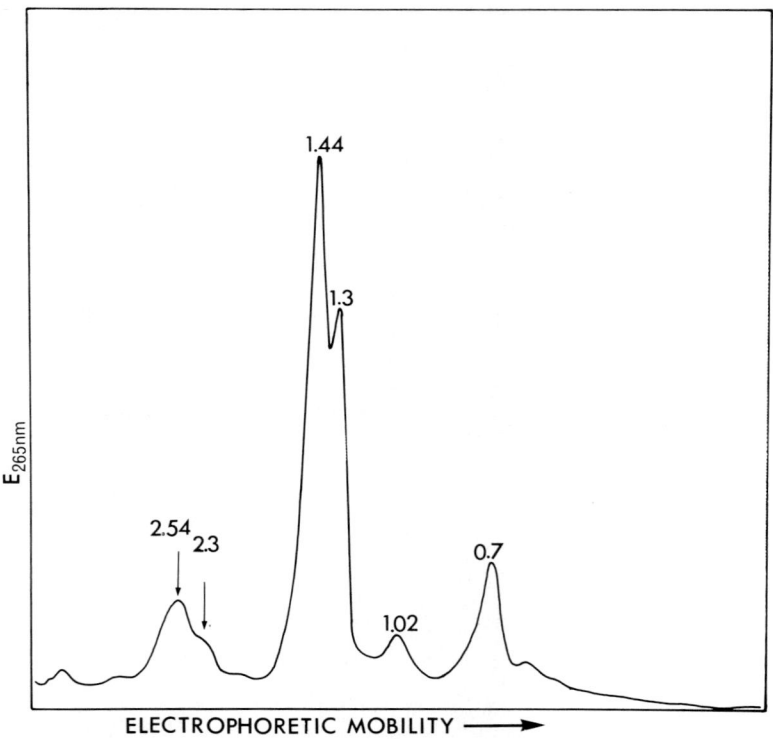

FIGURE 8.   Polyacrylamide gel electrophoresis of nucleic acids from purified mung bean (*Phaseolus aureus*) nucleoli. (From Thompson and Grierson, D., unpublished data).

high molecular weight component of 0.9×10^6 has also been observed to accumulate in carrot discs and may also be a processing intermediate in the synthesis of the 0.7×10^6-mol wt rRNA. The similarity in the kinetics of accumulation of the 2.8 and 2.2×10^6 molecules make it unlikely that the larger RNA is converted to the smaller one, and the possibility exists that both may be initial transcription products. An essentially similar situation in which there are two high molecular weight precursors has been reported in cultured sycamore (*Acer pseudoplanatanus*) cells.[33,34] These workers also showed that methylation of the rRNA occurs during synthesis of the largest precursor molecules (molecular weight 3.4×10^6 and 2.4×10^6), as is known to be the case in mammalian tissues. However, in contrast to the mammalian system, there also appears to be a significant amount of direct methylation of the 1.4×10^6-mol wt precursor rRNA.

Direct evidence that the rapidly labeled high molecular weight RNA (31S, 2.2 to 2.4 $\times 10^6$ mol wt) is in fact a precursor to the mature cytoplasmic rRNAs (25S, 1.3×10^6 and 18S, 0.7×10^6 mol wt) has been provided[35] in germinating rye (*Secale cereale*) embryos. The authors showed by fingerprint analysis of T1-ribonuclease digests that all the large oligonucleotides in the 25S and 18S rRNA are present in the 31S species.

A generalized scheme for synthesis of cytoplasmic rRNA in plant tissues, based on observations with a number of plant species and from different tissues, can be summarized:

$$rDNA ---> 2.3\text{-}2.4 \times 10^6 \nearrow 1.4 \times 10^6 \rightarrow \underline{1.3 \times 10^6} \ (25S)$$
$$\searrow \underline{5.4 \times 10^3} \ (5.8S)$$
$$(31S) \searrow 0.9\text{-}1.0 \times 10^6 \rightarrow \underline{0.7 \times 10^6} \ (18S)$$

B. Synthesis and Processing of Chloroplast rRNA

Chloroplast from a range of higher plants have been shown to contain between 10 to 30 copies of a covalently closed circular DNA molecule with an estimated molecular weight of between 85×10^6 and 96×10^6. It is now established that chloroplast DNA codes for chloroplast rRNA, that the rRNA genes are closely linked (as in *E. coli*), and that they are represented twice per genome. Restriction enzyme analysis coupled with *in situ* hybridization with labeled chloroplast rRNAs suggest that the sequence of rRNA genes in the chloroplast DNA is probably 16S - 23S - 5S, analogous to the sequence of rRNA genes in bacteria.[37] Evidence is now accumulating to suggest that in chloroplasts of both algae and higher plants, the mature chloroplast rRNAs are synthesized from precursors of 10 to 20% greater molecular weight. The pattern of synthesis and maturation appears to follow that already established for bacteria[38] with the exception that several workers have now demonstrated a common precursor of both mature chloroplast rRNAs.[39-41] This precursor may be analogous to the 30S molecule which has only been detected in strains of *E. coli* deficient in RNase III.[42]

Several workers have investigated the synthesis of chloroplast rRNA by pulse labeling of young leaves with [^{32}P]orthophosphate, followed by polyacrylamide gel electrophoresis of the purified chloroplast RNA.[11,39-41] In short-term labeling they detected radioactive peaks which correspond to RNAs of molecular weight 2.7×10^6, 1.2×10^6, 0.65×10^6 and 0.47×10^6. After longer periods of incubation, radioactivity accumulated in the 1.05×10^6 and 0.56×10^6 mature chloroplast rRNAs. The kinetics of labeling are consistent with the 2.7×10^6 mol wt species being a precursor to the 1.2×10^6 and 0.65×10^6-mol wt RNAs, which are in turn processed to the 1.05×10^6 and 0.56×10^6-mol wt mature rRNAs.

The existence of the immediate precursors to the mature chloroplast rRNAs has now been confirmed in a range of higher plants and algae (see Figure 9). The rapidly labeled 2.7 to 2.9×10^6-mol wt RNA, which could be a possible candidate for the initial transcription product of the rRNA genes (i.e., a molecule which contains a copy of both 1.05×10^6 and 0.56×10^6-mol wt rRNAs) has also been reported in *Phaseolus aureus* chloroplasts[11] and in chloroplasts from chloramphenicol-treated *Spirodela oligorhiza*.[43] The only direct evidence that this molecule is a precursor of the chloroplast rRNAs comes from the recent work of Hartley et al.[40] and Bohnert et al.[44] They isolated chloroplasts from young spinach plants, allowed them to synthesize RNA in the presence of [^{3}H]uridine or [^{32}P] orthophosphate and isolated the labeled putative precursor molecules. They then investigated the extent of sequence homology between the 2.7×10^6-mol wt product and mature chloroplast rRNAs by DNA-RNA competition hybridization. Their results suggest that the 0.56×10^6 and 1.05×10^6 rRNAs are present in an approximately 1:1 molar ratio in the 2.7×10^6-mol wt RNA. From this, and other hybridization data obtained with in vivo labeled chloroplast rRNAs, they have proposed the following scheme for the synthesis of chloroplast rRNA in spinach (*Spinacia oleracea*):

$$\text{Chloroplast} \dashrightarrow 2.7 \times 10^6 \rightarrow ? \left\langle \begin{array}{l} 1.2 \times 10^6 \longrightarrow \begin{array}{l} \underline{1.05 \times 10^6} \quad (23S) \\ \underline{3.3 \times 10^3} \quad (4.5S) \end{array} \\ \\ 0.7 \times 10^6 \quad \underline{0.56 \times 10^6} \quad (16S) \end{array} \right.$$

Additional preliminary hybridization experiments by Bohnert et al.[44] suggest that the 2.7×10^6-mol wt rRNA also contains regions homologous to the chloroplast-specific 5S and 4.5S rRNAs. The nature of the rapidly labeled 0.47 to 0.49×10^6-mol wt RNA component is still unknown. It could well be a metastable processing end product, although the work of Spiers[45] suggests that it is composed of three classes of RNA that together display mRNA activity in an in vitro protein synthesizing system from *E. coli*.

The pattern of synthesis and maturation of the rRNAs of blue-green algae appear to be similar to that reported for chloroplasts and bacteria.[46,47] However, there has been no report of a single high molecular weight precursor analogous to the 2.7×10^6-mol wt RNA found in chloroplasts.

C. Synthesis of Mitochondrial rRNA

The only reported study on the synthesis and maturation pathways of mitochondrial rRNA is by Kuriyama and Luck[48] on *Neurospora crassa*. They showed that the initial precursor has a molecular weight of 2.4×10^6 and is processed via intermediates of 1.6×10^6 and 0.9×10^6 mol wt to the mature rRNAs of 1.28×10^6 and 0.72×10^6 mol wt.

D. Ribosomes and RNA Synthesis in *Euglena gracilis*

Early experiments with *E. gracilis* lead to conflicting estimates of the size and composition of both chloroplast and cytoplasmic ribosomes in this organism. However, once problems of cell breakage and ribonuclease activity had been overcome, there was general agreement that cytoplasmic ribosomes were unusually large, 87S, with subunits of 67S and 47S. The larger subunit appears to contain rRNA of 1.4×10^6 mol wt, along with a large amount of a breakdown product of 1.1×10^6 mol wt, while the smaller subunit has an rRNA of 0.9×10^6 mol wt.[5,6,49] These unusual sizes may relate to the unique phylogenetic position of *E. gracilis* between the plant and animal kingdoms. The pathway of synthesis of these rRNAs is postulated as:[50,51]

$$4.3 \times 10^6 \left\langle \begin{array}{l} \underline{0.9 \times 10^6} \\ \\ 2.34 \times 10^6 \rightarrow 1.4 \times 10^6 \longrightarrow (1.1 \times 10^6) \end{array} \right.$$

The intermediate of 2.34×10^6 mol wt can be up to 3% of the total cellular RNA.

In contrast, the chloroplast ribosomes of *E. gracilis* are analagous to other chloroplast systems, and the 70S ribosome has 50S and 30S subunits containing rRNA of molecular weight 1.1×10^6 and 0.56×10^6.[5,6,49] This rRNA is processed from precursors of 1.16×10^6 and 0.64×10^6 mol wt with no evidence for a single high molecular weight precursor containing both mature rRNA sequences.[51]

V. LOW MOLECULAR WEIGHT RNAs OF PLANT RIBOSOMES

In addition to high molecular weight RNA, the large subunit of plant ribosomes also contain low molecular weight RNAs. Dyer and Payne[52] have recently reviewed their own important contribution to this field and, together with results of other workers, have considered the structure, function, and evolution of these molecules.

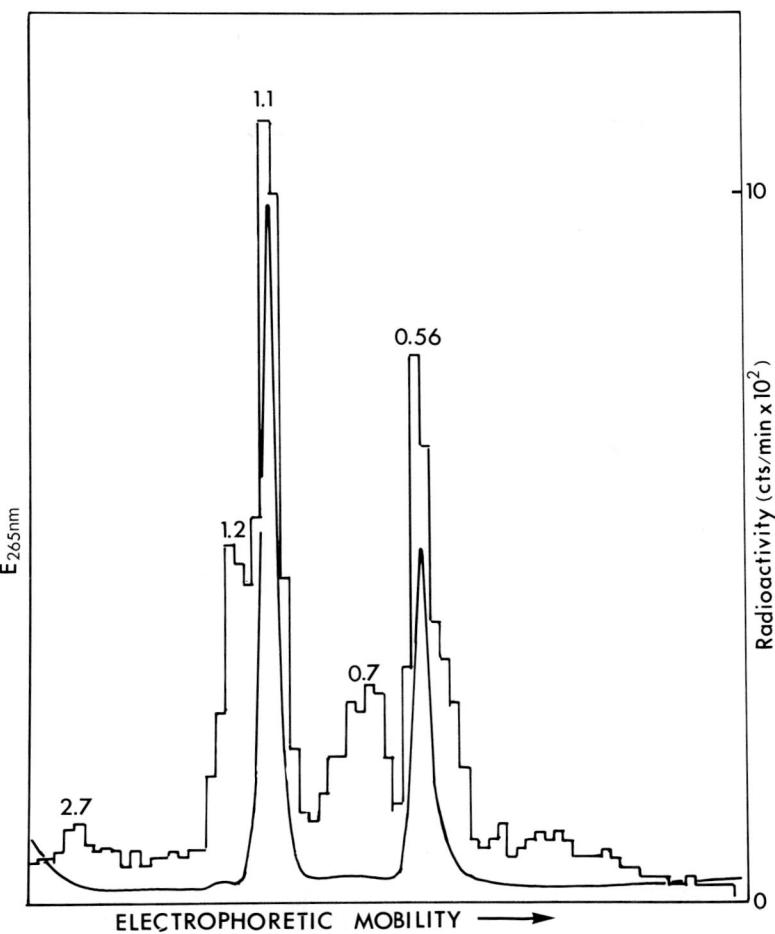

FIGURE 9.   Polyacrylamide gel electrophoresis of nucleic acid extracted from chloroplasts prepared from spinach (*Spinacia oleracea*) leaves labeled for 5 hr with [^{32}P] orthophosphate. 14-day-old plants were labeled via the roots with [^{32}P] orthophosphate (60 μCi/ml). Chloroplast extraction and nucleic acid preparation were performed as described in Reference 22. Electrophoresis was for 5 hr at 3 mA/9 cm gel at a temperature of 5°C (—)E$_{265nm}$ (⌐⌐) radioactivity per 0.5 mm slice. RNA components are referred to as 10^{-6} × molecular weight.

A. 5S rRNA

A point of similarity between the cytoplasmic, chloroplast, and mitochondrial ribosomes of higher plants has been the demonstration that each contains one molecule of a low molecular weight RNA with a sedimentation coefficient of 5S. Although the 5S rRNAs from the chloroplast, mitochondria, and cytoplasm are very similar in size, they can be separated by gel electrophoresis (Figures 4 and 10) and methylated albumin kieselguhr chromatography.[12] Their electrophoretic mobilities relative to those of *E. coli* 5S RNA (120 nucleotides) and plant 4S RNA (tRNA) (78 nucleotides) suggest that the plant cytoplasmic 5S RNA molecule contains 118 nucleotides and has a molecular weight of 37,5000. The chloroplast component is slightly larger and contains about 122 nucleotides, corresponding to a molecular weight of 39,400[12] (See Table 4).

The demonstration that plant mitochondrial ribosomes contain a 5S rRNA (approximately 120 nucleotides and molecular weight 38,750) has now been confirmed independently by two different laboratories.[53,54] This is in contrast to the mitochondrial

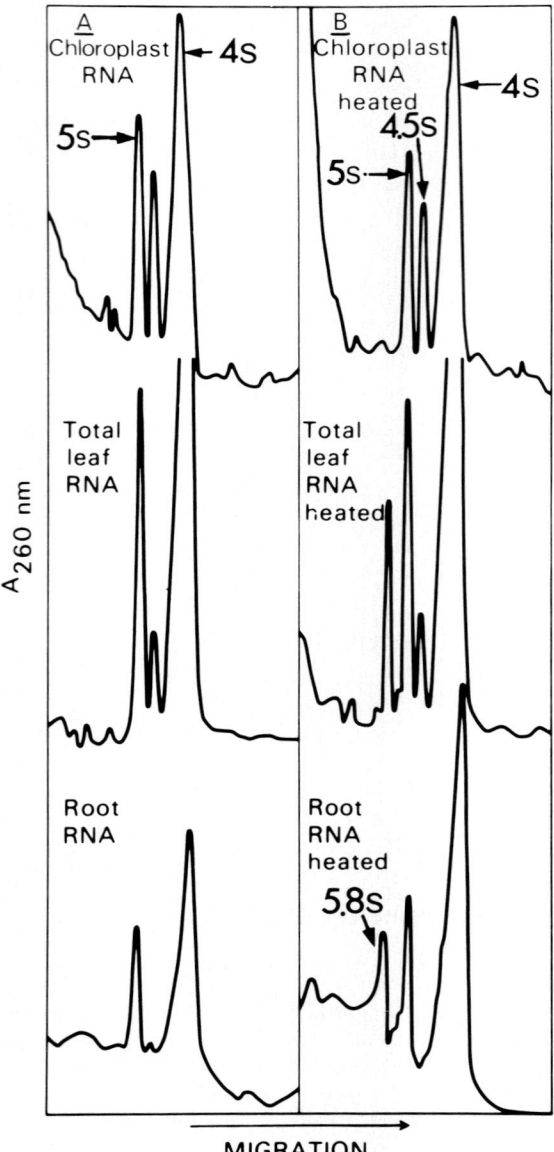

FIGURE 10.   Polyacrylamide gel electrophoresis of low molecular weight RNA from spinach (*Spinacia oleracea*) chloroplasts, total leaf and root. RNA samples in (*A*) were unheated, while those in (*B*) were heated to 65°C for 5 min followed by rapid cooling and electrophoresis on 10% polyacrylamide gels. (Modified from Whitfeld, P. R., Leaver, C. J., Bottomley, W. and Aitchison, B. A., in *Nucleic Acid and Protein Synthesis in Plants,* Bogorad, L. and Weil, J. H., Eds., Centre National de la Recherche Scientifique, Paris, 1977, 255. With permission.)

ribosomes of animals and fungi in which such a component has not been detected.[26] This failure to find a 5S RNA in most mitochondrial ribosomes is surprising, particularly as the presence of such a highly conserved molecule in all other translation systems suggests that this RNA species plays a fundamental role.

The small size of 5S RNA and its relative ease of purification has enabled several workers to carry out nucleotide sequence analysis.[52,55] It appears to be a highly conserved molecule not only in size, but in base composition (it contains no modified bases) and primary and secondary structure.

Dyer and colleagues have compared the sequences of cytoplasmic 5S RNA from widely separated taxonomic groups, including both monocotyledons and dicotyledons.[52] They found only eight base differences between the most dissimilar groups studied. In flowering plants, these workers have shown[55] that while the sequence of chloroplast 5S rRNA is obviously very different from that of cytoplasmic 5S rRNA, certain sequences of nucleotides in the molecule are very similar to those in the 5S rRNA of the blue-green algae, *Anacystis nidulans*. This does not necessarily indicate that blue-green algae as such were the progenitors of chloroplasts (as suggested in the endosymbiont theory for the origin of chloroplasts), but that the ancestors of the chloroplasts may have been photosynthetic prokaryotes.[52]

The cytoplasmic 5S rRNA is apparently transcribed in the nucleoplasm, separately from the high molecular weight rRNAs. There is an indication for bacteria, blue-green algae, and *Xenopus laevis* that 5S RNA is synthesized in a precursor form with a few extra nucleotides at the 5′- and 3′-ends respectively, which are removed before or during incorporation of 5S RNA into the large ribosomal subunit precursor. Treatment of the large subunit with either formamide or EDTA causes the release of 5S RNA in the form of a ribonucleoprotein complex. In higher plant cytoplasmic ribosomes, a single protein remains attached to 5S rRNA. It has a molecular weight of approximately 38,000 and is similar in size to the protein which remains attached to 5S rRNA when mammalian ribosomes are disrupted by similar treatments.[52] The complex from both animal and bacterial ribosomes has GTPase activity, and it is postulated that in the intact ribosome it may be involved in the translocation step of protein synthesis. In plants, Azad and Lane[56] have shown that cytoplasmic 5S rRNA complexes preferentially with 18S rRNA when heated together to temperatures which only partially disrupt hydrogen bonding. They go on to propose that 5S rRNA may serve as a bridge to mediate reversible association between the small and large ribosomal subunits.

B. 5.8S rRNA

A further difference between 70S and 80S ribosomes is the presence of a 5.8S RNA component which is hydrogen-bonded to the high molecular weight RNA of the large ribosomal subunit of animal, plant, and fungal cytoplasmic ribosomes.[57] A comparable RNA has not been detected in chloroplast, mitochondrial, or prokaryotic ribosomes (Figures 4 and 10).

It has been shown that there is one molecule of 5.8S rRNA per ribosome with a calculated molecular weight of 50,400. It therefore contains about 157 nucleotide residues[57] (see Table 4).

In vivo labeling of rRNA from artichoke suggests that the 5.8S and 25S rRNAs are initially synthesized as part of a common precursor molecule with a molecular weight of 1.39×10^6. During the final stage of processing of the rRNA gene transcript to the mature 25S rRNA, the 5.8S rRNA is specifically cleaved from the 3′ end of the 1.39×10^6 precursor. This observation could perhaps explain the finding that 5.8S rRNA contains modified (probably methylated) nucleotides, as do the high molecular weight rRNAs which is in contradistinction to the other low molecular weight RNAs.

Sequence analysis of 5.8S rRNA from several flowering plants suggests only a limited homology with sequences in yeast and hepatoma 5.8S rRNAs. The composition of the flowering plant oligonucleotides demonstrates that the molecule is very different from 5S rRNA, and probably the two molecules did not have a common ancestor.

TABLE 4

Estimated Molecular Weights and Nucleotide Contents of Low Molecular Weight RNA from the Large Subunits of Plant Ribosomes

RNA species	Cytoplasm		Chloroplast		Mitochondria	
	Molecular Weight	Number of nucleotides	Molecular weight	Number of nucleotides	Molecular weight	Number of nucleotides
5.8 S	50,400	157	—	—	—	—
5.0 S	37,500	118	39,400	122	38,750	120
4.5 S	—	—	25,000 or 33,000	80 or 103	—	—

From Dyer, T. A., Bowman, C. M., and Payne, P. I., in *Nucleic Acids and Protein Synthesis in Plants*, Bogorad, L. and Weil, J. H., Eds., Plenum Press, New York, 1976, 121. With permission.

C. 4.5S rRNA

In the last few years, several groups have independently reported that the chloroplast ribosomes of angiosperms contain an apparently unique, low molecular weight ribosomal RNA in addition to 5S rRNA.[44,52,59] This RNA is present in approximately equimolar amounts with respect to 5S RNA and is a component of the 50S ribosomal subunit.[59] This molecule has been referred to as 4.5S rRNA although it does not appear to be of similar size in all angiosperms. It normally migrates between 5S and 4S RNAs on SDS-acrylamide gel electrophoresis (Figure 10) and has a molecular weight between 25,000 and 33,000 which corresponds to between 80 and 103 nucleotides, depending on plant species (Table 4). The smaller form of this molecule (80 nucleotides) has been found in *Vicia faba;* it migrates to the same position as tRNA and can easily be mistaken for it.[52]

The 4.5S rRNA can be released from the chloroplast ribosome by treatment with ethylenediamine tetraacetic acid (EDTA). In this regard it is similar to 5S RNA and distinct from the 5.8S RNA of the cytoplasmic 60S ribosomal subunit which is only dissociated from 25S rRNA by conditions which disrupt base pairing. Dyer et al.[55] have recently shown by oligonucleotide fingerprinting that the 4.5S rRNA molecule is essentially homogenous and that it is completely different from any of the other low molecular weight rRNAs or tRNAs studied so far.

At this stage, one can only speculate about the nature of the 4.5S rRNA. It has been suggested that it could be the chloroplast equivalent of cytoplasmic 5.8S rRNA, but in contrast to this molecule, it contains no methylated bases. A likely possibility is that the 4.5S rRNA is a piece of transcribed spacer, perhaps with no particular role in the ribosome. This suggestion is supported by recent kinetic studies on the radioactive labeling of the 4.5S rRNA molecule in spinach leaves, which indicate that it originates from the same precursor molecule as the 23S rRNA.[44,44a]

VI. CONCLUSION

The synthesis and processing of cytoplasmic rRNAs has been described in several species of plant, and it is now generally accepted that the mature cytoplasmic rRNAs are transcribed from rDNA as a large precursor molecule processed to yield rRNA characteristic of the particular organism. We know very little about the selective processing and methylation of precursor rRNA or the coordination of synthesis and assembly of ribosomal protein and rRNA.

The early descriptive phase of research on plant rRNAs is coming to an end, and of more importance in the future will be an understanding of the regulation of ribosome production in relation to the developmental and physiological status of the plant. An example of a development phase which may be amenable to this type of study is seed germination. This represents a dramatic change from the dormant embryo to the rapidly growing seedling, which is associated with a rapid change in the rate of synthesis and accumulation of ribosomal RNA.

The synthesis and processing of chloroplast rRNA appears to be similar to the situation in prokaryotes although the suggestion that the mature chloroplast rRNAs are initially transcribed as a single high molecular weight precursor molecule requires confirmation. Also of potential interest is an understanding of the coordination of synthesis and assembly in time and space of the nuclear-coded, cytoplasmically synthesized ribosomal proteins with the structural rRNAs coded and synthesized in the chloroplast.

Our present knowledge of mitochondrial ribosomes and rRNA in higher plants is at the purely descriptive phase, and we await detailed studies on their synthesis and function. While the discovery of a 4.5S rRNA component of unknown function in the chloroplast ribosome as recently as 1975 suggests that there may be minor, but nevertheless important, rRNA molecules yet to be described.

It has become the vogue in recent years to study plant growth and development in terms of changing populations of messenger RNAs and enzymes, while neglecting the regulation of synthesis and turnover of the protein synthetic machinery. Perhaps it is now the time to reexamine this important area with the new knowledge and techniques available.

ACKNOWLEDGMENTS

The author wishes to acknowledge financial support from the Science Research Council and The Royal Society.

REFERENCES

1. Leaver, C. J. and Pope, P. K., The biosynthesis of plant mitochondrial proteins in *Nucleic Acids and Protein Synthesis in Plants*, Bogorad, L. and Weil, J. H., Eds., Plenum Press, New York, 1976, 213.
2. Attardi, G. and Amaldi, F., Structure and synthesis of ribosomal RNA, *Annu. Rev. Biochem.*, 39, 183, 1970.
3. Loening, U. E., Molecular weights of ribosomal RNA in relation to evolution, *J. Mol. Biol.*, 38, 355, 1968.
4. Maden, B. E. H., The structure and formation of ribosomes in animal cells, *Prog. Biophys. Molec. Biol.*, 22, 5, 1971.
5. Rawson, J. R., and Stutz, E., Isolation and characterization of *Euglena gracilis* cytoplasmic and chloroplast ribosome and their ribosomal RNA components, *Biochim. Biophys. Acta*, 190, 368, 1969.
6. Scott, N. S., Munns, R., and Smillie, R. M., Chloroplast and cytoplasmic ribosomes in *Euglena gracilis*, *FEBS Lett.*, 10, 149, 1970.
7. Amaldi, F., Nonrandom variability in evolution of base compositions of ribosomal RNA, *Nature (London)*, 221, 95, 1969.
8. Faumann, M., Rabinowitz, M., and Getz, G. S., Base composition and sedimentation properties of mitochondrial RNA of *Saccharomyces cerevisiae*, *Biochim. Biophys. Acta*, 182, 355, 1969.
9. Oshio, Y. and Hase, F., Studies on nucleic acids in chloroplasts isolated from *Chlorella protothecoides*, *Plant Cell Physiol.*, 9, 69, 1968.
10. Rossi, L. and Gualerzi, C., Nonrandom differences in the base composition of chloroplast and cytoplasmic ribosomal RNA from sone higher plants, *Life Sci.*, 9, 1401, 1970.

11. **Grierson, D. and Loening, U. E.,** Ribosomal RNA precursors and the synthesis of chloroplast and cytoplasmic ribosomal ribonucleic acid in leaves of *Phaseolus aureus, Eur. J. Biochem.,* 44, 501, 1974.
12. **Payne, P. I. and Dyer, T. A.,** Characterization of cytoplasmic and chloroplast 5S ribosomal ribonucleic acid from broad bean leaves, *Biochem. J.,* 124, 83, 1971.
13. **Payne, P. I. and Dyer, T. A.,** Characterization of the ribosomal ribonucleic acids of blue-green algae, *Arch. Mikrobiol.,* 87, 29, 1972.
14. **Stanley, W. M., Jr., and Bock, R. M.,** Isolation and physical properties of the ribosomal ribonucleic acid of *Escherichia coli, Biochemistry,* 4, 1302, 1965.
15. **Cammarano, P., Felsani, A., Gentile, M., Gualerzi, G., Romeo, A., and Wolf, G.,** Formation of active hybrids 80-S particles from subunits of pea seedlings and mammalian liver ribosomes, *Biochim. Biophys. Acta,* 281, 625, 1972.
16. **Lyttleton, J. W.,** Isolation of ribosomes from spinach chloroplasts, *Exp. Cell. Res.,* 26, 312, 1962.
17. **Whitfeld, P. R.,** Chloroplast RNA, in *The Ribonucleic Acids,* Stewart, P. R. and Letham, D. S., Eds. Springer-Verlag, Berlin, 1973, 179.
18. **Ellis, R. J. and Hartley, M. R.,** Nucleic acids of chloroplasts, in *Biochemistry of Nucleic Acids,* Burton, K., Ed., Butterworth and University Park Press, London, 6(1), 11, 1974.
19. **Loening, U. E. and Ingle, J.,** Diversity of RNA components in green plants, *Nature,* (London), 215, 363, 1967.
20. **Ingle, J.,** Synthesis and stability of ribosomal RNAs in radish, *Plant Physiol.,* 43, 1448, 1968.
21. **Leaver, C. J. and Ingle, J.,** The molecular integrity of chloroplast ribosomal ribonucleic acid, *Biochem. J.,* 123, 235, 1971.
22. **Leaver, C. J.,** Molecular integrity of chloroplast ribosomal ribonucleic acid, *Biochem. J.,* 135, 237, 1973.
23. **Doolittle, W. F.,** Postmaturational cleavage of 23S ribosomal ribonucleic acid and its metabolic control in the blue-green alga *Anacystic nidulans, J. Bacteriol.,* 113, 1256, 1973.
24. **Leaver, C. J.,** The biogenesis of plant mitochondria, in *The Chemistry and Biochemistry of Plant Proteins,* Harborne, J. B. and Van Sumere, C. F., Eds., Academic Press, London, 137, 1975.
25. **Pring, D. R.,** Maize mitochondria: Purification and characterization of ribosomes and ribosomal ribonucleic acid, *Plant Physiol.,* 53, 677, 1974.
26. **Borst, P.,** Mitochondrial nucleic acids, *Annu. Rev. Biochem.,* 41, 33, 1972.
27. **Buetow, D. E., Kissel, M. S. and Zablen, L.,** Phylogenetic origin of chloroplast 16S rRNA, in *Nucleic Acids and Protein Synthesis in Plants,* Bogorad, L. and Weil, J. H., Eds., Centre National de la Recherche Scientifique, Paris, 641, 1976.
28. **Cunningham, R. S., Gray, M. W., Doolittle, W. F., and Bonen, L.,** The procaryotic nature of wheat embryo mitochondrial 18S ribosomal RNA, in *Nucleic Acids and Protein Synthesis in Plants,* Bogorad, L. and Weil, J. H., Eds., Centre National de la Recherche Scientifique, Paris, 243, 1976.
29. **Ingle, J.,** The regulation of ribosomal RNA synthesis, in *Biosynthesis and its Control in Plants,* Milborrow, B. V., Ed., Academic Press, London, 69, 1973.
30. **Rungger, D. and Crippa, M.,** The primary ribosomal DNA transcript in eucaryotes, *Prog. Biophys. Molec. Biol.,* 31, 247, 1977.
31. **Perry, R. P., Cheng, T. Y., Freed, J. J., Greenberg, J. R., Kelly, D. E., and Tartof, D. D.,** Evolution of the transcription unit of ribosomal RNA, *Proc. Natl. Acad. Sci. U.S.A.,* 65, 609, 1970.
32. **Leaver, C. J. and Key, J. L.,** Ribosomal RNA synthesis in plants, *J. Mol. Biol.,* 49, 671, 1970.
33. **Cox, B. J. and Turnock, G.,** Synthesis and processing of ribosomal RNA in cultured plant cells, *Eur. J. Biochem.,* 37, 367, 1973.
34. **Rogers, M. E., Loening, U. E. and Fraser, R. S. S.,** Ribosomal RNA precursors in plants, *J. Mol. Biol.,* 49, 681, 1970.
35. **Cecchini, J. P. and Miassod, R.,** Ribosomal cistrons in higher plant cells, *Biochem. Biophys. Acta,* 418, 104, 1976.
36. **Sen, S., Payne, P. I., and Osborne, D. J.,** Early RNA synthesis during the germination of rye (*Secale cereale*) embryos and the relationship to early protein synthesis, *Biochem. J.,* 148, 381, 1975.
37. **Whitfeld, P. R., Aitchison, B. A., Bottomley, W. and Leaver, C. J.,** Analysis of the coding capacity of EcoRI restriction fragments of spinach chloroplast DNA, in *Genetics and Biogenesis of Chloroplasts and Mitochondria,* Bücher, Th., Neupert, W., Sebald, W., and Werner, S., Eds., North-Holland, Amsterdam, 1976, 361.
38. **Pace, N. R.,** The structure and synthesis of ribosomal RNA of procaryotes, *Bacteriol. Rev.,* 37, 562, 1973.
39. **Hartley, M. R. and Ellis, R. J.,** Ribonucleic acid synthesis in chloroplasts, *Biochem. J.,* 134, 249, 1973.
40. **Hartley, M. R., Head, C. W., and Gardiner, J.,** The synthesis of chloroplast RNA, in *Nucleic Acid and Protein Synthesis in Plants,* Bogorad, L. and Weil, J. H., Eds., Centre National de la Recherche Scientifique, Paris, 1977, 419.

41. Detchon, P. and Possingham, J. V., Chloroplast ribosomal ribonucleic acid synthesis in cultured spinach leaf tissue, *Biochem. J.*, 136, 829, 1973.

42. Nikolaev, N., Schlessinger, D., and Wellauer, P. K., 30S preribosomal RNA of *E. coli* and products of cleavage by ribonuclease II. Length and molecular weight, *J. Mol. Biol.*, 86, 741, 1974.

43. Rosner, A., Posner, H. B., and Gressel, J., Synthesis and processing of RNA in *Lemna, Plant Cell Physiol.*, 14, 555, 1973.

44. Bohnert, H. J., Driesel, A. J. and Hermann, R. C., Characterization of the RNA components synthesized by isolated chloroplasts, in *Genetics and Biogenesis of Chloroplasts and Mitochondria*, Bücher, Th., Neupert, W., Sebald, W., and Werner, S., Eds., North-Holland, Amsterdam, 1976, 629.

44a. Hartley, M. R., unpublished data.

45. Spiers, J., Studies on rapidly labelled RNA species synthesized in chloroplast of spinach, in *Nucleic Acid and Protein Synthesis in Plants*, Bogorad, L. and Weil, J. H., Eds., Centre National de la Recherche Scientifique, Paris, 425, 1977.

46. Grierson, D. and Smith, H., The synthesis and stability of ribosomal RNA in blue-green algae, *Eur. J. Biochem.*, 36, 280, 1973.

47. Doolittle, W. F., Ribosomal ribonucleic acid synthesis and maturation in the blue-green algae *Anacystis nidulans, J. Bacteriol.*, 111, 316, 1972.

48. Kuriyama, Y. and Luck, D. J. L., Ribosomal RNA synthesis in mitochondria of *Neurospora crassa, J. Mol. Biol.*, 73, 425, 1973.

49. Mendiola, L. R., Kovacs, A., and Price, C. A., Separation and partial characterization of chloroplast and cytoplasmic ribosomes from *Euglena gracilis, Plant Cell Physiol*, 11, 335, 1975.

50. Brown, R. D. and Haselkorn, R., Synthesis and maturation of cytoplasmic ribosomal RNA in *Euglena gracilis, J. Mol. Biol.*, 59, 491, 1971.

51. Scott, N. S., Precursors of chloroplast ribosomal RNA in *Euglena gracilis, Phytochemistry*, 15, 1207, 1976.

52. Dyer, T. A., Bowman, C. M., and Payne, P. I., The low-molecular-weight rRNAs of plant ribosomes; their structure, function and evolution, in *Nucleic Acids and Protein Synthesis in Plants*, Bogorad, L. and Weil, J. H., Eds., Plenum Press, New York, 121, 1976.

53. Leaver, C. J. and Harmey, M. A., Higher-plant mitochondria contain a 5S ribosomal ribonucleic acid component, *Biochem. J.*, 157, 275, 1976.

54. Cunningham, R. S., Bonen, L., Doolittle, W. F., and Gray, M. W., Unique species of 5S, 18S and 26S ribosomal RNA in wheat mitochondria, *FEBS Lett.*, 69, 116, 1976.

55. Dyer, T. A. and Bowman, C. M., A sequence analysis of low-molecular-weight rRNA from chloroplasts of flowering plants, in *Genetics and Biogenesis of Chloroplasts and Mitochondria*, Bücher, Th., Neupert, N., Sebald, W., and Werner, S., Eds., North-Holland, Amsterdam, 645, 1976.

56. Azad, A. A. and Lane, B. G., A possible role for 5S rRNA as a bridge between ribosomal subunits, *Can. J. Biochem.*, 51, 1669, 1973.

57. Payne, P. I. and Dye, T. A., Plant 5.8S RNA is a component of 80S but not 70S ribosomes, *Nature (London), New Biol.*, 235, 145, 1972.

58. Hepburn, A.G. and Ingle, J., Origins of 7S ribosomal RNA in artichoke, *Phytochemistry*, 14, 1157, 1975.

59. Whitfeld, P. R., Leaver, C. J., Bottomley, W. and Aitchison, B. A., Low-molecular weight (4.5s) ribonucleic acid in higher-plant chloroplast ribosomes, *Biochem. J.*, 175, 1103, 1978.

PLANT MESSENGER RNA

T. C. Hall

TABLE OF CONTENTS

I. INTRODUCTION

The involvement of RNA in protein synthesis was clear at the turn of the century;[1] in 1954, Zamecnik and Keller[2] demonstrated amino acid incorporation using a cell-free system. Astrachan and Volkin[3] found that RNA which was rapidly synthesized and degraded in *Escherichia coli* following infection by bacteriophage T2[4] had a base distribution closely resembling that of the infecting phage DNA. The correct interpretation — that this RNA played the role of a messenger between the inserted phage DNA and the appearance of viral proteins in the bacterial cell — was provided by Jacob and Monod.[5] Towards this understanding of information flow from DNA through RNA to proteins, the bacterial ribosomal system[6] for cell-free amino acid incorporation into protein was essential. This system was inhibited by DNase,[7] and it was realized that its activity was dependent on RNA transcribed from the DNA contaminating the ribosomal preparation. Restoration of amino acid incorporating activity to the DNase-treated system was subsequently shown by Nirenberg and Matthaei[8] in experiments where polyphenylalanine was synthesized in response to the addition of the synthetic polyribonucleotide poly(U). This observation provided the "Rosetta stone" for understanding the translation of sequences of genome nucleic acid into amino acid sequences of protein.[9] Using polynucleotide templates, the three-base (triplet) nature of each amino acid codon has subsequently been determined.[10]

The exciting results with synthetic polynucleotide templates needed to be confirmed with biological messengers. The discovery, in 1956, that RNA extracted from tobacco mosaic virus (TMV) was infective[11,12] provided a clue that RNA could carry information for protein synthesis, although at that time a direct role in the synthesis of coat and other viral proteins was not seriously considered. However, a striking stimulation of amino acid incorporation was obtained by addition of TMV RNA to the *E.coli* cell-free system.[13,14] It turned out that TMV RNA was an unfortunate choice because processing of this template is necessary (see Section III, Volume II, Chapter by Davies) before products identifiable as being similar to those made in response to viral infection (notably the coat protein) can be translated in vitro.[15]

The discovery of small RNA viruses (phages) which infected *E. coli* (notably f2,[16] and relatives such as MS2, R17 and Qβ), and whose RNA was readily translated by cell-free extracts of the host organisms to yield recognizable polypeptide products,[17] put the new science of molecular biology firmly — indeed, almost exclusively — into the hands of microbiologists for over a decade. Without doubt, the ease of genetic and biochemical manipulation intrinsic to prokaryotes dictated that microbial systems should, and for several years to come will still, be the prime candidates for elucidating gene expression at the molecular level. Nevertheless, understanding regulation of important eukaryotic proteins requires that messenger RNAs (mRNAs) be isolated from higher organisms. The type of information following from such isolation is detailed in Sections III,D,3 and VI. Suffice it to say here that possession of a specific mRNA provides an approach to detecting the transcriptional availability of specific cistrons of the genome DNA at carefully regulated developmental stages, in specific tissues and in response to diverse triggers such as hormones and environmental events.

After several controversial reports, convincing evidence for a eukaryotic mRNA (for mouse hemoglobin) was provided in 1971 by Lockard and Lingrel.[18] Since then, many animal mRNAs that function as templates for protein synthesis in vitro have been isolated, and these are providing exciting probes towards elucidating processes regulating gene expression.

There are now several papers describing mRNA or polysome-directed synthesis of plant proteins.[19-53] A summary of these systems is compiled as Table 1. Considerable biological importance is attached to many of the proteins for which these plant messengers serve as templates, and some are discussed in this chapter.

In addition to the current interest in plant mRNAs, it is pleasing that plant extracts have become widely accepted as very useful systems for the translation of a wide range of mRNAs. Great credit is due to Abraham Marcus and co-workers who pioneered the wheat embryo system for protein synthesis in vitro.[54] The demonstration that addition of plant viral messengers such as satellite tobacco necrosis virus RNA[54a] and brome mosaic virus RNA [54b] to the wheat embryo system resulted in synthesis of accurate translation products greatly stimulated the use of this material. A further stimulus was the demonstration that systems using wheat germ, a readily available commercial preparation of wheat embryos, also supported very active protein synthesis.[55,55a] Details for preparation and use of these systems and techniques for the isolation of plant viral mRNAs can be found in Section III, Volume II, Chapters by Zaitlin and Davies. In this chapter, current techniques used for isolating higher plant mRNAs and the biochemical characteristics of those which have been studied in some depth are described. Since the availability of plant mRNAs now permits studies which were not practicable until very recently, the directions in which they can be used to understand gene expression in plants are reviewed.

II. ISOLATION OF PLANT mRNAs

A. Preparation and Stability of Polysomes
1. Preparation of Polysomes

A polysome consists of mRNA to which several ribosomes are attached, each bearing a nascent polypeptide chain whose length depends on how much of the cistronic template has been read during translation (Figure 1). On the molecular scale, polysomes are very large structures, and they can be readily separated from other cellular components by physical techniques. For plant materials, the conditions for polysome isolation described by Verma et al.[34] are very useful and are detailed in Tables 2 and 3.

It is frequently thought (usually correctly) that the best polysome preparations are those with a high ratio of large polysomes, as judged by sucrose density gradient separation, because these presumably have suffered the least degradation.[57] However, each mRNA is of finite length, and ribosomes occupy discrete sections of the mRNA as they pass along it. Indeed, the presence of a ribosome protects a segment (about 30 nucleotides[58]) of the mRNA from nuclease digestion; this phenomenon has been cleverly exploited in obtaining specific regions of mRNAs, especially initiation sites.[59] The distance between the centers of two adjacent ribosomes on a polysome varies, but appears to be approximately 80 bases at maximum utilization of the mRNA.[60] Consequently, an mRNA for a small polypeptide is unlikely to bear as many ribosomes as a polysome engaged in synthesis of a high molecular weight polypeptide. However, factors such as the rate of polypeptide chain initiation,[61] the frequency of codons for certain amino acids (related to the presence of specific isoaccepting tRNAs and their aminoacylation enzymes in the tissues),[62] and perhaps the tertiary structure of the mRNA can influence the number of ribosomes attached to it. Although the rate of reading of mRNAs is often assumed to be constant for a particular tissue, future studies may well show that the speed of ribosome progression along the mRNA in fact varies (see Section V.F). With the effects mentioned above, this would help explain variations in length of polysomes engaged in the synthesis of similar sized polypeptides. Evidence has been advanced that polysomes synthesizing the large subunit (LS) of RuBPCase* (55,000 mol wt) bear only three ribosomes,[48] although with the minimum

* Ribulose 1,5-bisphosphate carboxylase.

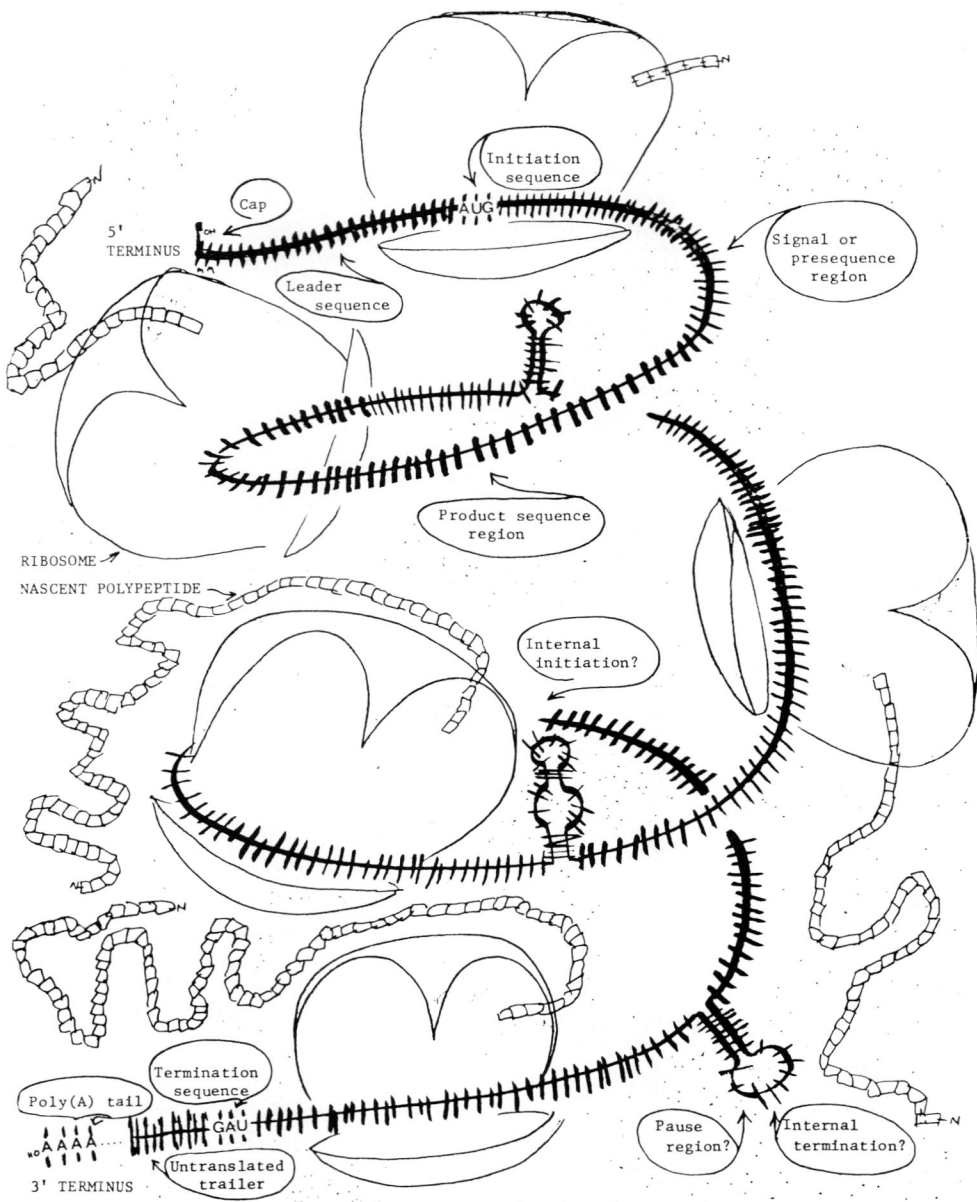

FIGURE 1.   Polysome structure and diagrammatic representation of mRNA regions. The 5-mer polysome shown would be about the size expected for the synthesis of a polypeptide of 11,000 daltons. Each
ribosome protects about 30 bases; and at maximum utilization of mRNA, ribosomes are typically approximately 80 bases apart. The regions of mRNA noted are discussed in Section IV.

spacing between adjacent ribosomes noted above, one could expect as many as 18
ribosomes per LS RuBPCase mRNA of some 1500 nucleotides.

2. Problems Arising from Nucleases

Among the many hydrolytic enzymes released on homogenization of plant tissues
are active ribonucleases (RNases). A useful classification is included in the recent review of plant nucleases by Wilson,[63] but the specific character of individual plant

TABLE 1

Plant Polysome and mRNA Preparations Having Amino Acid Incorporating Activity In Vitro

Plant and tissue	Poly-somes	mRNA	Cell-free system	Label	Activity, product characterization, notes	Ref.
Seed						
Avena sativa groat	9-mers	no	WG	^{3}H-leu, val, phe	36,107 cpm/0.32 A_{260} units RNA, possibly one-third following new initiation. 56,000, 31,000 and 21,000 dalton polypeptides made; cyanogen bromide peptides suggested some similarity to native oat globulin.	19
Glycine max soaked seed	yes	"Heavy"	misc.	^{14}C-leu	13,560 cpm/mg ribosomal protein. Low levels of incorporation with endogenous system or with heterologous system (mainly *E. coli*)	20
G. max ripening cotyledon	yes	poly(A)	WG	^{3}H-leu	30,000 cpm/0.0067 A_{260} units RNA. Both poly(A) RNA and polysome-directed reactions gave a small proportion of products that co-electrophoresed with peptides of soybean 7S and 11S proteins.	20a
Hordeum vulgare aleurone layer	yes	poly(A)	WE	^{35}S-met	SDS gel of α-amylase immunoprecipitate material. System responded in 3 hr to gibberellic acid GA addition.	21
H. vulgare endosperm	yes	no	WG	^{3}H-leu	SDS gel of products soluble in 55% propan-2-ol: putative hordeins.	22
Phaseolus vulgaris ripening cotyledon	yes	16S poly(A)	WG	^{3}H-leu ^{35}S-met	Polysomes direct synthesis of 53,000, 47,000 and 43,000 dalton G1 protein polypeptides. 16S mRNA directs (10,000 cmp ^{35}S met/0.0025 A_{260} units) synthesis of 47,000 and 43,000 dalton subunits only. Products identified by SDS gel mobility, immunoprecipitation and peptide mapping.	23,24, 24a, text
P. vulgaris ripening cotyledon	yes	yes	endogenous	^{35}S-met tRNA	RNA from polysomes bound to millipore filter, 690 cpm met-puromycin/0.4 A_{260} "messenger" RNA. Some initiation claimed.	25
Pisum sativum ripening cotyledon	yes	poly(A)	WG	^{14}C amino acids	SDS gel: 35 bands with polysomes; 31 with mRNA. Some similarities with storage protein tryptic map.	26
Secale cereale dry seed	no	poly(A)	WG	^{35}S-met	Multiple bands, none comparable to storage protein. mRNA degraded on imbibition. Dry seed of *Brassica napus*, *Pisum sativum*, *Vicia faba* also studied.	27
Triticum vulgare	no	total RNA	WE	^{3}H-amino acids	11,000 cpm/10 μg RNA.	28
Zea mays endosperm	MB and free	no	WG	^{14}C-leu ^{14}C-lys	35,358 cpm/40—80 μg polysome RNA. SDS gel mobility of 70% ethanol sol. products and leu/lys ratio similar to zein polypeptides.	29
Z. mays endosperm	MB	poly(A)	WG	^{14}C-leu	143,000 cpm/1.7 μg mRNA (540,000 daltons); Zeins of 21,800 and 19,000 mol wt synthesized	30
Z. mays endosperm	from protein body	yes	maize WG	^{3}H-leu ^{14}C-leu	18,000 dpm ^{14}C-leu/2 μg mRNA (416,000 daltons). Products soluble in 70% ethanol show electrophoretic similarities to zein polypeptides. mRNA-directed products were 24,500 and 21,100 daltons; probably each contains a transport presequence.	31,31a
Seedling						
Phaseolus vulgaris seedlings	yes	poly(A)	WG	^{35}S-met	SDS gel fluorography. Products gave many bands, pattern varying with light treatment, possibly phytochrome effect.	32
Pharbitis nil epicotyl	yes	poly(A)	E. coli	^{3}H-leu	7000 cpm/5 μg RNA.	33
Pisum sativum epicotyl	9-mers	poly(A) + and −	WG	^{3}H-leu	Buffer-soluble cellulase (SDS gels of immunoprecipitate product) synthesized preferentially on membrane-bound polysome-derived mRNA in response to auxin	34,35
Zea mays	yes	no	endogenous	^{14}C-leu	Some indication of stimulation by kinetin.	36

TABLE I (continued)

Plant Polysome and mRNA Preparations Having Amino Acid Incorporating Activity In Vitro

Plant and tissue	Poly-somes	mRNA	Cell-free system	Label	Activity, product characterization, notes	Ref.
Seed						
Root						
Glycine max root nodule	yes	poly(A)	WG	^{3}H-leu	38,000 cpm/4-8 A_{260}. SDS gels of leghemoglobin immunoprecipitated product. Appears to be made on membrane bound ribosomes.	37,38, 85a
Petroselinum sativum callus culture of root explant.	8-mers	no	WG	^{3}H-leu ^{35}S-met	SDS gel of product gave some 20 bands. Aurintricarboxylic acid (ATA) did not inhibit (no initiation); not sensitive to Mg^{++} concentration.	39
Leaf						
Lemna gibba plants	no	poly(A)	WG	^{35}S-met	SDS gel fluorography. Difference in culture in light/dark apparent for 8 bands, especially one of 18,000 mol wt.	40
Phaseolus vulgaris greening seedling leaf	yes	no	endogenous	^{14}C-leu	Activity stimulated by rat liver supernatant low with bean supernatant Sephadex® G-100 chromatography of RuBCase SS immunoprecipitate.	41
Phaseolus vulgaris greening seedling leaf	yes	poly(A)	WG	^{35}S-met	400,000 cpm/μg mRNA. Many bands, but 34,000, 32,000, and 25,000 seen in illumination with white light.	42
Tortula ruralis hydrated and desiccated moss	yes	poly(A)	WG	^{14}C-leu	9,175 cpm/10 μg. Polysomes dissociate on drying, re-associate on rehydration. Evidence for long-lived mRNA.	43,44
Triticum vulgare greening seedling leaf	yes	no	WG	^{3}H-leu	128,000 cpm/A_{260}. Estimated 20—40 amino acid residues added per chain. Immunoprecipitate and SDS gel of product suggested some RuBCase SS formed.	45
Chloroplast						
Chlamydomonas reinhardtii	MB	no	endogenous	^{3}H-amino acid mix	1 nmol amino acid/mg RNA/hr. SDS gel suggested some partial synthesis of chloroplast membrane peptide.	46,47
C. reinhardtii	yes	14S RNA	*E. coli*	^{3}H-arg	100,000 cpm/μg RNA; no evidence for poly(A) tract; SDS gel of immunoprecipitate RuBCase LS with *E. coli* system, none with WG.	48
C. reinhardtii	no	total, poly(A)	WG	^{35}S-met	1 A_{260} unit mRNA/mℓ; carboxylase SS immunoprecipitate and identified on gels. See Section III.C.	49
Euglena gracilis	no	poly(A) + and −	WG	^{35}S-met	With poly(A) RNA 4×10^6 cpm/μg; multiple bands, no RuBCase LS; with poly(A)-RNA 2×10^5 cpm, LS seen by SDS gel fluorography	50
Spinacea oleracea	no	total	*E. coli* WG	^{35}S-met	Evidence by SDS gel autoradiography that *E. coli* system translated only plastid RNA, and WG translated only cytoplasmic RNA. Also evidence that DNA transcribed and translated by this system.	51
S. oleracea	no	total	*E. coli*	^{35}S-met	132,000 cpm/500 μg chloroplast RNA RuBCase LS (52,000 mol wt) immunoprecipitate on SDS gel; chymotryptic map.	53

Note: Plastid systems and systems coded by viral RNAs have been omitted. Systems are grouped according to part of plant. Several publications from one group may be included in one entry. Direct comparison of activity of the various systems is difficult; consult references for details. WE = wheat embryo; WG = wheat germ; MB = membrane bound.

RNases has received surprisingly little attention. While ribosomes do protect regions of mRNA from degradation by low concentrations of RNase, there is ample space between ribosomes for RNase to cleave the mRNA. Higher concentrations of RNase can degrade the ribosomes themselves. The presence of these RNases in plants is a great hinderance to the isolation of polysomes, and tissues having low endogenous levels are superior starting materials for obtaining mRNA.

Metal ions such as zinc and copper, chelators such as ethylenediamine tetraacetic

TABLE 2

Scheme for Isolation and Initial Characterization of Polysomes, Polysomal RNA, and Poly(A) Messenger RNA[a]

Plant tissue (approximately 1 g/4 ml) extracted in TE buffer with Polytron homogenizer for 1 min; spin 23,000 × g for 10 min.

Pellet = membrane bound polysomes, still bound.

Homogenize in MD buffer with Dounce® homogenizer; spin 23,000 × g for 10 min.

Supernatant = Free polysomes

Supernatant = Membrane-bound polysomes (released)

Pellet (discard)

Wash by pelleting through 3 ml SC buffer by centrifugation at 105,000 × g for 90 min.

Pellet = Washed polysomes   Spt. discard.

Suspend with Dounce homogenizer into a small volume of PB, spin at 10,000 × g for 10 min. Supernatant is polysomes ready for addition to cell-free system. Profile may be analyzed after adding 100 μg to a 15 to 40% linear sucrose gradient made in PB and spinning at 105,000 × g in a Spinco® SW-41 rotor for 85 min.

Suspend in PD buffer; make to 1% (w/v) with SDS at room temperature. Extract RNA by vigorous shaking with an equal volume of a 50:50:1 mixture of chloroform:phenol:isoamyl alcohol[21] for 10 min. Chill to 5°C; spin 12,000 × g for 10 min. Take the aqueous (upper) phase and re-extract with phenol mixture. Make the second aqueous phase to 2% with Na acetate pH 5.5 and precipitate with 2 vol ethanol at −20°C overnight. Collect RNA by spinning at 12,000 × g for 20 min. Wash the pellet two times with a 2:1 mixture of ethanol:0.2 M NaCl = polysome RNA

Apply in RB to an oligo(dT) cellulose column and wash with RB minus SDS to remove nonadsorbed RNA. Elute with 10 mM Tris-acetate pH 7.6 to obtain poly(A) RNA. Two or three cycles on the oligo(dT) column may be required to remove contaminating nonpoly(A)-RNA. Finally, make to 2% with K acetate pH 5.5 and precipitate with 2.5 vol ethanol at −20°C.

Centrifuge to pellet poly(A) RNA = messenger RNA which may be dissolved in water. The presence of poly(A) may be checked by hybridization to radioactive poly(U).

[a]   This is essentially the method of Reference 34; details of buffer compositions are given in Table 3.

---

acid (EDTA) and EGTA,*[64] detergents such as sodium dodecyl sulfate (SDS) and Extran,®[65] alkylating agents such as diethylpyrocarbonate,[66] absorbants such as Bentonite and various other compounds including heparin and polyvinylsulfate have been employed to inhibit RNases during polysome and mRNA extraction. Careful use of EDTA or EGTA, high pH and high ionic strength,[57,64] has proven to be effective for polysome isolation from many plant tissues. The sucrose used in polysome purification should be free of RNase; sometimes filtration through activated charcoal[67] helps remove both ultraviolet absorbing materials and endogenous RNase. A complex of 2 mM guanosine and 0.2 mM vanadyl sulfate has also been used to inhibit RNase.[68]

Relatively simple precautions, but ones often neglected, are those which will inactivate or remove RNase from apparatus to be used in polysome or mRNA isolation. Glassware should be heated to 225°C for 2 hr, and other materials (e.g., plastics, tubing, cuvettes) can be decontaminated by soaking in 1% (w/v) SDS at 60°C for 15

*   Ethyleneglycol-bis(β)-aminoethyl ether)-N,N′-tetraacetic acid.

TABLE 3

Buffers Used in the Isolation of Polysomes and mRNA[a]

Concentration, references

Buffer	Composition	34 (1975)	57 (1972)	48 (1976)	41 (1974)
Tissue extraction (TE)	Tris-HCl, pH 8.2	50 mM			
	KCl	500 mM			
	Mg acetate	5 mM			
Membrane-dissociation (MD)	Tris-acetate, pH 8.5	150 mM			
	Mg acetate	10 mM			
	Nonidet P40	0.5%			
Tissue extraction/ membrane dissociation (TEMD)	Tris	150 mM acetate, pH 8.5	200 mM acetate, pH 8.5	50 mM chloride pH 7.6	100 mM chloride pH 7.5
	sucrose				500 mRzsM
	KCl	200 mM	200 mM		50 mM
	Mg	50 mM	60 mM		
		20 mM acetate	30 mM chloride	2.5 mM chloride	10 mM chloride
	mercaptoethanol	4 mM			5 mM
	Na triisopropylna- phthalene sulfonate			2%	
	NaCl			1%	
	Vanadyl sulfate				0.6%
Sucrose cushion (SC)	Tris	50 mM acetate pH 8.5	40 mM chloride pH 8.5		100 mM
	sucrose	1.5 M	1.5 M		1.0 M
	KCl	20 mM	20 mM		50MM
	mercaptoethanol				5 mM
Polysome buffer (PB)	Tris	50 mM acetate pH 8.5	40 mM chloride pH 8.5		
	KCl	20 mM	20 mM		
	Mg	10 mM acetate	10 mM chloride		
Polysome-dissociation (PD)	Tris-acetate, pH 9.0	100 mM			
	NaCl	100 mM			
	Na$_2$ EDTA	2 mM			
	SDS	1%			
mRNA buffer (RB)	Tris-acetate, pH 7.6	10 mM			
	NaCl	400 mM			
	SDS	0.5%			

[a] The buffers used in Table 2 correspond to those of Reference 34.

min, followed by rinsing with sterile distilled water. Fingerprints contain active pancreatic RNase, and use of disposable gloves borders on the mandatory. Laboratory visitors should be discouraged from touching materials! After taking the above precautions, mRNA fractions freed by RNase are stable, and most purification steps can safely be done at room temperature.

B. Isolation of mRNA from Polysomes

Since most tissues synthesize many proteins at any given time, the isolated polysomes contain a variety of mRNAs. Therefore, either the specific mRNA to be isolated must be present in relatively large amounts or some procedure for selecting polysomes bearing this mRNA must be undertaken. Schimke et al.[69] used immunoadsorbtion of polysomes synthesizing ovalbumin prior to isolation of mRNA for this protein. Several other mRNAs have been isolated in this way, but the antibody must be highly selective and must not react with ribosomes.

Phenolization in the presence of SDS (Table 2) is most frequently used to extract RNA from polysomes, but contamination with large quantities of ribosomal RNA (rRNA) hinders subsequent purification of the mRNA. Therefore, until techniques for separating poly(A) containing mRNA from rRNA (see Section II.D) were developed, methods for releasing mRNA from the polysome without dissociation of the ribosomes were favored. One such method is chelation of magnesium with (2 to 200 mM) EDTA, the mRNA then being separated from ribosomes by sucrose density gradient centrifugation. High salt concentrations (e.g., 0.88 M KCl) are also effective in stripping ribosomal subunits from polysomes.[70] Using a high salt buffer of 0.5 M KCl, 5 mM MgCl$_2$ in 50 mM Tris-HCl (pH 7.5) plus 1 mM puromycin (all concentrations final), Blobel[71] separated the large and small subunits of reticulocyte ribosomes and globin mRNA. While the procedures would seem to be of general application, no reports of successful plant mRNA isolation by means of high salt or puromycin release have appeared.

C. Preparation of Total Nucleic Acid

Since prokaryotic and plastid mRNAs typically do not have a poly(A) sequence at the 3′-terminus, the methods given later for eukaryotic mRNAs cannot be used. However, total RNA extracted from leaves or isolated chloroplasts have stimulated amino acid incorporation in cell-free systems, and some RuBPCase large subunit has been synthesized (see Section III. C).[52,53] Procedures used to prepare the total RNA usually involve thorough mixing of an equivolume mixture of phenol and buffered detergent, e.g., 10 mM Tris-HCl, pH 7.4 (containing 1% tri-*iso*propylnaphthalene sulfonate, 6% *p*-aminosalicylate, 50 mM NaCl and 10 mM MgCl$_2$) and freshly distilled phenol (saturated with 10 mM Tris-HCl, pH 7.4, and containing 10% *m*-cresol and 1% 8-hydroxyquinoline).[72] Chloroform is usually added,[50] as this helps to chase the RNA into the aqueous (upper) layer.[73] For RNAs bearing poly(A) tracts, chloroform plus the use of pH 9 extraction buffers has been found to protect the poly(A) sequence, and to help in partitioning the RNA into the aqueous phase.[73a] RNA is obtained from the aqueous layer by the addition of 2 volumes of ethanol, usually in the presence of 0.5 to 1 M NaCl or acetate to help precipitation, followed by storage at −20°C for several hours.

Direct extraction of total RNA can also be useful as the first step in isolating eukaryotic mRNAs, followed by fractionation into poly(A) and nonpoly(A) containing fractions by oligo(dT)-cellulose chromatography,[50] or further purification by sucrose density gradient centrifugation.[74] Procedures for separation by preparative electrophoresis, especially on formamide[75,76] and agarose[77] gels, show considerable promise for yielding pure mRNA species.

Extraction of total RNA has proven an attractive alternative to polysomal techniques in the isolation of globulin mRNA from developing bean seeds.[77a] Cotyledons excised from maturing French bean pods were stored at −70°C after initial freezing in liquid N$_2$. Total nucleic acid was extracted with a hot buffered SDS solution, followed by phenolization and oligo(dT)-cellulose chromatography.[78] We have also found that digestion of protein with proteinase K is a useful alternative to phenol extraction. This procedure is detailed in Table 4.

Total nucleic acid extraction has also been used in the preparation of mRNA for the small subunit of RuBPCase.[49] *Chlamydomonas reinhardtii* cells were suspended in 50 mM Tris-HCl (pH 7.5), 150 mM NaCl, 5 mM EDTA, proteinase K at 40 μg/ml and 2% SDS. After stirring at room temperature for 15 min, the RNA was extracted twice with phenol-chloroform-*iso*amyl alcohol, precipitated with ethanol and chromatographed on oligo(dT)-cellulose.

D. Separation of mRNA Dependent on a Poly(A) 3′-Tract

Many eukaryotic mRNAs have a 3′-terminal sequence of adenosine residues. The length of this poly(A) tract has been shown to vary during the lifetime of the messenger (see Section V. H.3). Newly synthesized mouse globin mRNA was shown to have a tract some 150 nucleotides in length, while that in reticulocytes is approximately 50

TABLE 4

Isolation and Oligo(dT) Cellulose Chromatography of Bean Cotyledon mRNA

To 5 g of 12 to 14 mm long, ice-cold bean cotyledons, 13 ml of hot (100°C) buffer (containing 1% SDS, 30 mM EGTA, and 5 mM dithiothreitol in 200 mM Na borate buffer, pH 9 with NaOH) were added, and the mixture ground in a Polytron® homogenizer for about 1 min. Then 6.5 mg Proteinase K was added,[a] and the mixture incubated at 37°C for 1 hr. After addition of 1 ml of 2 M KCl (to precipitate SDS), the extract was centrifuged at 10,000 rpm in a Beckman® JA 21 rotor for 10 min.

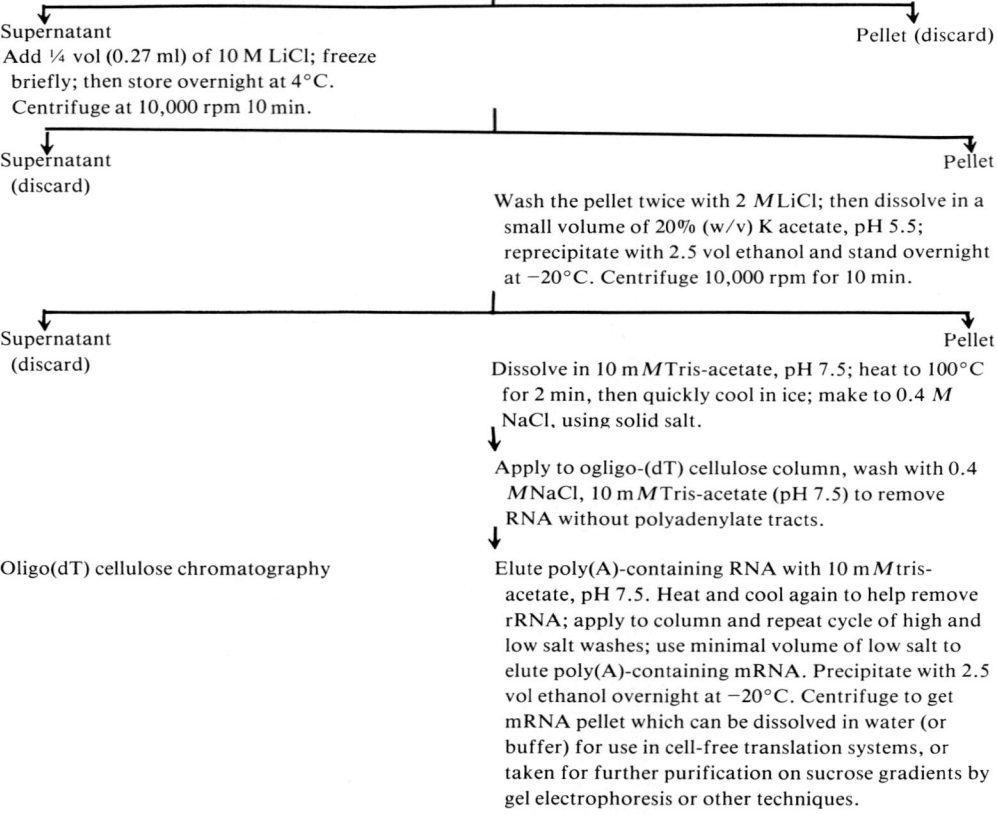

Supernatant
Add ¼ vol (0.27 ml) of 10 M LiCl; freeze briefly; then store overnight at 4°C. Centrifuge at 10,000 rpm 10 min.

Pellet (discard)

Supernatant (discard)

Pellet
Wash the pellet twice with 2 M LiCl; then dissolve in a small volume of 20% (w/v) K acetate, pH 5.5; reprecipitate with 2.5 vol ethanol and stand overnight at −20°C. Centrifuge 10,000 rpm for 10 min.

Supernatant (discard)

Pellet
Dissolve in 10 mM Tris-acetate, pH 7.5; heat to 100°C for 2 min, then quickly cool in ice; make to 0.4 M NaCl, using solid salt.

Apply to ogligo-(dT) cellulose column, wash with 0.4 M NaCl, 10 mM Tris-acetate (pH 7.5) to remove RNA without polyadenylate tracts.

Oligo(dT) cellulose chromatography

Elute poly(A)-containing RNA with 10 mM tris-acetate, pH 7.5. Heat and cool again to help remove rRNA; apply to column and repeat cycle of high and low salt washes; use minimal volume of low salt to elute poly(A)-containing mRNA. Precipitate with 2.5 vol ethanol overnight at −20°C. Centrifuge to get mRNA pellet which can be dissolved in water (or buffer) for use in cell-free translation systems, or taken for further purification on sucrose gradients by gel electrophoresis or other techniques.

[a]   Proteinase K may be preincubated at pH 7.5 in order to digest any RNase contamination present.

residues long.[79] RNA from both a lower plant (*Euglena gracilis*) and from a higher plant (*Vicia faba*) was found to have poly(A) tracts 150 to 250 nucleotides long.[80] Histone messengers have been convincingly shown to lack poly(A) 3′ sequences;[81] several plant viral mRNAs, e.g., from brome mosaic virus[82] and eggplant mosaic virus[83] do not have terminal poly(A), and there are several reports of nonpolyadenylated plant mRNAs.[50,84] Prokaryotic mRNAs typically are not polyadenylated, and this may be related to the fact that they are often translated as they are being transcribed and do not have to traverse a nuclear membrane.

For those mRNAs that have a sizable poly(A) sequence, techniques utilizing hydrogen bonding to inert matrices such as poly(U) Sepharose® or oligo(dT) cellulose make mRNA isolation relatively easy.[78,85] In practice, rRNA often shows an unfortunate capacity to stick to the mRNA (or, to the oligo(dT) cellulose column) at high salt conditions, and special care must be taken in removal of this contaminant from mRNA preparations. It is, perhaps, significant that very few plant mRNA preparations have been shown by electrophoretic analysis to be free from ribosomal RNA contamination, although it is acknowledged that, in many cases, the very small amounts of mRNA available presently preclude use of this analytical technique. Zein mRNA,[30] G1 globulin mRNA (see Figure 4), and leghemoglobin mRNA[85a] are probably, to date, the only plant mRNAs obtained essentially free of rRNA. Procedures designed to reduce or eliminate rRNA binding to cellulose columns include the use of dimethyl sulfoxide (DMSO) and heat.[86] Heating to 90°C for 2 min, followed by quenching in ice immediately prior to the second cycle of purification of seed globulin mRNA on an oligo(dT) cellulose column was effective in removing rRNA contamination (Table 4 and Figure 5).

Thus, the development of effective techniques for separating poly(A) containing RNA from other RNAs has greatly simplified isolation of mRNA. As noted above, rapid procedures for extraction of total RNA from plant material can be expected to yield active mRNA. Some problems with this approach remain: high quantities of DNA may be encountered which require removal by spooling, filtration through glass wool, or by DNase digestion. In any case, the starting material should be enriched with the mRNA of interest.

III. SPECIFIC MESSENGERS

A. Leghemoglobin mRNA

The roots of leguminous plants develop small (approximately 2-mm diameter nodules in response to symbiotic infection with nitrogen fixing bacteria. The association appears to be quite specific since infection by different bacterial strains results in markedly different efficiency of nitrogen fixation. The formation of the nodules also appears to be a well-defined and integrated result of the bacterial plant symbiosis. On exposure to air, a cut root nodule can appear to bleed. This is due to the presence of leghemoglobin, a myoglobin-like protein thought to function as an efficient oxygen carrier in the maintenance of an optimal partial pressure of oxygen for both ATP synthesis and nitrogen fixation.[87]

The uptake of fixed nitrogen by leguminous plants is critically important to agriculture. While the use of chemically synthesized nitrogenous fertilizers has become very important in modern agriculture, these are energy-expensive; and improvement of the efficiency of biological nitrogen fixation is a very worthwhile ambition. A detailed understanding of the molecular processes involved is a necessary approach to this aim. The studies of Verma and colleagues[37,38] with mRNA for leghemoglobin are a good example of how investigations at the molecular level can resolve questions which have proven intractable by other approaches.

In a landmark paper, Verma et al.[37] described the isolation of mRNA which, in a wheat embryo cell-free system, served as a template for the synthesis of polypeptide products identifiable as the slow (15,900 mol wt) and fast (15,600 mol wt) chains of leghemoglobin. This paper was the first to show convincing evidence of a functional nuclear-coded (as opposed to viral) mRNA to be isolated from plant tissue. As in most successful isolations of mRNAs specific for a defined protein, the nodule tissue had a high content of the protein of interest. Clearly, a high content (or the rapid synthesis) of a single type of protein over a defined developmental stage requires an enrichment of mRNA template in the cells. Another advantage is a low endogenous RNase level.

The first step in leghemoglobin mRNA isolation was preparation of a polysome fraction. The high pH, high ionic strength buffer used was very similar to that developed by Davies, Larkins, and Knight for pea polysome isolation (see Table 3).[57] Total polysomal RNA was dissociated with chloroform-phenol-*iso*amyl alcohol, and fractionated by oligo(dT) cellulose chromatography. The isolated mRNA was found to be 9S.

A major question left by this study was whether leghemoglobin mRNA actually derives from the plant tissue or from bacteria in the nodule. There has been considerable debate as to the site of synthesis of leghemoglobin. Studies based on the detection of the heme moiety[88,89] suggested that the leghemoglobin was located within the membrane sacs surrounding the bacteria within the nodule; thus, if it were excreted from the bacterium, the protein would actually be of prokaryotic origin. However, the polysomes from which leghemoglobin mRNA was isolated were 80S, not 70s,[38] and therefore were eukaryotic rather than prokaryotic. Verma and Bal[38] also showed that leghemoglobin synthesis appears to be associated with free cytoplasmic polysomes rather than membrane-bound polysomes, and that ferritin-conjugated antibodies to leghemoglobin were localized exclusively with the plant cell cytoplasm, and adjacent to the outer surface of the membrane surrounding the bacteroids. Preliminary evidence suggested the cDNA, synthesized from the 9S leghemoglobin mRNA hybridized with plant DNA, essentially confirming that leghemoglobin is indeed a gene for a plant protein.

B. Cellulase mRNA

Higher plant cellulases are found in specific locations at particular stages of tissue development. Regulatory roles in cell expansion, leaf abscission, pollen tube elongation, fruit ripening, and vascular differentiation have been suggested. Byrne et al.[35] reported the purification of two cellulases which differed from each other in molecular weights, amino acid composition, and solubility in 20 mM phosphate buffer. Cellulase activity was increased almost 100-fold by treatment of pea epicotyls with the herbicide (2-4-dichlorophenoxy)acetic acid, an analog of the plant hormone auxin (indoleacetic acid). Using a wheat embryo system,[90] Verma and colleagues[34] found that mRNA obtained by oligo(dT) cellulose chromatography of total RNA isolated from hormone-treated epicotyl tissue yielded a product which could be immunoprecipitated with antisera to the buffer soluble cellulase. No product corresponding to the buffer insoluble cellulase was observed; and it was concluded that the mRNA for this protein had no poly(A) tract (hence was not retained on the oligo(dT) cellulose) or that buffer insoluble peptides were synthesized, but being insoluble in the medium, were removed along with nascent peptides and ribosomes.

Separation of membrane-bound and free polysomes revealed preferential synthesis of the buffer soluble cellulase on membrane-bound polysomes. Further, the ability to isolate functional mRNA revealed that while there was a lag in the appearance of buffer soluble cellulase activity following (2,4-dichlorophenoxy)acetic acid treatment, there was no lag in mRNA synthesis. No evidence for a preexisting, nonfunctional mRNA was obtained. Thus, regulation of buffer soluble cellulase synthesis appeared to be at two levels. Transcriptional control is evidenced by the lack of mRNA formation until stimulated by hormone treatment. Translational control also appears to occur as mRNA was found to be present before presence of cellulase could be detected in the tissue.

Verma's paper[34] is a veritable mine of information for those interested in polysome and mRNA isolation, cell-free translation, and polypeptide product characterization.

C. Ribulose 1,5-bisphosphate Carboxylase (RuBPCase) mRNA

RuBPCase has been found in all organisms using carbon dioxide as a carbon source.[91] While its best known function is in carboxylation, it is also an oxygenase and is thus involved in both photosynthesis and photorespiration. Because of its central role in plant metabolism, this enzyme has been studied in depth (for review, see References 92 to 94). Very large amounts of RuBPCase are present in leaves of higher plants, and Ellis[95] noted that it may well be the most abundant protein in nature. It is not clear why such large quantities are synthesized in leaves; perhaps it is a relatively inefficient enzyme, and considerable amounts are rquired for adequate fixation of carbon dioxide. This enzyme may also function as a temporary storage protein, available in adverse conditions.

The native protein (approximately 500,000 mol wt) is located in the stroma of chloroplasts; when the plastids are broken during aqueous extraction, it is found as a highly soluble protein in the supernatant fraction. Because of its ubiquity, this protein was readily separated from the other leaf proteins and was originally called Fraction I protein by Wildman and Bonner.[96] This name is still used synonymously with RuBPCase. Dissociation by means of urea, SDS, or alkaline pH yields a large subunit (LS) of molecular weight 5.2 to 6.0×10^4 and a small subunit (SS) of 1.2 to 2.4×10^4. Isoelectric focusing on acrylamide gels reveals several bands for each subunit,[97] and there is currently much interest in evolutionary relationships revealed by differences in these peptides among plant genera.[97,98] A mutation in the LS was found to be inherited via chloroplastic DNA, but allelic forms of the SS segregated as expected for control by a nuclear gene.[99-101] These findings were intriguing since they suggested that the native molecule results from protein synthesis directed by mRNA deriving from both the plastid DNA and nuclear DNA.[92] Experimental evidence for differential sites of biosynthesis has now been obtained. Blair and Ellis[102] incubated isolated pea chloroplasts in [^{35}S]met or [^{14}C]leu and found a highly labeled polypeptide product which migrated coincident with the large subunit of RuBPCase. Maps of tryptic peptides made from the in vitro (plastid)-synthesized material were coincident with those from authentic LS. Incorporation of amino acids into the LS was inhibited some 80% by addition of D-threochloramphenicol, consistent with synthesis by a plastid (prokaryotic) system. Conversely, Criddle et al.[103] found that incorporation of amino acids into the small subunit was preferentially inhibited by cycloheximide, an inhibitor of cytoplasmic protein synthesis systems. Polysomes isolated from greening bean (*Phaseolus vulgaris*) leaves synthesized a product which could be immunoprecipitated with antisera to SS;[41,45] ribosomes derived from these polysomes were found to be at least 95% cytoplasmic.

Recently, Dobberstein et al.[49] compared products translated from poly(A) containing mRNA and cytoplasmic polysomes from *Chlamydomonas reinhardtii* by a wheat germ system. The mRNA gave a 40- to 60-fold stimulation of incorporation and product analysis showed a prominent band with a molecular weight of 20,000 which immunoprecipitated with monovalent antibody to SS of RuBPCase. Polysome incorporation gave a 30-fold stimulation, and a polypeptide of 16,500 daltons (identical in mobility to authentic SS) could be immunoprecipitated from the labeled products. Incubation of mRNA in the presence of increasing amounts of postribosomal supernatant yielded increasing amounts of 16,500 and decreasing amounts of 20,000 immunoprecipitable product, suggesting that endoproteolytic cleavage of a fragment of 3500 daltons occurred. A peptide of this size was found by analysis of products resulting from incubation of the 20,000-dalton in vitro product with postribosomal supernatant; when this supernatant was heated to 95°C prior to incubation, no cleavage to 16,500 and 3500 peptides occurred. The 20,000-dalton polypeptide was, therefore, termed a putative biosynthetic precursor of SS. The 3500 fragment is probably a presequence (see Section V.E), the first to be identified in plants.

Since it appears that a constant ratio of large and small subunits of RuBPCase is synthesized, some coordinating regulatory system is necessary. Further, if the SS mRNA is transcribed and translated in the cytoplasm, processes for membrane transport and association with the LS are required.[95,104] Dobberstein et al.[49] have suggested that the 3500 fragment of the putative biosynthetic precursor of SS may be involved in the transport of SS across the double chloroplast membrane. They drew an analogy to diptheria toxin where a precursor polypeptide is cleaved to yield disulfide-bound α and β chains, of which only the α subunit traverses the eukaryotic plasma membrane.

While the studies referred to above strongly suggest a differential origin (transcription) of the mRNAs for large and small subunit, they do not rule out the possibility that the LS is translated from mRNA transported across the plastid membrane from its original site of transcription. Some light on this aspect was shed by Bottomley and colleagues[51] who found that chloroplast mRNA could be translated by a 70S system from *E. coli* but not by the 80S system from wheat germ. Conversely, spinach cytoplasmic mRNA was translated by the wheat germ system but not by the prokaryotic system. Low levels of [^{35}S]met were incorporated into material identified (by chymotryptic peptide chromatography) as LS by an *E. coli* S30 system dependent on RNA derived from spinach chloroplasts. This RNA lacked poly(A).[53] Large subunit mRNA from *Euglena* also lacks a significant poly(A) tract, but has been reported to be faithfully translated by wheat germ S30 extracts.[105] Plastid mRNA from duckweed (*Spirodela*) coding for a membrane-bound protein was also translated by 80S ribosomes;[106] thus, the claim for differential recognition (or lack thereof) of chloroplast and cytoplasmic mRNAs by 70S and 80S ribosomal systems needs careful confirmation.

Following up previous evidence for identification of LS mRNA in whole cell extracts of *Chlamydomonas reinhardtii*[74] and its synthesis on polysomes bearing only 2 to 5 ribosomes,[48,107] Gelvin et al.[108] have reported the conclusive demonstration that the chloroplast genome is the coding site for LS mRNA. This conclusion appears likely, and the techniques used were appropriate; however, because of its importance, the evidence presently available should be critically evaluated. ^{32}P-labeled LS mRNA was obtained by eluting sections of gels cut from a position corresponding to that expected for LS mRNA. Incubation of a mixture of labeled mRNA with whole cell 7 to 15S RNA (from sucrose gradients) or with RNA from small polysomes in a cell-free system from *E. coli*, resulted in low levels of [^{3}H]arg incorporation. Although 26.1% of the acid precipitable counts could be immunoprecipitated (by a double immune system[107]), there was only a 1.2-fold stimulation of incorporation by this putative mRNA preparation (17,600 cpm over a background of 14,500). The labeled mRNA was then used to locate *Eco*RI restriction fragments of *C. reinhardtii* chloroplast DNA, and was hybridized to [^{3}H]adenine-labeled chloroplast DNA (in the presence of excess of the unlabeled chloroplast and cytoplasmic rRNAs). Assuming that the radioactive mRNA was indeed pure, it was calculated that there are approximately 75 LS genes per chloroplast or per cell, and one LS gene per chromosome.

D. Seed Storage Protein mRNAs
1. Introduction
While RuBPCase may be the most abundant protein,[95] the highest protein concentration in plants is found in the seeds; as much as 40% of the dry weight of soybeans is protein. The deposition of protein in the seed typically occupies a relatively short period in the life cycle of the plant, and usually the storage protein comprises only a few molecular species. Thus, the seed is very attractive for studies on protein synthesis. The rapid accumulation of a tissue-specific protein over a defined developmental period is especially useful in that molecular processes of control (as in the turning on and off of specific genes) and regulation (of the relative amounts of polypeptides in polymeric molecules, and of the total amount of protein synthesized) are amenable to study.

Some of the earliest successful studies on plant protein synthesis in vitro were those of Mans and Novelli using maize kernels.[109] The use of wheat seeds to study cellular mRNA existing prior to imbibition during germination[110] led to development of cell-free systems which are probably the most active known and certainly the easiest to prepare (see also Volume II, Section III by Davies). An early, exciting, and probably prophetic study was the use of a complex transcription-translation system whereby pea cotyledon chromatin gave rise to a radioactive product identified immunochemically as storage globulin.[111] Other early work on cell-free synthesis of seed proteins (e.g., wheat endosperm[112]) was marred by microbial contamination of the preparations.[113]

2. Long-lived mRNA

One question raised by the early studies of Marcus and Feeley[110] is that of the existence of long-lived mRNA. That mRNA species are transcribed prior to maturation of flowering plant seeds appears reasonable since a whole new spectrum of proteins arises as the seed imbibes water and starts to germinate. An excellent review of the many studies related to the existence of long-lived mRNA in plants is that by Payne.[114]

Extensive work on carboxypeptidase and isocitrate lyase in cotton has indicated that the mRNAs were transcribed in immature seeds (probably between 30 and 35 days after flowering[115]) and stored in the dry seed until required in germination.[116,117] Unfortunately, almost all the evidence rests on the specific inhibition of mRNA synthesis by actinomycin D. As Payne explains in detail,[114] the methodology used allows for other explanations, and the specificity of actinomycin D inhibition is now in serious doubt.[118]

Preliminary evidence for the existence of mRNA in dry seeds of pea and other plants capable of directing storage protein synthesis in vitro has been advanced.[119] While a certain proportion of the very large amount of storage protein mRNA present during seed ripening may remain intact and functional during dehydration of the maturing cotyledons, there seems to be little ground for an important role of these mRNAs during germination.

Some of the most convincing evidence for long-lived mRNA in seeds to date is that provided by Spiegel and Marcus,[120] who used cordycepin and α-amanitin to show that polysome formation in early germination stages (immediately on imbibition) of wheat embryos was independent of transcription or polyadenylation of mRNA. While it is possible that some mRNAs essential to germination processes are stored in dry seeds as ribonucleoprotein complexes or particles,[121] it appears that only a very small amount, if any, mRNA of critical importance in germination is stored in the mature seed.

Many mosses occupy environmental niches where they are subject to rapid desiccation and rehydration at, and for, unpredictable intervals. Polysomes run-off during desiccation of the gametophyte of *Tortula ruralis* were found to be reestablished within 2 hr,[43] even in the presence of actinomycin D and cordycepin (alone or in combination).[44] Profiles of radioactivity incorporated prior to dehydration into the RNA (prelabeled RNA) did not change in the rehydrated moss, nor was newly-incorporated [³H]uridine seen in the polysomes formed early in rehydration. By extraction of poly(A) containing RNA from prelabeled tissue, Dhinsda and Bewley[44a] have now demonstrated that mRNA in the dried moss remains labeled and active in protein synthesis. It is difficult to tell if this mRNA is actually used in polysomes formed early on rehydration of slowly dried moss, since free, radioactive, uridine is present from prelabeling. However, the lack of incorporation of fresh isotopic label into polysomal RNA for the first hour or two of rehydration, together with the activity in dried moss, would seem to confirm the contention that mRNA is conserved during dehydration phases in this plant.

3. Practical Applications of Seed mRNA Studies

In addition to basic scientific knowledge likely to derive from studies of seed protein synthesis, the potential exists for using this information towards increasing food supplies. Isolation of specific mRNAs is the first step in the elucidation of systems controlling their transcription from the cellular genome.[122] A knowledge of ways to induce earlier onset and delayed termination of seed protein synthesis and accumulation, even if only by a small period, would dramatically enhance crop yields. The isolation of specific mRNAs for edible seed proteins would permit the identification, by hybridization techniques, of the chromosomal location and reiteration frequency of the genes for these important proteins. They would also permit evaluation of the possibility that intervening sequences occur in seed protein genes, as they do in several animal protein genes (see Section V. A).

Another possibility arising from mRNA isolation is that of cDNA synthesis. If seed storage protein mRNA could be successfully used as the template for reverse transcriptase, the cDNA obtained would effectively represent an isolated food protein gene. Techniques are now available for the integration of such isolated genetic information into other organisms, although there is considerable question as to whether the integrated materials will be faithfully transcribed and translated. Several segments of eukaryotic DNA have been inserted into prokaryotes and transcribed at least in part. A major problem has been obtaining full-length DNA copies of mRNA, but recently high-fidelity reverse transcripts have been obtained.[123] A geneticist's appraisal of the role for biochemical approaches to modification of crops plants in the future is given in Volume II, Section IV by Bingham; and current progress with crown gall, a promising vector for genetic engineering experiments, is reviewed in Volume II, Section IV by Schell.

4. Zein mRNA

Zein is a class of macromolecules, heterogeneous in size and charge, which constitute the major part of the storage protein of corn (*Zea mays*) endosperm.[124] They are hydrophobic proteins soluble in 70% ethanol, high pH, high concentrations of urea, and several organic solvents. They are, therefore, readily separated from other proteins. Additionally, the low lysine and tryptophane content further aids in their identification. On SDS electrophoresis, four major polypeptide chains can be resolved, two of which, having molecular weights of 23,000 and 19,000, are the major components. Burr and Burr[31] clearly showed the retention of polysomes on the surface of zein-containing protein bodies in corn. These polysomes stimulated amino acid incorporation into protein when added to a cell-free system with products synthesized by polysomes from corn shoots, and on SDS gel electrophoresis of in vitro products soluble in 70% ethanol a peak co-incident with ^{14}C-alkylated zein was obtained.

More recently, Burr et al.[31a] prepared protein bodies from maize endosperm collected 15 to 17 days after fertilization, obtained total RNA by SDS-phenol extraction, and then isolated poly(A)-containing RNA by oligo(dT)-cellulose chromatography. RNA having messenger activity was freed of ribosomal RNA contamination by sucrose density gradient centrifugation in the presence of dimethyl sulfoxide to provide denaturing conditions. This RNA was, on a weight basis, three times as active as was unfractionated brome mosaic virus RNA when translated in a wheat germ system. Two major bands with molecular weights of 24,500 and 20,100 were detected on analysis of polypeptide products by gel electrophoresis. These products were approximately 2000 and 1000 daltons larger than the authentic zein polypeptides, and it is thought that they contain presequences associated with membrane transport into the developing protein body. Analysis of the size of the mRNA preparation by electrophoresis under totally denaturing conditions (in 2.2 M formaldehyde) yielded an estimate of 1104±58 nucleotides, and electron microscopic measurement gave a value of 1230 nucleotides.

These values correspond to a molecular weight of approximately 416,000, which is considerably smaller than that obtained by Larkins et al.[30] under conditions that were not totally denaturing; if correct, they exclude the possibility that the major zein polypeptides are coded by a dicistronic mRNA containing no overlapping sequences.

Larkins and Dalby[29] added polysomes from corn endosperm to the much more active wheat germ system and obtained definitive evidence for the synthesis of zein in vitro. In a later study, it was clearly shown that membrane-bound, rather than free ribosomes, were the sites of zein synthesis.[30] Free poly(A) containing mRNA was isolated from these membrane-bound polysomes and shown to direct the cell-free synthesis of zein polypeptides. Thus, both electron microscopic and molecular evidence strongly suggest that zein is synthesized on the surface of the protein bodies.

As in the case of RuBPCase small subunit, the question remains as to how the zein polypeptides cross membranes. The in vitro translation products of several eukaryotic mRNAs for secretory proteins have been found slightly larger than the functional proteins.[78] They bear an N-terminal extension of some 20 amino acid residues, and it has been theorized that such presequences play a role in membrane transport (see Section V. E).[125] It is interesting, in this connection, that the isolated zein mRNA was found to sediment at 13.2S on sucrose gradients, but that its mobility on SDS gels corresponded to a molecular weight (540,000) expected of a 16S mRNA. As shown in Section III. D, we have found G1 mRNA also to run between the 25S and 18S rRNAs on SDS gels, but to sediment slower than the 18S subunit on sucrose gradients. However, while the G1 polypeptides are 43,000, 47,000, and 53,000; the major zein polypeptides are approximately 21,800 and 19,000 daltons. Thus, the mRNA isolated by Larkins and colleagues[30] appears either to be a dicistronic messenger or to contain a leader sequence (see Section V. C) of considerable length; in the latter case, there would be ample information for a polypeptide presequence.

5. Globulin mRNA

Legume cotyledons contain large quantities of protein; in soybean (*Glycine max*) protein represents some 40% of the seed dry weight. Some 60 to 80% of the total protein is globulin (protein requiring salt for solubility). Typically, the globulin fraction can be separated into two fractions. One requires appreciably more salt for solubility than the other. The molecular structure of the seed globulins is rather complex, and is a current topic of debate (see Reference[126]). Studies in my laboraaory have concentrated on the globulin fraction from the French bean (*Phaseolus vulgaris*) which requires high salt (approximately 0.2 M NaCl) for solubility. We have termed this the G1 fraction as it is the first globulin fraction to precipitate on dilution of a 0.5 M NaCl extract (pH 3.5 with HCl) of the cotyledons with 5 to 8 volumes of distilled water.[127] In the cultivar Tendergreen, there are three polypeptide subunits with molecular weights of 53,000, 47,000, and 43,000; examination of other seed lines revealed several modified types, including a modification of the largest subunit to give a 50,500 mol wt species which was inherited as a simple Mendelian trait.[128,129] Depending on the cultivar, G1 globulin accounts for 34 to 51% of the total protein in bean seeds.[130]

After sedimentation of the G1 fraction, the supernatant yields a second precipitate, the globulin-2 (G2) fraction, on overnight dialysis against water. The G2 fraction contains several polypeptides probably from several unrelated proteins.

Electrophoresis of proteins from developing seeds under dissociating conditions clearly shows the changing profile of the G1 and G2 polypeptides (Figure 2). Each of the three G1 subunits can first be seen when the cotyledons are approximately 8 mm long, and they remain in the same proportion throughout subsequent development. The several G2 polypeptides are not synchronized in appearance or disappearance; they therefore give the impression of being present in different proportions at various developmental stages. The dramatic appearance of the G1 subunits at the 8 mm stage has been confirmed immunochemically.[130a]

Polysomes from 12 to 15 mm long cotyledons (15 to 20 days after anthesis) yielded polypeptide products which co-electrophoresed with G1 subunits in cell-free reactions containing wheat germ S23.[23] Electron microscopy of the bean cotyledon shows the polysomes to be grouped on endoplasmic reticulum proximal to protein bodies surrounded by a single membrane (Figure 3). This contrasts with the situation in corn, in which zein is synthesized on polysomes at the membrane surface of protein bodies.[31] Larkins et al.[30] found that products of in vitro synthesis using membrane-bound polysomes from corn endosperm contained much higher amounts of zein polypeptides than did those synthesized on free polysomes. We find approximately equal quantities of G1 polypeptides in products synthesized by free or by membrane-bound polysomes from bean cotyledons.

Extraction and oligo(dT) cellulose chromatography of RNA from bean cotyledons, using procedures detailed in Table 4, yields material which is very active in directing amino acid incorporation in the wheat germ system. Even without recourse to immunoprecipitation, autoradiography of [^{35}S]met-labeled products using this mRNA shows major products migrating at positions corresponding to the 47,000 and 43,000 dalton subunits of authentic G1 protein (Figure 4). The immunoprecipitated material is even cleaner, providing further evidence for the accurate translation of this globulin mRNA.[131] Recently, peptide mapping has been used to confirm that the 47,000 and 43,000 dalton products are faithfully translated from G1 mRNA preparation.[24a] Electrophoresis of the poly(A) containing mRNA fraction shows it to be only slightly contaminated by rRNA (Figure 5), and I am confident that the remaining rRNA will be eliminated by sucrose gradient centrifugation or elution from gels in the near future.

Since there appears to be approximately equal amounts of 53,000 and 47,000 mol wt subunits in the G1 monomer, the question arises as to whether mRNA for all three subunits is present. If the 53,000 subunit mRNA lacks a 3'-poly(A) tract, it would not be bound to the oligo(dT) cellulose column, and hence absent from our mRNA fraction. Alternatively, the mRNA could be present, but initiation of translation occurs at low efficiency. Another explanation for the lack of the 53,000 subunit in the cell-free products could be that it derives from the second cistron of a dicistronic messenger (wheat germ S23 is thought to initiate poorly at internal sites of polycistronic messengers, e.g., phage RNA). However, the gels of Figure 5 suggest a molecular weight of 700,000 to 750,000 for the G1 mRNA (capable of coding peptides up to 75,000), and there is no trace of an RNA big enough for a 100,000 (or two 50,000) molecular weight polypeptide. Although a remote possibility, it cannot be excluded at this point that the 53,000 product is that of an overlapping gene such as recently discovered for phage $\phi\chi$ 174,[132] and postulated for some plant viruses.[133]

Further purification of G1 mRNA will permit determination of the relative amounts of mRNA for each subunit, and of the initiation, elongation, and termination characteristics. Also of interest is the mechanism controlling the transcription of these mRNAs in certain tissues at certain times, and whether specificity for addition of sugar residues (G1 is a glycoprotein[129]) lies solely in the amino acid sequence, or is in some way a function of the mRNA. As discussed elsewhere (Section III. D.3), questions such as how many copies exist of the genes for each subunit may also be approached once adequate quantities of mRNA are available.

6. Amylase mRNA

The aleurone layers of germinating barley seeds have been shown to be the site of

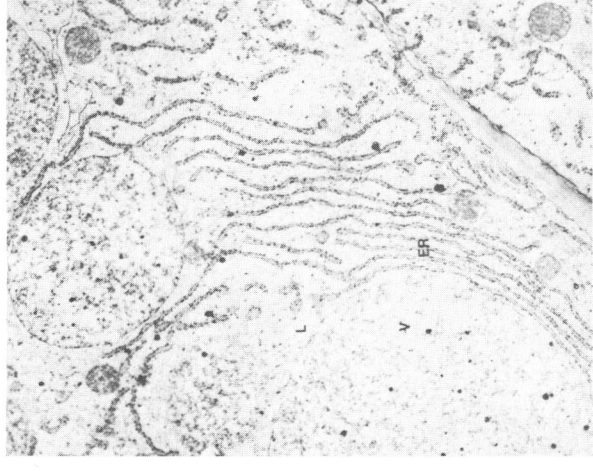

FIGURE 3. Ultrastructure of developing french bean cotyledon. The endoplasmic reticulum (ER), heavily laden with membrane-bound ribosomes appears to be surrounding the vacuole (V). At this stage of development (14 mm), active synthesis of the storage globulins G1 and G2 is taking place. These are being deposited in the vacuoles which are, in fact, in transition to becoming protein bodies. A few lipid droplets (L) can be seen. Magnification is 7750 times.

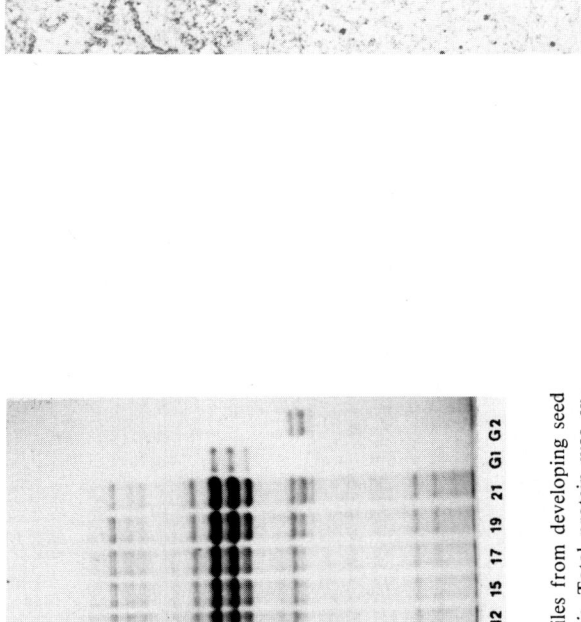

FIGURE 2. Polypeptide profiles from developing seed cotyledons of *Phaseolus vulgaris*. Total protein was extracted from cotyledons of increasing length (6 to 21 mm, as denoted below the lanes) and, after dissociation, subjected to SDS gel electrophoresis. Authentic samples of globulin fractions G1 and G2 were run in the outer lanes. Note that while the intensity of G1 polypeptides increased as the cotyledons grew from 9 to 21 mm, the relative staining for each of the three subunits (53,000; 47,000 and 43,000 daltons) remains constant. The polypeptides in the G2 fraction are seen to be in different proportions at different stages of development (From Sun, S. M., Buchbinder, B. U., and Hall, T. C., *Plant Physiol.*, 61, 920, 1978. With permission.)

A   B   A B C D

FIGURE 4. 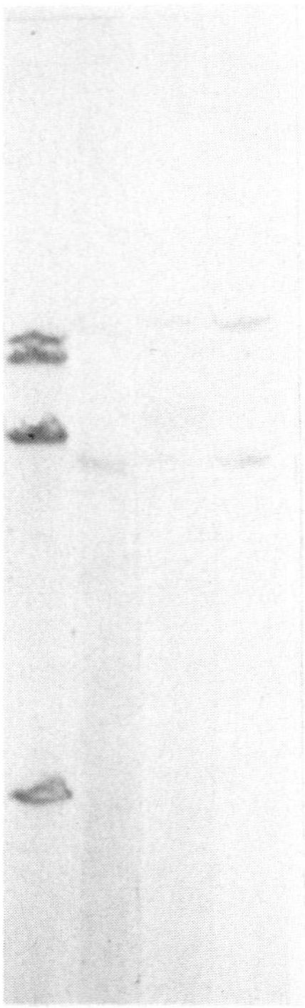   FIGURE 5.

FIGURE 4.   Autoradiography of seed mRNA-directed products synthesized in the wheat germ cell-free system. Lane A shows 10 μl of a 50 μl reaction containing 1 μg of mRNA extracted from 14 mm bean cotyledons by procedures detailed in Table 4. The 47,000 and 43,000 subunits of G1 globulin (see caption to Figure 2) can be seen clearly, but only the faintest trace of the 53,000 mol wt subunit can be detected. There is evidently some contamination by G2 fraction mRNAs, as shown by the synthesis of the two polypeptides seen lower in the gel. The bright, diffuse, band represents partial translation products. A comparable amount of a reaction mixture from which seed mRNA was omitted was added to lane B; clearly little endogenous template activity was present in the wheat germ S23 preparation used.

FIGURE 5.   Gel electrophoresis of bean seed mRNA. Total RNA was extracted from developing seeds (see Section, 3.D.5), and poly(A)-containing RNA isolated on an oligo(dT) cellulose colum, using the techniques detailed in Table 4. Some 26S and 16S rRNA contamination can be seen together with the heterodisperse mRNA (gel C); the rRNAs were also run as markers in gels B and D. Gel A contained brome mosaic virus RNAs, which have molecular weights ($\times$ 10^{-6}) of 1.1, 0.99, 0.7, and 0.28.

synthesis of α-amylase in response to the hormone gibberellic acid (GA$_3$).[134] Detailed studies provided evidence that GA$_3$ enhanced incorporation of labeled ribonucleotides into poly(A)-containing RNA in these layers.[135,136]

Higgins et al.[21] extracted total RNA from hormone-treated and untreated tissue. Using oligo(dT) cellulose chromatography, they separated total RNA extracts into poly(A)-containing and poly(A)-lacking RNA fractions and used them as templates in a wheat germ system.[137] Autoradiography of the products showed that only the poly(A)-containing template directed the synthesis of a 45,000 mol wt polypeptide, and this was selectively precipitated from other products in the postribosomal supernatant by a double antibody technique[138] using α-amylase antiserum produced in rabbits together with goat anti-rabbit gammaglobulin serum. Further microdensitometry[139] of gel autoradiographs of products synthesized using templates extracted up to 16 hr after GA$_3$ application showed a great increase in translatable mRNA for α-amylase between 8 and 12 hr after hormone addition.

IV. MISCELLANEOUS MESSENGERS

Since many, if not most, eukaryotic mRNAs have a 3'-poly(A) tract, the technique of oligo(dT) cellulose chromatography provides a ready procedure for their isolation. Consequently, there has recently been a rapid increase in the number of reports of plant mRNA fractions isolated. In many cases, there has been no attempt at detailed characterization of the products. A summary of plant mRNAs isolated to date, including these discussed in detail above, is presented in Table 1.

V. STRUCTURE AND FUNCTION

A. Introduction

It is now clear that mRNAs have considerable information built into their primary and secondary structure in addition to that of being templates for protein synthesis. Using examples from eukaryotic, prokaryotic, and viral messengers, the composite picture given in Figure 1 may be derived from current information. It is highly likely that other structural features remain to be discovered. For example, an area of intensive research has arisen from the finding that the structural gene sequences coding for several animal proteins contain intervening sequences that are not represented in the mature (polysomal) mRNA. The natural ovalbumin gene contains seven such intervening sequences.[139a] The discovery of intervening sequences followed from experiments such as those in which cloned ovalbumin DNA and purified ovalbumin mRNA were hybridized. Electron microscopy clearly showed that, instead of forming a uniform double-stranded hybrid, there were internal regions (R loops) of DNA which contained no sequence homology with the mature ovalbumin mRNA. It is not known if intervening sequences may be "skipped" during transcription of mRNA from the genome, but in the case of the mouse β-globin gene (which contains a 550-base pair intervening sequence) it has been shown that a 15S β-globin precursor mRNA is transcribed which contains both the coding and intervening sequences.[139b] This finding demands the existence of systems for the removal of internal RNA sequences and subsequent rejoining of internal RNA sequences. Such a "splicing" system opens up many possibilities, for example, it may be conjectured that mRNAs can be formed which will result in the synthesis of families of polypeptides that contain common sequences that are covalently bound to unique sequences. Initial evidence has been advanced for avian sarcoma virus which suggests that a common 5'-terminal nucleotide sequence is somehow spliced into the individual species of viral-specific mRNAs present in infected cells.[139c]

Since no plant mRNA has been obtained in sufficient amounts and with the high degree of purity required, the features described below, in many cases, of necessity relate to mRNAs that are not transcripts of plant genomic information. Nevertheless, almost all of these features have been shown to be recognized by wheat germ translation systems. It is, therefore, not unrealistic to assume that they will eventually be identified in plant mRNAs.

2. The 5'-terminal Cap

In 1974, it was shown that the 5'-terminus of reovirus and several tumor virus mRNAs was a 7-methylguanosine, linked through the 5'-carbon of its ribose to the 5'-carbon of the second nucleotide. This 5'-5' linked m⁷GpppX was termed a "cap," and it has since been found at the 5'-terminus of most (but not all) viral and eukaryotic mRNAs.[140] The second and third bases of many eukaryotic mRNAs are also methylated. Perry and Kelley[141] showed that unprocessed messenger in the nucleus, hnRNA, was frequently of the type m⁷GpppXᵐpYp... (called a cap I structure), while processed messengers from the cytoplasm had an additional 2'-0-methylated nucleotide in the third position. The latter were termed cap II structures, m⁷GpppXᵐpYpᵐ...; it was noted that the Y base was usually a pyrimidine. Methylation of this third base appears to be a secondary modification, occurring in the cytoplasm and possibly after or as the mRNA is incorporated into polysomes.

Studies with reovirus suggested that the cap was essential for translation,[142] but this now appears not to be the case for all mRNAs as several plant viral RNAs lacking a cap, or from which the cap has been removed, have been translated in vitro (see Volume II, Section III by Davies).[143]

C. Leader Sequences

Studies with plant viral and prokaryotic mRNAs have revealed that there may be as few as ten bases preceding the initiation codon, as in brome mosaic virus (BMV) RNA4,[144] or many, as in MS2 RNA, where the A-protein genome follows a 129-nucleotide leader sequence.[62]

Dasgupta et al.[144] showed that wheat embryo ribosomes did not bind significantly to the cap, but did bind efficiently to BMV RNA 4 fragments containing the first 22 nucleotides from the 5'-terminus. Shine and Dalgarno[145] noted that coliphage ribosome binding sequences were short, and all contained at least part of the purine-rich sequence 5'-AGGAGGU-3'. This sequence is complementary to a region at the 3'-end of *E. coli* 16S rRNA.[145,146] Thus, prokaryotic and eukaryotic ribosome binding sites are probably short, but appear to differ in sequence (although purine-rich, BMV RNA 4 contains only one G in the first 22 nucleotides). If RuBPCase LS mRNA contains the prokaryotic ribosome binding sequence, while the nuclear-coded SS mRNA does not, this could explain the differential translation of these mRNAs by bacterial and wheat germ cell-free systems noted above (see Section III, C). Since only a short sequence appears necessary for ribosome binding, it seems likely that the long leader sequences, when present, may play an additional role or roles.

An unsolved mystery is that of if, and how, specific messengers are selected. While eukaryotic systems, such as that from wheat germ, translate both eukaryotic and prokaryotic mRNAs with high fidelity (albeit with different efficiencies), it seems likely that, in vivo, considerable messenger selection is exercised. For example, it is known that in the reticulocyte system the rate constant for ribosome attachment is lower for α-globulin mRNA than for β-globulin mRNA.[147] Such effects make it necessary to exercise great caution in confirming that stimulation of translation of a specific mRNA is directly related to addition of an initiation factor. Whether or not leader sequences play any part towards recognition of mRNAs by ribosomes or in the interaction with modulating factors in certain tissues is entirely speculative.

D. Initiation Sites

Eukaryotic mRNAs studied to date have been largely monocistronic; and although the exceptions include several plant viral mRNAs, many of these appear to use devices such as partial transcription so as to present themselves as if they were monocistronic in the host cell (see Chapter 11 for a detailed discussion of this phenomenon). The evidence is, however, not conclusive, and polycistronic mRNAs having internal initiation sites may yet be found to occur frequently in higher organisms. Even the existence of overlapping genes[132] (read from the same RNA template, but with different frames) cannot be excluded.[133]

In both eukaryotic and prokaryotic mRNAs, the initiation codon is typically that for methionine, AUG, but the 5'-terminal cistron coding for the A-protein of bacteriophage MS2 has a GUG initiation codon.[62] In this mRNA, incorrect reading frames are blocked by nonsense codons. Since the same AUG triplet is used for insertion of methionine into internal positions of polypeptides, some system for distinguishing between initiation and elongation methionine codons must exist. Certainly, they are recognized by different isoaccepting tRNAs; that for prokaryotic initiation ($tRNA_f^{Met}$) permits the addition of a formyl residue to the amino group of methionine after esterification, while that for elongation ($tRNA_m^{et}$) does not. Although eukaryotic $tRNA_f^{Met}$ is not formylated in the cytoplasm of eukaryotic cells, it can be formulated by the bacterial enzymes.

Some evidence for initiation factors specific for certain mRNAs has been advanced, e.g., eIF-3 which appears to direct specifically the synthesis of myosin or myoglobin.[148] Several other factors are involved in initiation steps of protein synthesis, and these have been reviewed in detail by Weissbach and Ochoa.[149] Evidence for modification of eukaryotic initiation factors by phosphorylation has been advanced, and the role of cyclic or noncyclic nucleotides in this reaction is currently under debate.[150] In plants, the most detailed studies on initiation factors and their functions have been done with wheat germ extracts.[151,152]

E. Presequences

The N-terminal methionine of many proteins is cleaved or modified (e.g., acetylated) after translation, and often several amino acid residues are removed. An intriguing finding is the existence of presequences (sometimes termed "signal" sequences) of some 20 amino acid residues on many secretory proteins.[153] These are believed to play a role in the vectoral discharge and sequestering of the nascent polypeptide chains,[125] being removed from the rest of the molecule as it passes through a membrane boundary. It would not be surprising if seed proteins stored in membrane-bound protein bodies contained such sequences, and the exciting observation of a presequence on RuBPCase SS mRNA has already been mentioned (Section III. C).[49]

F. Product Sequences

The tertiary structure of the mRNA sequence coding for the polypeptide product may be very important towards regulation of translation. Initial interpretations of the structure of MS2 phage RNA (the first organism for which the entire genetic information has been sequenced) have provided fascinating glimpses of these types of control.

It is evident that many regions of MS2 mRNA are capable of hydrogen bonding with other regions, and it appears that the bonding is temporarily altered as ribosomes pass along during translation. That the ribosome binding site for the A-protein cistron is available only during translation of the other cistrons appears well established, as is the fact that expression of the replicase gene is controlled by translation of the coat protein gene. Fiers et al.[62] have also suggested that the frequency of certain code words may well have a bearing on the frequency of translation of (i.e., modulate) a given cistron.

Although it is usually assumed that the rate of translation of a given mRNA is constant, it is certainly possible that the observations have only revealed the average reading rate, while ribosomes may in fact traverse different regions of the template at varying speeds. Thus, I find it tempting to speculate that at loops or, alternatively, in regions which can hydrogen-bond, the reading may be slowed. Ribosomes might even be expected to pause at certain sites before continuing. This effect could help to explain the multiple minor bands seen in electrophoretograms of cell-free products, as at the moment of dissociation ribosomes would be paused at these locations, and would bear unfinished products of discrete length. These minor bands are typically reproducible from one experiment to the next. Further, ribosomes at pause regions might be more susceptible to release from the messenger, in part accounting for the phenomenon of premature termination.

Evidence for termination of less than full length polypeptides has been adduced recently by Van Tol and Van Vloten-Doting[153a] during translation of RNA component 1 of alfalfa mosaic virus (AMV) by a rabbit reticulocyte lysate system. At low RNA concentrations (5μg/mℓ) a single polypeptide of 115,000 daltons was the major translation product. As the concentration of AMV RNA 1 was increased to 150 μg/mℓ the proportion of two shorter polypeptides (mol wt 62,000 and 58,000) increased, a sharp transition in product profile occurring between RNA concentrations of 20 and 40 μg/mℓ. Since only one ribosome-binding site is thought to be present on AMV RNA 1 and because all three polypeptides were shown to have the same N terminal amino acid sequence, it was deduced that the codon occuring immediately to the 3′ side of the AMV RNA 1 sequences coding for the 58,000 and 62,000 dalton polypeptides is recognized only by tRNAs that are present in low concentration. The shortage of these tRNAs (specific-isoaccepting tRNAs or suppresor tRNAs) would consequently cause a pause at the position of insertion of the amino acid donated by these tRNAs. As noted by the authors,[153a] an intriguing question is if each or all of the three polypeptide products translated from this mRNA have (similar or different) biological function. Other roles for pause regions could be to allow the nascent polypeptide to assume a certain tertiary configuration, to bind to a membrane, or to undergo modification (e.g., glycosylation, demethionylation, acetylation).

G. Termination Sequences

The codons UAA, UAG, and UGA are not recognized by tRNAs, but by specific proteins, the release factors. Examples of the use of each of these codons have been found, and in some instances two stop signals occur in succession. In eukaryotes, GTP appears to be involved in polypeptide chain release; in prokaryotes, no absolute requirement for GTP has been established.

Presumably, the presence of the ribosome together with its nascent polypeptide is required for the binding of release factors. Otherwise, these proteins could bind to many out-of-frame triplets which could be seen as termination codons. In the case of MS2 RNA, the termination signals for both the A-protein and replicase genes are UAG preceded by an arginine codon; they are located at the top of hairpin bends.[154] If internal initiation sites exist in eukaryotic mRNAs, then internal termination sites would also be expected. Such situations permit the occurrence of "read through" proteins; it seems possible that some of the in vitro products of TMV RNA translation are such proteins. Relatively little work has been done towards understanding termination and release of polypeptide chains, perhaps because this happens all too easily during protein synthesis in vitro!

H. 3'-Terminal Structures

1. Nontranslated Trailer Sequences

As at the 5'-terminus, so at the 3'-terminus, there may be a considerable nucleotide sequence which is not translated and whose functions have yet to be fully understood. Using MS2 RNA once again as an example, there is a 174-nucleotide-long untranslated segment following the UAG termination signal of the replicase gene.[154] This segment, like the 5'-leader sequence, appears stringently dependent on its primary structure[155] — there being essentially no nucleotide changes (compared with approximately one change for every 40 nucleotides for the coat protein genes) between this sequence in the related phages R17 and MS2. It seems probable that the configuration of this structure is important for its recognition by enzymes, especially the RNA replicase (as transcription proceeds in a 3'→5' direction). Another specific instance of enzyme recognition follows.

2. Aminoacylatable Sequences

The demonstration by Pinck et al.,[156] that TYMV RNA could function like a tRNA in the acceptance of an amino acid, was entirely unexpected. Subsequently, mRNAs from several plant viruses have been found to have this property,[157] and it is discussed in greater detail in Volume II, Section III by Lane. For the present discussion, however, it is only necessary to point out that the 3'-terminal structures of these RNAs must encode similar recognition properties for the aminoacyl-tRNA synthetases as do tRNAs, despite the marked differences in sequence and configuration.[82] Like MS2 RNA, the sequence of the untranslated segment of each of the four RNA components of BMV (over 160 nucleotides long) appears to be highly conserved.[158]

Perhaps even more curious than aminoacylation of plant viral mRNAs is the situation in mengovirus and encephalomyocarditis virus, both of which infect animals, where the genomic RNA can accept histidine and serine, respectively.[159,160] These mRNAs must undergo some form of cleavage prior to aminoacylation as they typically have poly(A) tracts at their 3'-termini. It will be interesting to determine any nuclear-coded mRNAs can undergo similar modification and aminoacylation reactions.

3. The Poly(A) 3'-Tract

The practical application of the 3'-terminal poly(A) sequence present on many eukaryotic mRNAs towards their isolation has been detailed above (Section II, D). Polyadenylation of mRNA generally appears to be a posttranscriptional event, but may be genome-coded in some instances; virion mRNAs of cowpea mosaic virus have a poly(A) 3'-terminus of 100 to 200 residues.[161] Addition of poly(A) usually takes place in the nucleus, and the suggestion has been made that it plays a role in transport of messenger sequences from nucleus to cytoplasm.[162] However, some polyadenylation, or turnover, may occur in the cytoplasm,[163] and poly(A) polymerase has even been isolated from wheat chloroplasts.[164] Despite the lack of poly(A) tracts on mRNAs of prokaryotes or their viruses, poly(A) can be enzymatically added; bacteriophage RNA to which a poly (A) sequence was added was infectious, but the progeny RNAs were not adenylated.[165]

Early reports of poly(A)-containing RNA in plants[166,167] gave length estimates of from 100 to 250 nucleotides. It is established that the length of the poly(A) tract decreases as the messenger ages. Varying lengths of poly(A) on globin mRNA did not affect its template efficiency in wheat germ,[168] but poly(A) may help to protect the 3'-terminus from exonuclease degradation; nonpolyadenylated mRNAs were destroyed more rapidly than polyadenylated mRNAs when injected into frog oocytes.[169] Interestingly, new data shows that mouse globin precursor mRNA, when first transcribed (as heterogeneous nuclear, hnRNA), contains sequences at least seven times (27S) the mRNA length. This is processed in several steps and a poly(A) tail appears only after cleavage to a 15S molecule.[170]

VI. DISCUSSION: REGULATION OF mRNA

A. Transcriptional Control

The transcription of mRNA from the genome by DNA-dependent RNA polymerases (see Volume I, Section II by Becker) is clearly the primary site of regulation of mRNA function. Frequently, the synthesis and accumulation of enzymes and other proteins (e.g., seed storage proteins, Section III. D) is restricted to certain developmental periods and specific tissues. It is unlikely that the mRNAs for these proteins are transcribed at other times and in other cells; the ways in which genome expression is repressed and expressed are only just now becoming amenable to study. Doubtless, histones and phosphorylated proteins[171] are involved, but their role is far from being elucidated. Hopefully, the availability of plant mRNAs for cell-free studies and for obtaining radioactive cDNA probes for examining chromosomal function and organization (see Sections III. D. 3, IV. B and Volume I, Section I by Walbot and Goldberg) will prove a powerful aid in understanding transcriptional control.

Very little evidence has been advanced for the presence of untranslated mRNA in the eukaryotic cytoplasm, and that for factors controlling the rate of translation of specific mRNAs has been questioned.[147] Thus, although the processes of transcription and translation are not so closely associated in eukaryotes as in prokaryotes (where the mRNA is often translated as it is being transcribed), understanding transcriptional control remains the key study area towards understanding control of gene expression. Nevertheless, other factors do influence mRNA function, and some are briefly discussed below.

B. Translational Control

Although mRNAs present in the eukaryotic cytoplasm are usually thought of as being immediately translated, it is possible — if polycistronic, nuclear-coded mRNAs exist — that the availability of initiation sites may greatly influence translation of the individual cistrons, as discussed for MS2 phage RNA (Section V.D).

In some tissues, e.g., cotyledons of germinating soybean,[172] it has been established that the spectrum of tRNAs changes during development. This, and the evidence for preference for certain codewords for translation of MS2 phage coat protein,[62] strongly suggest that the translation of certain mRNAs can be restricted (or enhanced) by the presence of isoaccepting tRNAs and their synthetases (Volume I, Section II by Weil).

C. Selectivity of Cell-free Systems

It is important to remember that much of our present knowledge of mRNA function, not only for plants but also for many eukaryotic mRNAs, comes from using the wheat germ system. Despite its excellence, one must remain cautious in accepting all results at face value. A very real question is why the cell-free system from wheat (and some other cereal) embryos is capable of translating heterologous mRNAs, while extracts from other plants are essentially inactive. Legume cotyledons are very active in protein synthesis, yet S23 preparations made in a similar way to that for wheat germ are not capable of translating supplied viral RNA. They can, however, be made very active in elongation of polypeptide chains.[173] Many animal and a few plant (especially viral-coded[174]) messengers have been translated using the reticulocyte system, as modified by micrococcal nuclease treatment to remove endogenous globin mRNA activity.[175] A recent paper[176] on light-induced increases of phenylalanine ammonia-lyase mRNA in parsley (*Petroselenum hortense*) cell cultures showed that in wheat germ, mainly small (about 25,000 dalton) products were translated, while an 83,000 dalton product (co-electrophoresing with authentic protein) was obtained using a reticulocyte lysate system. The comparison is obscured, however, by the fact that the reticulocyte products shown were of immunoprecipitated material while those from wheat germ were not. Also, the wheat germ system used[177] was admittedly deficient in the synthesis of large proteins. The criticism made above — that the wheat germ system is essentially unique among plants — can to some extent be leveled at the reticulocyte system in its respect to animals. Thus, it will be worthwhile to compare translation products made in different systems (see Section III, Volume II Chapter by Davies). Evidence has been published to support the view that there is no efficient heterologous translation of mRNA by prokaryotic and eukaryotic systems.[177a] Microinjection of mRNA into frog oocytes[178] is a powerful technique whose use has been inadequately explored for plant mRNAs. It has been used in demonstrating the translation of large polypeptides from TMV RNA,[179] and for preliminary examination of proteins implicated in incompatibility studies on *Petunia*.[180]

D. Chemical and Environmental Regulation

The interesting stimulation of amylase synthesis in response to gibberellic acid (Section III.D.6) and cellulase synthesis in response to a growth substance (Section III.B) are examples of the role of plant hormones in controlling gene expression. The understanding of the involvement of kinetins and other plant growth regulators towards gene expression will surely be assisted as the pathways of transcription and translation in plants become experimentally approachable.

In plants, the environment produces rapid physiological and biochemical responses. In seeds, the cessation of protein synthetic activity during desiccation and resumption during imbibition and germination have been alluded to several times. Studies on the changes in glyoxysomal enzymes and their mRNAs promise to provide fascinating insight to the interrelated control systems operating in this early phase of plant growth.[181,182] The effects of desiccation and rehydration on mRNA and protein synthesis have also been noted in connection with the moss *Tortula ruralis,* where polysomes run-off during drying of the gametophyte, but are restored within 2 hr of wetting,[44] quite possibly using conserved mRNA (Section III.D.2).

Many changes in plants are induced by daylength and season. Complex interactions of light quality and temperature are involved in the processes by which plants acclimate to very cold winter temperatures.[183] Many of these processes are physical, but are accompanied by changes in the spectrum of proteins (and hence protein synthesis); during the spring deacclimation of woody twigs, these changes[184] are not dissimilar to germination events. Clearly, light has major effects on plants; the example of mRNA stimulation for phenylalanine ammonia-lyase[176] is a demonstration of a direct response of mRNA to light. Of greater significance is the regulation of synthesis of mRNAs for the nuclear- and plastid-coded subunits of RuBPCase (Section III.C) a study of the enzyme which has made life as we know it possible on this planet and one which will doubtless play a major role in revealing evolutionary trends in mRNA structure, regulation, and function.

ACKNOWLEDGMENTS

I appreciate the useful suggestions made by my colleagues, especially S. M. Sun and B. U. Buchbinder. Support for various aspects of this chapter from the National Science Foundation, the National Institutes of Health, The Herman Frasch Foundation; Centre National de la Recherche Scientifique (France), the Graduate School and The College of Agricultural and Life Sciences of the University of Wisconsin-Madison, is gratefully acknowledged.

REFERENCES

1. Holloway, B. W. and Ripley, S. H., Nucleic acid content of reticulocytes and its relation to uptake of radioactive leucine *in vitro, J. Biol. Chem.,* 196, 695, 1952.
2. Zamecnik, P. C. and Keller, E. B., Relation between phosphate energy donors and incorporation of amino acids into proteins, *J. Biol. Chem.,* 209, 337, 1954.
3. Astrachan, L. and Volkin, E., Properties of amino acid turnover in T2-infected *Escherichia coli, Biochim. Biophys. Acta,* 29, 536, 1958.
4. Hershey, A. D., Nucleic acid economy in bacteria infected with bacteriophage T2. II. Phage precursor nucleic acid, *J. Gen. Physiol.,* 37, 1, 1953.
5. Jacob, F. and Monod, J., Genetic regulatory mechanisms in the synthesis of proteins, *J. Mol. Biol.,* 3, 318, 1961.
6. Lamborg, M. R. and Zamecnik, P. C., Amino acid incorporation into protein by extracts of *E. coli, Biochim. Biophys. Acta,* 42, 206, 1960.
7. Tissieres, A., Schlessinger, D., and Gros, F., Amino acid incorporation into proteins by *Escherichia coli* ribosomes, *Proc. Natl. Acad. Sci. USA.,* 46, 1540, 1960.
8. Nirenberg, M. W. and Matthaei, J. H., The dependence of cell-free protein synthesis in *E. coli* upon naturally occurring or synthetic polyribonucleotides, *Proc. Natl. Acad. Sci. USA.,* 47, 1588, 1961.
9. Crick, F. H. C., The recent excitement in the coding problem, in *Progress in Nucleic Acid Research,* Davidson, J. N. and Cohn, W. E., Eds., Academic Press, New York, 1, 163, 1963.
10. Watson, J. D., *Molecular Biology of the Gene,* 3rd ed., W. A. Benjamin, Inc., Menlo Park, California, 1976, 356.
11. Gierer, A. and Schramm, G., Infectivity of ribonucleic acid from tobacco mosaic virus, *Nature,* London, 177, 702, 1956.
12. Fraenkel-Conrat, H., The role of nucleic acid in the reconstitution of active tobacco mosaic virus, *J. Am. Chem. Soc.,* 78, 882, 1956.
13. Matthaei, J. H., Jones, O. W., Martin, R. G., and Nirenberg, M. W., Characteristics and composition of RNA coding units, *Proc. Natl. Acad. Sci. USA.,* 48, 666, 1962.
14. Tsugita, A., Fraenkel-Conrat, H., Nirenberg, M. W., and Matthaei, J. H., Demonstration of the messenger role of viral RNA, *Proc. Natl. Acad. Sci. USA.,* 48, 846, 1962.
15. Hunter, T. R., Hunt, T., Knowland, J., and Zimmern, D., Messenger RNA for the coat protein of tobacco mosaic virus, *Nature,* London, 260, 759, 1976.
16. Loeb, T. and Zinder, N. D., A bacteriophage containing RNA, *Proc. Natl. Acad. Sci. USA.,* 47, 282, 1961.
17. Nathans, D., Notani, G., Schwartz, J. H., and Zinders, N. D., Biosynthesis of the coat protein of coliphage f2 by *E. coli* extracts, *Proc. Natl. Acad. Sci. USA.,* 48, 1424, 1962.
18. Lockard, R. E. and Lingrel, J. B., Identification of mouse haemoglobin messenger RNA, *Nature,* London, 233, 204, 1971.
19. Luthe, D. S. and Peterson, D. M., Cell-free synthesis of globulin by developing oat (*Avena sativa*) seeds, *Plant Physiol.,* 59, 836, 1977.
20. Mori, T. and Yokoyama, Z., Ribonucleic acids with messenger activities in the cotyledons of soybean seeds, *Agric. Biol. Chem.,* 39, 785, 1975.
20a. Beachy, R. N., Thompson, J. F., and Madison, J. T., Isolation of polyribosomes and messenger RNA active in *in vitro* synthesis of soybean seed proteins, *Plant Physiol.,* 61, 139, 1978.

21. Higgins, T. J. V., Zwar, J. A., and Jacobsen, J. V., Gibberellic acid enhances the level of translatable mRNA for α-amylase in barley aleurone layers, *Nature,* London, 260, 166, 1976.
22. Fox, J. E., Pratt, H. M., Shewry, P. R., and Miflin, B. J., The *in vitro* synthesis of hordein with polysomes from normal and high lysine varieties of barley, *Colloq. Int. C.N.R.S.,* 261, 501, 1977.
23. Sun, S. M., Buchbinder, B. U., and Hall, T. C., Cell-free synthesis of the major storage protein of the bean, *Phaseolus vulgaris* L., *Plant Physiol.,* 56, 780, 1975.
24. Hall, T. C., Bliss, F. A., Ryan, D. S., and Sun, S. M., The subunit structure and cell-free synthesis of the major storage protein from bean (*Phaseolus vulgaris* L.), seeds, *Colloq. Int. C.N.R.S.,* 261, 335, 1977.
24a. Hall, T. C., Ma, Y., Buchbinder, B. U., Pyne, J. W., Sun, S. M., and Bliss, F. A., Messenger RNA for G1 protein of French bean seeds: cell-free translation and product characterization, *Proc. Natl. Acad. Sci. USA.,* 75, 3196, 1978.
25. Wells, G. N. and Beevers, L., Protein synthesis in cotyledons of *Pisum sativum* L. II. The requirements for initiation with plant messenger RNA, *Plant Sci. Lett.,* 1, 281, 1973.
26. Higgins, T. J. V., and Spencer, D., Cell-free synthesis of pea storage proteins, *Colloq. Int. C.N.R.S.,* 261, 327, 1977.
27. Payne, P. I., Gordon, M. E., Dobrzanska, M., Parker, M. L., and Barlow, P. W., The long-lived messenger RNA of dry seeds: Evidence for its existence, intracellular location and role in germination, *Colloq. Int. C.N.R.S.,* 261, 487, 1977.
28. Schultz, G. A., Chen, D., and Katchalski, E., Localization of a messenger RNA in a ribosomal fraction from ungerminated wheat embryos, *J. Mol. Biol.,* 66, 379, 1972.
29. Larkins, B. A. and Dalby, A., *In vitro* synthesis of zein-like protein by maize polyribosomes, *Biochem. Biophys. Res. Commun.,* 66, 1048, 1975.
30. Larkins, B. A., Jones, R. A., and Tsai, C. Y., Isolation and *in vitro* translation of zein messenger ribonucleic acid, *Biochemistry,* 15, 5506, 1976.
31. Burr, B. and Burr, F. A., Zein synthesis in maize endosperm by polysomes attached to protein bodies, *Proc. Natl. Acad. Sci. USA.,* 73, 515, 1976.
31a. Burr, B., Burr, F. A., Rubenstein, I., and Simon, M. N., Purification and translation of zein messenger RNA from maize endosperm protein bodies, *Proc. Natl. Acad. Sci. USA.,* 75, 696, 1978.
32. Giles, A. B., The mobilization of stored messenger RNA of *Phaseolus vulgaris* induced by light, *Colloq. Int. C.N.R.S.,* 261, 603, 1977.
33. Yoshida, K., RNA containing poly-A sequence in *Pharbitis nil* cotyledons and its template activity *in vitro, Plant and Cell Physiol.,* 15, 441, 1974.
34. Verma, D. P. S., Maclachlan, G. A., Byrne, H., and Ewings, D., Regulation and *in vitro* translation of messenger ribonucleic acid from auxin-treated pea epicotyls, *J. Biol. Chem.,* 250, 1019, 1975.
35. Byrne, H., Christou, N. V., Verma, D. P. S., and Maclachlan, G. A., Purification and characterization of two cellulases from auxin-treated pea epicotyls, *J. Biol. Chem.,* 250, 1012, 1975.
36. Klämbt., D., Cytokinin effects on protein synthesis of *in vitro* systems of higher plants, *Plant and Cell Physiol.,* 17, 73, 1976.
37. Verma, D. P. S., Nash, D. T., and Schulman, H. M., Isolation and *in vitro* translation of soybean leghemoglobin mRNA, *Nature,* London, 251, 74, 1974.
38. Verma, D. P. S. and Bal, A. K., Intracellular site of synthesis and localization of leghemoglobin in root nodules, *Proc. Natl. Acad. Sci. USA.,* 73, 3843, 1976.
39. Pfisterer, J. and Kloppstech, K., Free and membrane bound polysomes from plant cell cultures, *Colloq. Int. C.N.R.S.,* 261, 279, 1977.
40. Tobin, E. M. and Klein, A. O., Isolation and translation of plant messenger RNA, *Plant Physiol.,* 56, 88, 1975.
41. Gray, J. C. and Kekwick, R. G. O., The synthesis of the small subunit of ribulose 1,5-bisphosphate carboxylase in the French bean, *Phaseolus vulgaris, Eur. J. Biochem.,* 44, 493, 1974.
42. Giles, A. B., Grierson, D., and Smith, H., *In vitro* translation of messenger-RNA from developing bean leaves. Evidence of stored messenger-RNA and its light-induced mobilization into polyribosomes, *Planta,* 136, 31, 1977.
43. Tucker, E. B. L., and Bewley, J. D., Plant desiccation and protein synthesis III. Stability of cytoplasmic RNA during dehydration, and its synthesis during rehydration of the moss *Tortula ruralis, Plant Physiol.,* 57, 564, 1976.
44. Bewley, J. D. and Dhindsa, R. S., Stability of components of the protein synthesizing complex of a plant during desiccation, in *Translation of Natural and Synthetic Polynucleotides,* Legocki, A., Ed., University of Agriculture in Poznan, Poland, 1977.
44a. Bewley, J. D., and Dhindsa, R. S., personal communication.
45. Roy, H., Patterson, R., and Jagendorf, A. T., Identification of the small subunit of ribulose 1,5-bisphosphate carboxylase as a product of wheat cytoplasmic ribosomes, *Arch. Biochem., Biophys.,* 172, 64, 1976.

46. **Margulies, M. M. and Michaels, A.,** Biosynthesis of chloroplast membrane proteins in *Chlamydomonas reinhardtii, Colloq. Int. C.N.R.S.,* 261, 395, 1977.

47. **Michaels, A. and Margulies, M. M.,** Amino acid incorporation into protein by ribosomes bound to thylakoid membranes: formation of discrete products, *Biochim. Biophys. Acta,* 390, 352, 1975.

48. **Howell, S., Heizmann, P., and Gelvin, S.,** Properties of the mRNA and localization of the gene coding for the large subunit of ribulose bisphosphate carboxylase in *Chlamydomonas reinhardtii, Colloq. Int. C.N.R.S.,* 261, 313, 1977.

49. **Dobberstein, B., Blobel, G., and Chua, N.-H.,** *In vitro* synthesis and processing of a putative precursor for the small subunit of ribulose-1,5-bisphosphate carboxylase of *Chlamydomonas reinhardtii, Proc. Natl. Acad. Sci. USA.,* 74, 1082, 1977.

50. **Sagher, D., Grosfeld, H., and Edelman, M.,** Large subunit ribulosebisphosphate carboxylase messenger RNA from *Euglena* chloroplasts, *Proc. Natl. Acad. Sci. USA.,* 73, 722, 1976.

51. **Bottomley, W., Higgins, T. J. V., and Whitfeld, P. R.,** Differential recognition of chloroplast and cytoplasmic messenger RNA by 70S and 80S ribosomal systems, *FEBS Lett.,* 63, 120, 1976.

52. **Bottomley, W., Higgins, T. J. V., Whitfeld, P. R., and Leaver, C. J.,** The products of *in vitro* protein synthesizing systems programmed by chloroplast and cytoplasmic RNA and chloroplast DNA, *Colloq. Int. C.N.R.S.,* 261, 413, 1977.

53. **Hartley, M. R., Wheeler, A., and Ellis, R. J.,** Protein synthesis in chloroplasts V. Translation of messenger RNA for the large subunit of fraction I protein in a heterologous cell-free system, *J. Mol. Biol.,* 91, 67, 1975.

54. **Marcus, A., Efron, D., and Weeks, D. P.,** The wheat embryo cell-free system, in *Nucleic Acids and Protein Synthesis, Methods Enzymol.,* Vol. 30, Moldave, K. and Grossman, L., Eds., Academic Press, New York, 1974.

54a. **Klein, W. H., Nolan, C., Lazar, J. M., and Clark, J. M., Jr.,** Translation of satellite tobacco necrosis virus ribonucleic acid. I. Characterization of *in vitro* procaryotic and eucaryotic translation products, *Biochemistry,* 11, 2009, 1972.

54b. **Shih, D. S. and Kaesberg, P.,** Translation of brome mosaic viral ribonucleic acid in a cell-free system derived from wheat embryo. *Proc. Natl. Acad. Sci. USA.,* 70, 1799, 1973.

55. **Roberts, B. E. and Patterson, B. M.,** Efficient translation of tobacco mosaic virus RNA and rabbit globin 9S RNA in a cell-free system from commercial wheat germ, *Proc. Natl. Acad. Sci. USA.,* 70, 2330, 1973.

55a. **Davies, J. W. and Kaesberg, P.,** Translation of virus mRNA: Synthesis of bacteriophage Qβ proteins in a cell-free extract from wheat embryo, *J. Virol.,* 12, 1434, 1973.

56. **Marcu, K. and Dudock, B.,** Characterization of a highly efficient protein synthesizing system from commercial wheat germ, *Nucleic Acids Res.,* 1, 1385, 1974.

57. **Davies, E., Larkins, B. A., and Knight, R. H.,** Polyribosomes from peas. An improved method for their isolation in the absence of ribonuclease inhibitors, *Plant Physiol.,* 50, 581, 1972.

58. **Takanami, M., Yan, Y., and Jukes, T. H.,** Studies on the site of ribosomal binding of f2 bacteriophage RNA, *J. Mol. Biol.,* 12, 761, 1965.

59. **Steitz, J. A.,** Polypeptide chain initiation: Nucleotide sequences of the three ribosomal binding sites on bacteriophage R17 RNA, *Nature,* 224, 957, 1969.

60. **Watson, J. D.,** *Molecular Biology of the Gene,* 3rd ed., W. A. Benjamin, Inc., Menlo Park, California, 1976, 340.

61. **Fan, H. and Penman, S.,** Regulation of protein synthesis in mammalian cells. II. Inhibition of protein synthesis at the level of initiation during mitosis, *J. Mol. Biol.,* 50, 655, 1970.

62. **Fiers, W., Contreras, R., Duerinck, F., Haegeman, G., Iserentant, D., Merregaert, J., Min Jou, W., Molemans, F., Raeymaekers, A., Van den Berghe, A., Volckaert, G., and Ysebaert, M.,** Complete nucleotide sequence of bacteriophage MS2 RNA: primary and secondary structure of the replicase gene, *Nature,* London, 260, 500, 1976.

63. **Wilson, C. M.,** Plant nucleases, *Annu. Rev. Plant Physiol.,* 26, 187, 1975.

64. **Larkins, B. A. and Davies, E.,** Polyribosomes from peas. III. Stimulation of polysome degradation of exogenous and endogenous calcium, *Plant Physiol.,* 52, 655, 1973.

65. **Zabel, P., Jongen-Neven, I., and Van Kammen, A.,** *In vitro* replication of cowpea mosaic virus RNA. II. Solubilization of membrane-bound replicase and the partial purification of the solubilized enzyme, *J. Virol.,* 17, 679, 1976.

66. **Weeks, D. P. and Marcus, A.,** Polyribosome preparation in the presence of diethylpyrocarbonate, *Plant Physiol.,* 44, 1291, 1969.

67. **Brakke, M. K.,** Density-gradient centrifugation, in *Methods in Virology,* Vol. 2. Maramorosch, K. and H. Koprowsky, Eds., Academic Press, New York, 1967, 93.

68. **Gray, J. C.,** The inhibition of ribonuclease activity and the isolation of polysomes from leaves of the French bean *Phaseolus vulgaris, Arch. Biochem. Biophys.,* 163, 343, 1974.

69. **Schimke, R. T., Sullivan, D., Kiely, M. L., Gonzales, C., and Taylor, J. M.,** Immunoabsorption of ovalbumin synthesizing polysomes and partial purification of ovalbumin mRNA, *Methods Enzymol.,* 30, 631, 1974.

70. **Hanna, N. and Godin, C.,** Comparison of the effect of various methods of dissociation on the protein composition of rat liver ribosomal subunits, *Can. J. Biochem.,* 53, 935, 1975.

71. **Blobel, G.,** Release, identification and isolation of messenger RNA from mammalian ribosomes, *Proc. Natl. Acad. Sci. USA.,* 68, 832, 1971.

72. **Leaver, C. J. and Ingle, J.,** The molecular integrity of chloroplast ribosomal nucleic acid, *Biochem. J.,* 123, 235, 1971.

73. **Penman, S.,** RNA metabolism in the HeLa cell nucleus, *J. Mol. Biol.,* 17, 117, 1966.

73a. **Eaton, B. T., and Faulkner, P.,** Heterogeneity in the poly(A) content of the genome of sindbis virus, *Virology,* 50, 865, 1972.

74. **Howell, S. H., Heizmann, P., Gelvin, S., and Walker, L. L.,** Identification and properties of the messenger RNA activity in *Chlamydomonas reinhardti* coding for the large subunit of D-ribulose-1,5-bisphosphate carboxylase, *Plant Physiol.,* 59, 464, 1977.

75. **Staynov, D. Z., Pinder, J. C., and Gratzer, W. B.,** Molecular weight determination of nucleic acids by gel electrophoresis in nonaqueous solutions, *Nature (London), New Biol.,* 235, 108, 1972.

76. **Kazazian, H. H., Snyder, P. G., and Cheng, T.-C.,** Separation of α- and β-globin messenger RNAs by formamide gel electrophoresis, *Biochem. Biophys. Res. Commun.,* 59, 1053, 1974.

77. **Rosen, J. M.,** Isolation and characterization of purified rat casein messenger ribonucleic acids, *Biochemistry,* 15, 5263, 1976.

77a. **Buchbinder, B. U. and Hall, T. C.,** Direct isolation of mRNAs from developing bean (*Phaseolus vulgaris*) seed cotyledons, *Plant Physiol.,* submitted.

78. **Aviv, H. and Leder, P.,** Preparation of biologically active globin messenger RNA by chromatography on oligothymidylic acid cellulose, *Proc. Natl. Acad. Sci. USA.,* 69, 1408, 1972.

79. **Merkel, C. G., Kwan, S.-P., and Lingrel, J. B.,** Size of the adenylic acid region of newly synthesized globin messenger RNA, *J. Biol. Chem.,* 250, 3725, 1975.

80. **Sagher, D., Edelman, M., and Jakob, K. M.,** Poly(A)-associated RNA in plants, *Biochim. Biophys. Acta,* 349, 32, 1974.

81. **Nemer, A.,** Developmental changes in the synthesis of sea urchin embryo messenger RNA containing and lacking polyadenylic acid, *Cell,* 6, 559, 1975.

82. **Dasgupta, R. and Kaesberg, P.,** Sequence of an oligonucleotide derived from the 3′-end of each of the four brome mosaic viral RNAs, *Proc. Natl. Acad. Sci. USA.,* 74, 4900, 1977.

83. **Briand, J. P., Richards, K. E., Bouley, J. P., Witz, J.,and Hirth, L.,** Structure of the amino acid accepting 3′-end of high-molecular-weight eggplant mosaic virus RNA, *Proc. Natl. Acad. Sci. USA.,* 73, 737, 1976.

84. **Wheeler, A. M. and Hartley, M. R.,** Major mRNA species from spinach chloroplasts do not contain poly(A), *Nature,* London, 257, 66, 1975.

85. **Krystosek, A., Cawthon, M. L., and Kabat, D.,** Improved methods for purification and assay of eukaryotic messenger ribonucleic acids and ribosomes, *J. Biol. Chem.,* 250, 6077, 1975.

85a. **Verma, D. P. S., Ball, S., Guerin, C., and Wanamaker, L.,** Leghemoglobin biosynthesis in soybean root nodules. Characterization of the nascent and released peptides and the relative rate of synthesis of the major leghemoglobins, *Biochemistry,* 18, 476, 1979.

86. **Bantle, J. A., Maxwell, I. H., and Hahn, W. E.,** Specificity of oligo(dT)-cellulose chromatography in the isolation of polyadenylated RNA, *Anal. Biochem.,* 72, 427, 1976.

87. **Bergerson, F. J. and Turner, G. L.,** Leghaemoglobin and the supply of O_2 to nitrogen-fixing root nodule bacteroids: Studies of an experimental system with no gas phase., *J. Gen. Microbiol.,* 89, 31, 1975.

88. **Dilworth, M. J. and Kidby, D. K.,** Localization of iron and leghaemoglobin in the legune root nodule by electron microscope antoradiography, *Exp. Cell. Res.,* 49, 148, 1968.

89. **Bergerson, F. J. and Goodchild, D. J.,** Cellular location and concentration of leghaemoglobin in soybean root nodules, *Aust. J. Biol. Sci.,* 26, 741, 1973.

90. **Weeks, D. P. and Marcus, A.,** Preformed messenger of quiescent wheat embryos, *Biochim. Biophys. Acta,* 232, 671, 1971.

91. **McFadden, B. A.,** Autotrophic CO_2 assimilation and the evolution of ribolose diphosphate carboxylase, *Bacteriol. Rev.,* 37, 283, 1973.

92. **Kawashima, N. and Wildman, S. G.,** Fraction I protein, *Annu. Rev. Plant Physiol.,* 21, 325, 1970.

93. **Ellis, R. J.,** Fraction I protein, *Commentaries in Plant Science,* Pergamon Press, Oxford, 4, 29, 1973.

94. **Jensen, R. G. and Bahr, J. T.,** Ribulose 1,5-*bisphosphate* carboxylase-oxygenase, *Annu. Rev. Plant Physiol.,* 28, 379, 1977.

95. **Ellis, R. J.,** The search for plant messenger RNA, in *Perspectives in Experimental Biology,* Vol. 2, 283. Sunderland, N., Ed., Pergamon Press, New York, 1976.

96. **Wildman, S. G. and Bonner, J.,** The proteins of green leaves. I. Isolation, enzymatic properties and auxin content of spinach cytoplasmic proteins, *Arch. Biochem.,* 14, 381, 1947.

97. **Sakano, K., Kung, S. D., and Wildman, S. G.,** Identification of several chloroplast DNA genes which code for the large subunit of *Nicotiana* Fraction I proteins, *Molec. Gen. Genet.,* 130, 91, 1974.

98. **Chen, K., Johal, S., and Wildman, S. G.,** Phenotype markers for chloroplast genes in higher plants and their use in biochemical genetics, in *Nucleic Acids and Protein Synthesis in Plants,* Bogorad, L. and Weil, J. H., Eds., Plenum Press, New York, 1977, 183.

99. **Chan, P. and Wildman, S. G.,** Chloroplast DNA codes for the primary structure of the large subunit of Fraction I protein, *Biochim. Biophys. Acta,* 277, 677, 1972.

100. **Kawashima, N. and Wildman, S. G.,** Studies on Fraction I protein. IV. Mode of inheritance of primary structure in relation to whether chloroplast or nuclear DNA contains the code for a chloroplast protein, *Biochim. Biophys. Acta,* 262, 42, 1972.

101. **Kung, S-D.,** Tobacco fraction I protein: a unique genetic marker, *Science,* 191, 429, 1976.

102. **Blair, G. E. and Ellis, R. J.,** Protein synthesis in chloroplasts. I. Light-driven synthesis of the large subunit of Fraction I protein by isolated pea chloroplasts, *Biochim. Biophys. Acta,* 319, 223, 1972.

103. **Criddle, R. S., Dau, B., Kleinkopf, G. E., and Huffaker, R. C.,** Differential synthesis of ribulose-diphosphate carboxylase subunits, *Biochem. Biophys. Res. Commun.,* 41, 621, 1970.

104. **Ellis, R. J.,** The synthesis of chloroplast proteins, in *Nucleic Acids and Protein Synthesis in Plants,* Bogorad, L. and Weil, J. H., Eds., Plenum Press, New York, 1977, 195.

105. **Edelman, M., Sagher, D.,N., and Reisfeld, A.,** *In vitro* translation of large subunit ribulosediphosphate carboxylase from *Euglena, Colloq. Int. C.N.R.S.,* 261, 562, 1977.

106. **Rosner, A., Reisfeld, A., Jacob, K. M., Gressel, J., and Edelman, M.,** Shifts in the RNA and protein metabolism of *Spirodela* (duckweed), *Colloq. Int. C.N.R.S.,* 261, 562, 1977.

107. **Gelvin, S. and Howell, S. H.,** Identification and precipitation of the polyribosomes in *Chlamydomonas reinhardi* involved in the synthesis of the large subunit of D-ribulose-1,5-bisphosphate carboxylase, *Plant Physiol.,* 59, 471, 1977.

108. **Gelvin, S., Heizmann, P., and Howell, S. H.,** Identification and cloning of the chloroplast gene coding for the large subunit of ribulose-1,5-bisphosphate carboxylase from *Chlamydomonas reinhardi, Proc. Natl. Acad. Sci. USA.,* 74, 3193, 1977.

109. **Mans, R. J. and Novelli, G. D.,** Stabilization of the maize seedling amino acid incorporating system, *Biochim. Biophys. Acta,* 80, 127, 1964.

110. **Marcus, A. and Feeley, J.,** Activation of protein synthesis in the imbibition phase of seed germination, *Proc. Natl. Acad. Sci. USA.,* 51, 1075, 1964.

111. **Bonner, J., Huang, R. C., and Gilden, R. V.,** Chromosomally directed protein synthesis, *Proc. Natl. Acad. Sci. USA.,* 50, 893, 1963.

112. **Morton, R. K. and Raison, J. K.,** The separate incorporation of amino acids into storage and soluble proteins catalyzed by two independent systems isolated from wheat endosperm, *Biochem. J.,* 91, 528, 1964.

113. **Wilson, C. M.,** Bacteria antibiotics and amino acid incorporation into maize endosperm protein bodies, *Plant Physiol.,* 41, 325, 1966.

114. **Payne, P. I.,** The long-lived messenger ribonucleic acid of flowering-plant seeds, *Biol. Rev.,* 51, 329, 1976.

115. **Ihle, J. N. and Dure, L. S.,** The temporal separation of transcription and translation and its control in cotton embryogenesis and germination, in *Plant Growth Substances 1972,* Carr, D. J., Ed., Springer-Verlag, Berlin, 1972, 216.

116. **Dure, L. S.,** Regulation of protein synthesis in cotton seed embryogenesis and germination, *Biochem. Soc. Symp.,* 38, 217, 1973.

117. **Dure, L. S.,** Seed formation, *Annu. Rev. Plant Physiol.,* 26, 259, 1975.

118. **Fraser, R. S. S.,** Studies on messenger and ribosomal RNA synthesis in plant tissue cultures induced to undergo synchronous cell division, *Eur. J. Biochem.,* 50, 529, 1975.

119. **Payne, P. I., Gordon, M. E., Dobrzanska, M., Parker, M. L., and Barlow, P. W.,** The long-lived messenger RNA of dry seeds: Evidence for its existence, intracellular location and role in germination, *Colloq. Int. C.N.R.S.,* 261, 487, 1977.

120. **Spiegel, S. and Marcus, A.,** Polyribosome formation in early wheat embryo germination independent of either transcription of polyadenylation, *Nature,* London, 256, 228, 1975.

121. **Walbot, V., Harris, B. and Dure, L. S.,** The regulation of enzyme synthesis in the embryogenesis and germination of cotton, in *Developmental Biology of Reproduction,* 33rd Symp. Soc. Dev. Biol., Markert, C. L., Ed., Academic Press, New York, 1974, 165.

122. **Boyer, H. W., Tait, R. C., McCarthy, B. J., and Goodman, H. M.,** Cloning of eukaryotic DNA as an approach to the analysis of chromosome structure and function, in *Genetic Improvement of Seed Proteins,* National Research Council, Natl. Acad. Sci., 1976, 359.

123. **Rothenberg, E., Smotkin, D., Baltimore, D., and Weinberg, R. A.,** *In vitro* synthesis of infectious DNA of murine leukaemia virus, *Nature,* London, 269, 122, 1977.

124. **Gianazza, E., Viglienghi, V., Righetti, P. G., Salamini, F., and Soave, C.,** Amino acid composition of zein molecular components, *Phytochemistry,* 16, 315, 1977.

125. **Blobel, G., and Dobberstein, B.,** Transfer of proteins across membranes. I. Presence of proteolytically processed and unprocessed nascent immunoglobulin light chains on membrane-bound ribosomes of murine myeloma, *J. Cell Biol.,* 67, 835, 1975.

126. **Derbyshire, E., Wright, D. J., and Boulter, D.,** Legumin and vicilin, storage proteins of legume seeds, *Phytochemistry,* 15, 3, 1976.

127. **Sun, S. M. and Hall, T. C.,** Solubility characteristics of globulins from *Phaseolus* seeds in regard to their isolation and characterization, *Agric. Food Chem.,* 23, 184, 1975.

128. **Romero, J., Sun, S. M., McLeester, R. C., Bliss, F. A., and Hall, T. C.,** Heritable variation in a polypeptide subunit of the major storage protein of the bean, *Phaseolus vulgaris* L., *Plant Physiol.,* 56, 776, 1975.

129. **Hall, T. C., McLeester, R. C., and Bliss, F. A.,** Equal expression of the maternal and paternal alleles for the polypeptide subunits of the major storage protein of the bean *Phaseolus vulgaris* L., *Plant Physiol.,* 59, 1122, 1977.

130. **Ma, Y. and Bliss, F. A.,** Seed proteins of common bean (*Phaseolus vulgaris* L.), *Crop Sci.,* 18, 431, 1978.

130a. **Sun, S. M., Mutschler, M. A., Bliss, F. A., and Hall, T. C.,** Protein synthesis and accumulation in bean cotyledons during growth, *Plant Physiol.,* 61, 918, 1978.

131. **Hall, T. C., Sun, S. M., Buchbinder, B. U., and Belozerskii, M. A.,** The translation of mRNA for storage globulin of the bean, *Phaseolus vulgaris,* in *Translation of Natural and Synthetic Polynucleotides,* Legocki, A., Ed., University of Agriculture in Poland, 1977, 217.

132. **Barrell, B. G., Air, G. M., and Hutchison, C. H.,** III. Overlapping genes in bacteriophage $\phi\chi$ 174, *Nature,* London, 264, 34, 1976.

133. **Beachey, R. N., Zaitlin, M., Bruening, G., and Israel, H. W.,** A genetic map for the cowpea strain of TMV, *Virology,* 73, 498, 1976.

134. **Filner, P. and Varner, J. E.,** A test for *de novo* synthesis of enzymes: density labeling with H_2O^{18} of barley α-amylase induced by gibberellic acid, *Proc. Natl. Acad. Sci. USA.,* 58, 1520, 1967.

135. **Jacobsen, J. V. and Zwar, J. A.,** Gibberellic acid causes increased synthesis of RNA which contains poly(A) in barley aleurone tissue, *Proc. Natl. Acad. Sci. USA.,* 71, 3290, 1974.

136. **Ho, D. T. H. and Varner, J. E.,** Hormonal control of messenger ribonucleic acid metabolism in barley aleurone layers, *Proc. Natl. Acad. Sci. USA.,* 71, 4783, 1974.

137. **Higgins, T. J. V., Goodwin, P. B., and Whitfeld, P. R.,** Occurrence of short particles in beans infected with the cowpea strain of TMV. II. Evidece that the short particles contain the cistron for coat-protein, *Virology,* 71, 486, 1976.

138. **Partington, G. A., Kemp, D. J., and Rogers, G. E.,** Isolation of feather keratin mRNA and its translation in a rabbit reticulocyte cell-free system, *Nature (London) New Biol.,* 246, 33, 1973.

139. **Adenski, M.,** Polyacrylamide gel electrophoresis of viral RNA, in *Methods in Virology,* Maramorosch, K. and Kóprowski, H., Eds., Vol. 5, Academic Press, New York, 1971, 125.

139a. **Dugaiczyk, A., Woo, S. L. C., Lai, E. C., Mace, M. L., Jr., McReynolds, L., and O'Malley, B. W.,** The natural ovalbumin gene contains seven intervening sequences, *Nature, (London,)* 274, 328, 1978.

139b. **Tilghman, S. M., Curtis, P. J., Tiemeier, D. C., Leder, P., and Weissmann, C.,** The intervening sequence of a mouse β-globin gene is transcribed within the 15S β-globin mRNA precursor, *Proc. Natl. Acad. Sci. USA.,* 75, 1309, 1978.

139c. **Krzyzek, R. A., Collett, M. S., Lau, A. F., Perdue, M. L., Leis, J. P., and Faras, A. J.,** Evidence for splicing of avian sarcoma virus 5'-terminal genomic sequences onto viral-specific RNA in infected cells, *Proc. Natl. Acad. Sci. USA.,* 75, 1284, 1978.

140. **Shatkin, A.J. and Both, G. W.,** Reovirus mRNA: Transcription and translation, *Cell,* 7, 305, 1976.

141. **Perry, R. P. and Kelley, D. E.,** Kinetics of formation of 5'-terminal caps in mRNA, *Cell,* 8, 433, 1976.

142. **Both, G. W., Banerjee, A.K., and Shatkin, A. J.,** Methylation-dependent translation of viral messenger RNAs *in vitro,* *Proc. Natl. Acad. Sci. USA.,* 72, 1189, 1975.

143. **Shih, D. S., Dasgupta, R., and Kaesberg, P.,** 7-methyl-guanosine and efficiency of RNA translation, *J. Virol.,* 19, 637, 1976.

144. **Dasgupta, R., Shih, D. S., Saris, C., and Kaesberg, P.,** Nucleotide sequence of a viral fragment that binds to eukaryotic ribosomes, *Nature,* 256, 624, 1975.

145. **Shine, J. and Dalgarno, L.,** The 3'-terminal sequence of *Escherichia coli:* 16S ribosomal RNA: Complementarity to nonsense triplets and ribosome binding sites, *Proc. Natl. Acad. Sci. USA,* 71, 1342, 1974.

146. **Steitz, J. A. and Jakes, K.,** How ribosomes select initiator regions in mRNA: Base pair formation between the 3' terminus of 16S rRNA and the mRNA during initiation of protein synthesis in *Escherichia coli, Proc. Natl. Acad. Sci. USA.,* 72, 4734, 1975.

147. **Lodish, H. F.,** Translational control of protein synthesis, *Annu. Rev. Biochem.,* 45, 39, 1976.

148. **Heywood, S. M., Kennedy, D. S., and Bester, A. J.,** Separation of specific initiation factors involved in the translation of myosin and myglobin messenger RNAs and the isolation of a new RNA involved in translation, *Proc. Natl. Acad. Sci. USA.,* 71, 2428, 1974.

149. **Weissbach, H. and Ochoa, S.,** Soluble factors required for eukaryotic protein synthesis, *Annu. Rev. Plant Physiol.,* 45, 191, 1976.

150. **Traugh, J. A., Tahara, S. M., Sharp, S. B., Safer, B., and Merrick, W. C.,** Factors involved in initiation of haemoglobin synthesis can be phosphorylated *in vitro, Nature,* London, 263, 163, 1976.

151. **Treadwell, B. V., and Robison, W. G.,** Isolation of protein synthesis initiation factor MP from the high-speed supernatant fraction of wheat germ, *Biochem. Biophys. Res. Commun.,* 65, 176, 1975.

152. **Seal, S. N., Giesen, M., Roman, R., and Marcus, A.,** Functional characterization of the initiation factors of wheat germ, in *Nucleic Acids and Protein Synthesis in Plants,* Bogorad, L. and Weil, J. H., Eds., Plenum Press, New York, 1977, 167.

153. **Chan, S. J., Keim, P., and Steiner, D. F.,** Cell-free synthesis of rat preproinsulins: Characterization and partial amino acid sequence determination, *Proc. Natl. Acad. Sci. USA.,* 73, 1964, 1976.

153a. **Van Tol, G. L. and Van Vloten-Doting, L.,** *Eur. J. Biochem.* 93, 461, 1979.

154. **Vandenberghe, A., Min Jou, W., and Fiers, W.,** 3'-Terminal nucleotide sequence (n = 361) of bacteriophage MS2 RNA, *Proc. Natl. Acad. Sci. USA.,* 72, 2559, 1975.

155. **Weissmann, D., Billeter, M. A., Goodman, H. M., Hindley, J., and Weber, H.,** Structure and function of phage RNA, 1972, in *RNA Viruses: Replication and Structure,* FEBS Proc., 8th meeting, Amsterdam, 1972.

156. **Pinck, M., Yot, P., Chapeville, F., and Duranton, H. M.,** Enzymatic binding of valine to the 3'-end of TYMV-RNA, *Nature,* London, 226, 954, 1970.

157. **Hall, T. C. and Wepprich, R. K.,** Functional possibilities for aminoacylation of viral RNA in transcription and translation, *Annu. Microbiol., (Paris),* 127, 143, 1976.

158. **Bastin, M., Dasgupta, R., Hall, T. C., and Kaesberg, P.,** Similarity in structure and function of the 3'-terminal region of the four brome mosaic viral RNAs, *J. Mol. Biol.,* 103, 1, 1976.

159. **Salomon, R. and Littauer, U. A.,** Enzymatic acylation of histidine to mengovirus RNA, *Nature,* 249, 32, 1974.

160. **Lindley, I. J. D., and Stebbing, N.,** Aminoacylation of encephalomyocarditis virus RNA, *J. Gen. Virol.,* 34, 177, 1977.

161. **El Manna, M. M. and Bruening, G.,** Polyadenylate sequences in the ribonucleic acids of cowpea mosaic virus, *Virology,* 56, 198, 1973.

162. **Lewin, B.,** Units of transcription and translation: the relationship between heterologous nuclear RNA and messenger RNA, *Cell,* 4, 11, 1975.

163. **Perry, R. P.,** Processing of RNA, *Annu. Rev. Biochem.,* 45, 605, 1976.

164. **Burkard, G. and Keller, E. B.,** Poly(A) polymerase and poly(G) polymerase in wheat chloroplasts, *Proc. Natl. Acad. Sci. USA.,* 71, 389, 1974.

165. **Devos, R., Gillis, E., and Fiers, W.,** The enzymic addition of poly(A) to the 3'-end of RNA using bacteriophage MS2 RNA as a model system, *Eur. J. Biochem.,* 62, 401, 1976.

166. **Sagher, D., Edelman, M., and Jakob, K.,** Poly(A)-associated RNA in plants, *Biochim. Biophys. Acta,* 349, 32, 1974.

167. **Key, J. L. and Silflow, C.,** The occurrence and distribution of poly(A) ribonucleic acid in soybean, *Plant Physiol.,* 56, 364, 1975.

168. **Bard, E., Efron, D., Marcus, A., and Perry, R. P.,** Translational capacity of deadenylated messenger RNA, *Cell,* 1, 101, 1974.

169. **Huez, G., Marbaix, G., Hubert, E., Laclercq, M., Nudel, U., Soreq, H., Salomon, R., Lebleu, B., Revel, M., and Littauer, U. A.,** Role of the polyadenylate segment in the translation of globin messenger in *Xenopus oocytes, Proc. Natl. Acad. Sci. USA.,* 71, 3143, 1974.

170. **Bastos, R. N., and Aviv, H.,** Globin RNA precursor molecules: Biosynthesis and processing in erythroid cells, *Cell,* 11, 641, 1977.

171. **Stein, G. S. and Kleinsmith, L. J.,** *Chromosomal proteins and their roles in the regulation of gene expression,* Academic Press, New York, 1975.

172. Bick, N. D., Liebke, H., Cherry, J. H., and Strehler, B. L., Changes in leucyl- and tyrosyl-tRNA of soybean cotyledons during plant growth, *Biochim. Biophys. Acta,* 204, 175, 1970.

173. Hall, T. C., Bliss, F. A., Ryan, D. S., and Sun, S. M., The subunit structure and cell-free synthesis of the major storage protein from bean (*Phaseolus vulgaris* L.) seeds, *Colloq. Int. C.N.R.S.,* 261, 335, 1977.

174. Pelham, H. R. B. and Stuik, E. J., Translation of cowpea mosaic virus RNA in a messenger-dependent cell-free system from rabbit reticulocytes, in *Proc. Coll. Nucleic Acids Protein Synthesis in Plants,* C.N.R.S., Paris, 691, 1977.

175. Pelham, H. R. B. and Jackson, R. J., An efficient mRNA-dependent translation system from rabbit reticulocyte lysates, *Eur. J. Biochem.,* 67, 247, 1976.

176. Schröder, J., Light-induced increase of messenger RNA for phenylalanine ammonia-lyase in cell suspension cultures of *Petroselinum hortense, Arch. Biochem. Biophys.,* 182, 488, 1977.

177. Schröder, J., Betz, B., and Hahlbrock, K., Light-induced enzyme synthesis in cell suspension cultures of *Petroselinum hortense, Eur. J. Biochem.,* 67, 527, 1976.

177a. Herrlich, P. and Schweinger, M., Discrimination of messenger RNA, *FEBS Lett.,* 87, 1, 1978.

178. Gurdon, J. B., Lingrel, J. B., and Marbaix, G., Message stability in injected frog oocytes: Long life of mammalian α and β globin messages, *J. Mol. Biol.,* 80, 539, 1973.

179. Knowland, J., Protein synthesis directed by the RNA from a plant virus in a normal animal cell, *Genetics,* 78, 383, 1974.

180. van der Donk, J. A. W. M., Translation of plant messengers in egg cells of *Xenopus laevis, Nature,* London, 256, 674, 1975.

181. Becker, W. M., Leaver, C. J., and Weir, E. M., Developmental changes in cotyledonary RNA and protein during germination of cucumber, *Plant Physiol.,* 59, S57, 1977.

182. Leaver, C. J., Weir, E. M., Walden, R., Wietgrefe, S. W., and Becker, W. M., Developmental changes in cytoplasmic and orangellar RNAs during germination of cucumber, *Plant Physiol.,* 59, S57, 1977.

183. Weiser, C. J., Cold resistance and injury in woody plants, *Science,* 169, 1269, 1970.

184. Hall, T. C., McLeester, R. C., McCown, B. H., and Beck, G. E., Enzyme changes during deacclimation of willow stem, *Cryobiology,* 7, 130, 1970.

Index

INDEX

D